STRATEGY IN TRANSITION

Strategic Management Society Book Series

The Strategic Management Society Book Series is a cooperative effort between the Strategic Management Society and Blackwell Publishing. The purpose of the series is to present information on cutting-edge concepts and topics in strategic management theory and practice. The books emphasize building and maintaining bridges between strategic management theory and practice. The work published in these books generates and tests new theories of strategic management. Additionally, work published in this series demonstrates how to learn, understand, and apply these theories in practice. The content of the series represents the newest critical thinking in the field of strategic management. As a result, these books provide valuable knowledge for strategic management scholars, consultants, and executives.

Published

Strategic Entrepreneurship: Creating a New Mindset
Edited by Michael A. Hitt, R. Duane Ireland, S. Michael Camp, and Donald L. Sexton

Creating Value: Winners in the New Business Environment
Edited by Michael A. Hitt, Raphael Amit, Charles E. Lucier, and Robert D. Nixon

Strategy Process: Shaping the Contours of the Field
Edited by Bala Chakravarthy, Peter Lorange, Günter Müller-Stewens, and Christoph Lechner

The SMS Blackwell Handbook of Organizational Capabilities: Emergence, Development and Change
Edited by Constance E. Helfat

Mergers and Acquisitions: Creating Integrative Knowledge
Edited by Amy L. Pablo and Mansour Javidan

Strategy in Transition
Edited by Richard A. Bettis

Strategy in Transition

Edited by

Richard A. Bettis

Blackwell
Publishing

BLACKWELL PUBLISHING
350 Main Street, Malden, MA 02148-5020, USA
108 Cowley Road, Oxford OX4 1JF, UK
550 Swanston Street, Carlton, Victoria 3053, Australia

First published 2005 by Blackwell Publishing Ltd

Library of Congress Cataloging-in-Publication Data

International Strategic Management Society Conference (21st : 2002 : San Francisco, Calif.)
    Strategy in transition / [edited by] Richard A. Bettis.
        p. cm. — (Strategic Management Society book series)
    "Papers from the Strategic Management Society International Conference held in San Francisco
in October of 2001"—Introd.
    Includes bibliographical references and index.
    ISBN 1-4051-1849-0 (hardcover : alk. paper)
    1. Strategic planning—Congresses. 2. Knowledge management—Congresses.
3. Technological innovations—Management—Congresses. 4. Organizational
change—Congresses. 5. Industrial management—Congresses. I. Bettis, Richard Allan.
II. Title. III. Series.
HD30.28.I555 2001
658.4′012—dc22
                                                                        2004004344

A catalogue record for this title is available from the British Library.

Set in 10/12pt Galliard
by Graphicraft Limited, Hong Kong
Printed and bound in the United Kingdom
by MPG Books, Bodmin, Cornwall

For further information on
Blackwell Publishing, visit our website:
www.blackwellpublishing.com

Contents

Contributors

Bettis, Richard A.
Kenan-Flagler Business School, University of North Carolina
e-mail: r_bettis@unc.edu

Boccardelli, Paolo
Luiss Guido Carli University
e-mail: pboccardelli@luiss.it

Bresnen, Michael
Warwick Business School, University of Warwick
e-mail: irobmb@wbs.warwick.ac.uk

Costanzo, Laura A.
The International Institute of Banking and Financial Services (IIBFS), Leeds University Business School
e-mail: l.a.costanzo@lubs.leeds.ac.uk

Durisin, Boris
Bocconi University
e-mail: boris.durisin@sdabocconi.it

Edelman, Linda F.
Bentley College
e-mail: ledelman@bentley.edu

Grandi, Alessandro
University of Bologna
e-mail: agrandi@mail.ing.unibo.it

Kowalczyk, Stanley J.
College of Business, San Francisco State University
e-mail: sjk@sfsu.edu

von Krogh, Georg
University of St. Gallen
e-mail: georg.vonkrogh@unisg.ch

Magnusson, Mats G.
Chalmers University of Technology
e-mail: matmag@mot.chalmers.se

Newell, Sue
Bentley College
e-mail: snewell@bentley.edu

O'Brien, Jonathan P.
Krannert Graduate School of Management, Purdue University
e-mail: obrienjp@purdue.edu

Oblój, Krzysztof
Leon Kozminski Academy of Management and Entrepreneurship
School of Management, University of Warsaw
e-mail: kobloj@wspiz.edu.pl

Oriani, Raffaele
University of Bologna
e-mail: oriani@economia.unibo.it

Pratt, Michael G.
University of Illinois at Urbana-Champaign
e-mail: mpratt@uiuc.edu

Salmenkaita, Jukka-Pekka
Nokia Research Center
e-mail: Jukka-Pekka.Salmenkaita@hut.fi

Scarbrough, Harry
Leicester University
e-mail: hs28@le.ac.uk

Stewart, Alice C.
Guilford College
e-mail: astewart@guilford.edu

Swan, Jacky
Warwick Business School, University of Warwick
e-mail: J.A.Swan@warwick.ac.uk

Weigelt, Carmen
The Fuqua School of Business, Duke University
e-mail: cbw4@mail.duke.edu

West, Joel
College of Business, San José State University
e-mail: Joel.West@sjsu.edu

Zheng, Henry Y.
The Ohio State University Graduate School, The Ohio State University
e-mail: zheng.8@osu.edu

Introduction to Strategy in Transition

Richard A. Bettis

This collection represents the outstanding papers from the Strategic Management Society International Conference held in San Francisco in October of 2001. Over 500 academics, consultants, and executives from all over the world met for three days to discuss the state-of-the-art in strategy. This was an extraordinary level of attendance, given that the horrific events of September 11 had occurred only the previous month. The purpose of this introduction is to explain the overall context in which the papers were written and presented, and to engage the reader in some of the issues that engaged the conference participants.

It was obvious to attendees that strategy was in an intellectual and practical transition. Less than two years before the conference, the nature of the transition seemed different. It was a transition between the "old" economy and the "new" economy. The old economy was, well, old. It was bricks, tired old guys in suits, return on assets, and outdated strategic and economic concepts. The new economy was built around the Internet, exciting young men and women, revenue growth rates, new approaches to strategy, and ever-increasing stock values. It featured irreverent managers, enormous stock options, and "sandals and six pack" corporate cultures. It was inevitable. It was "cool." It was the future. We never got there. We arrived at a much different place. The future has a funny way with the inevitable.

The conference took place near Silicon Valley a few months after the burst of the "dotcom bubble" and only a few weeks after the catastrophic events of September 11, 2001. Throughout the conference there was a sense that the world had changed in ways that demanded a thorough reexamination of strategy. Certainly the dotcom debacle brought many aspects of recently received strategy into question. To have been featured by a cover article in the business press during the year or two before the bust seemed, in retrospect, to have been a leading indicator of strategic problems or failure. The queen of the new economy firms, Enron, was stumbling. As events would quickly prove, Enron, rather than a prototype for the new energy firm and industrial giant of the future, would become a sad monument to over-hyped strategy, financial manipulation, empty promises, and unethical leadership. By contrast, the petroleum dinosaur of the 1990s, ExxonMobil, was actually the energy company of the future. The world had turned upside down for a second time in just a few years.

Coming on the heels of the dotcom debacle, the events of September 11, 2001 raised further issues for strategy. The war the terrorists were intent on waging was apparently aimed at the industrial and industrializing countries of the world. The threat of terrorism was likely to take a significantly increased share of strategic attention and hard cash in many firms. War against an economic and social system is, after all, war against the firms that comprise the core of that system. Vulnerability of firm infrastructure was suddenly a huge strategic issue. Many firms depended on complex information systems that could be easily disrupted. Concentration of significant assets in vulnerable locations like the World Trade Center or even the greater New York City, London, or Tokyo area suddenly seemed strategically imprudent. Global supply chains seemed to be much more tenuous than ever thought. Visibility in host countries invited terrorist acts. The risk of terrorism that could significantly disrupt a major industrial economy on the scale of 9–11 was suddenly a possible event that commanded strategic scenarios and serious contingency planning.

New modes of thinking, new models, and new processes were much discussed and debated at the conference. At the core of these discussions were a set of outstanding papers and presentations, some of which are included in this volume. Subsequent events have further amplified a major conclusion of those assembled for the conference – that strategy was at an important breakpoint and that strategy in the twenty-first century will develop in some unexpected ways – following neither the supposed trajectory of the "old" nor "new" economy. This collection is part of the foundation on which the intellectual part of the transition of strategy can build.

Some Important Issues Discussed at the Conference

The purpose of this section is to review some of the important issues discussed at the conference. This list is not the result of any formal survey or vote. I alone decided which issues to include, although it reflects discussion with many others during and immediately after the conference. Other important issues were undoubtedly discussed that do not appear here. However, the list is not arbitrary. As Co-Chair of the conference along with Derek Dean of McKinsey, I was in attendance at all of the plenary sessions, attended numerous paper sessions, and engaged in many informal discussions. Furthermore, I read more than fifty of the papers that were presented. I also took notes throughout the conference. After the conference I asked several colleagues to summarize for me the most interesting discussions they heard at the conference. The selection of topics is subjective, but hopefully useful and reasonable.

I established two criteria for including an issue in this discussion: (1) I and at least one of my post-conference interlocutors had to directly observe it being energetically discussed in formal paper or plenary sessions *and/or* in informal discussions (the informal discussions seemed more energetic in general); and (2) it had to seem relevant to the transition of strategy that is the topic of this volume. There were many topics that fit the first condition, but did not seem to me to fit the second (e.g., diversification and performance). There were also some issues that seemed relevant to the second condition, but did not receive the energetic discussion required by the first condition (e.g., increasing movement of high value added jobs from developed to developing countries). It should be kept in mind that I bring the perspective of an academic. The papers in this volume were primarily written by academics, as were the aggregate of papers presented at the conference. However, the speakers at the plenary sessions were largely consultants and senior executives. My first condition above seemed to me to ensure that the practitioner perspective is heavily factored into the selection of topics.

I have tried to stay away from obscure-sounding academic terms and technical reference lists in characterizing topics. Instead, I have tried to frame the topics in terms of the general questions that were being widely discussed. The final selection of issues included four issues: (1) What are we really getting for all these stock options? (2) Why do we keep deceiving ourselves? (3) How important are network effects? (4) If we cannot trust the stock market, what can we trust? Perhaps the thing that most surprised me was not that these issues were new or completely different (they were not), but rather that practical and theoretical perspectives on them seemed to be in flux. Conclusions that had once seemed reasonable were eroding or gone. The way was clear to try new theories, concepts, or approaches.

In the end, this selection of issues is obviously idiosyncratic to some degree. I hope that readers will find the selection interesting. Ultimately, I hope these topics will stimulate thoughtful discussion and research.

What are we really getting for all of these stock options?

In the past two decades there has been an explosion in corporate governance research. This has occurred in economics, finance, and accounting in addition to strategy. It was surprising, then, how fresh the discussion of stock options seemed. In the 1990s there seemed no more broadly accepted principle of corporate governance than the use of stock options to align manager interests with strategy and ultimately with shareholder wealth maximization. This was simply a settled matter. Progressive firms offered significant stock options to a broad segment of employees, and huge options to top executives. This was the nature of alignment and motivation. Stock options did not merely help facilitate alignment and motivation. Stock options *were* alignment and motivation. At the conference, the abuse and failure of stock options were glaringly obvious. The efficacy of stock options was once again a subject of intense debate. The motivational value of large options was widely questioned, as was the potential dilution of earnings.

Why do we keep deceiving ourselves?

The study of cognitive influences on strategy has been a key area of interest for at least two decades. More specifically, the set of issues involving how managers and organizations continue on a set trajectory, even when confronted with severe disconfirming evidence, has been widely studied and discussed. Words such as routines, dominant logic, inertia, and groupthink come to mind. Participants at the conference had to come to terms with perhaps the most widespread and dramatic example of organizational- and self-deception in recent industrial history: the dotcom bubble. Strategies that, in retrospect, seemed totally illogical and even patently foolish had been pursued by many firms, touted in the business press, and taught in classrooms. Interesting and intense discussions around these issues seemed the most common topic of intellectual discourse at the conference.

How important are network effects?

If belief in the "new" economy was a religion, then network effects were its main deity. Other related concepts of prominence to the "new" economy included increasing returns, positive feedback, standards, and tipping markets. These are all important, powerful, and useful strategic concepts, but they had been dragged into the spiritual realm of unquestioned belief and universal application. The extreme idea that competition in network or network-related industries was always an all or nothing proposition, based on early and aggressive acquisition of market share, was the operative strategic principle for many managers. The widely quoted examples of this involved Microsoft, VCRs, and Apple Computer. Many questioned this extreme form of market determinism, but it still flourished in the years leading up to the conference.

By the time the conference started, it was becoming obvious that network effects were not the entire story, nor, in many cases, even the major plot line. There was a lot more to an effective strategy in network industries than simply exploiting network effects in a brute-force fashion. Coffee breaks at the conference resonated with related discussions.

If we cannot trust the stock market, what can we trust?

Nothing is more important to strategy than the creation of superior performance. Most managers think of performance in terms of profits or the ratio of profits to some measure of the resources used to generate the profits (e.g., invested capital or revenues). But measures involving profits are just a convenient surrogate for shareholder value creation. Ultimately it is stock price performance that is the goal of effective strategy. Superior strategies create considerable shareholder value. Inferior strategies do not. Stock price *is* the important measuring stick.

But then we had the huge run-up of many stock prices in the 1990s. The trend reached preposterous levels in the technology sector. The distortions were huge. Yahoo.com was worth $125.04 billion on January 3, 2000 (with 1999 revenues of $588.6 million, and 1999 net income of $61.1 million), while on the same date Ford was worth $59.55 billion (with 1999 revenues of $162.6 billion, and 1999 net income of $7.2 billion). Near the peak, firms with relatively modest profits were often selling at P/E multiples of over 100 and were using their high valuations to acquire other, more profitable firms. In fact, calculation of how fast and how long some firms would have to grow to rationally justify their valuations resulted in nonsensically large numbers. Under such circumstances, it was easy for strategy and resource allocation to become highly distorted.

Throughout the conference, discussions of how to think about the relationship between strategy and stock value started over breakfast and could be heard in restaurants into the late evening.

CHAPTER 1

Incumbent's Adaptation and Capability Building on the Internet: Empirical Tests of Innovativeness in the US Banking Industry

Carmen Weigelt

Keywords: Internet, capability building, innovation, adaptation.

Abstract

This paper applies an evolutionary economics and dynamic capability perspective to investigate the adaptation and innovation process incumbents experience when adopting the Internet. Distinguishing different levels of adoption and initial (path-dependent) from emerging capabilities, this study investigates the influence of different types of capability building on firm innovativeness. The degree of emerging capability development reflected in the strength of internal capability building versus external skill acquisition is found to affect a firm's innovativeness on the Internet, in addition to early choices often hypothesized to affect later adoption opportunities. Moreover, firms with higher innovativeness are also more likely to measure the performance of their Internet activities. The hypotheses are tested using survey data on Internet banking activities of 175 US banks.

Introduction

How do organizations adapt to technological change such as the Internet? The dynamic capability view suggests that firms adapt to technological change by recombining, integrating, and reconfiguring internal and external competencies (Teece et al., 1997). Firms innovate by changing and recombining established routines and skill repertoires (Schumpeter, 1934; Nelson and Winter, 1982). Such innovations can, for instance, occur through reconfiguration and redeployment of resources in acquisitions (Karim and Mitchell, 2000) or through transfer of internal capabilities such as R&D to related applications (Henderson and Cockburn, 1996; Helfat, 1997). Firms' past accumulated stock of knowledge and capabilities has shown to strongly affect the direction into which firms innovate and evolve (Mitchell, 1989; Klepper and Simons, 2000). How strong, however, is the effect of accumulated past experience and capabilities on the direction of technological change relative to current capability-building efforts? To what degree can a firm directly influence its current innovativeness and to what degree is it predetermined by its past history?

This paper studies how path-dependent effects and current capability-building efforts influence the adoption of innovations. Distinguishing between initial (path-dependent) and emerging capabilities (developed during the implementation process of an innovation), this study emphasizes that although adaptation processes are path dependent, firms can influence the direction of adaptation through current capability-building efforts. Consequently, current managerial and resource allocation decisions matter as much as those made in the past.

Moreover, prior research on technological innovation has struggled with the paradox of improved technological applications failing to create increased revenue streams for firms (Brynjolfsson and Hitt, 1996). Hence, higher firm innovativeness does not necessarily translate into improved performance unless firms can successfully commercialize their innovations. Therefore, it is important for firms to measure the performance of their innovations.

I study the effects of initial versus emerging capabilities on firm innovativeness using survey data on the Internet banking activities of 175 US banks in 1997 and 1998. The Internet has presented adaptation challenges for incumbents in many industries, repeatedly raising the question of how established firms can leverage their existing capabilities online while at the same time developing new ones. How much does path dependence matter? The data set provides an opportunity for testing drivers of early Internet adoption in banking. Moreover, the fast-paced Internet environment allows the studying of adoption dynamics across different levels of an innovation within short time periods of only a few years.

The paper proceeds in the following way: First, I provide an overview of prior research on innovation and dynamic capabilities to position the study. Second, I discuss the theoretical framework and hypotheses. Third, I present the data, variable measurements, statistical analysis, and results. The paper concludes with a discussion section.

Prior Research on Innovation

Prior research on innovation is extensive, ranging from diffusion studies at the population level to determinants of organizational innovativeness at the firm level and innovation processes within organizations (Wolfe, 1994). Rogers (1995) defines innovation as an "idea, practice, or object that is perceived as new by an individual or other unit of adoption" (11). While diffusion research has assessed the rates and patterns of innovation adoption at the population level (Rogers, 1995), studies on organizational innovativeness have focused on how organizational factors, particularly structure, influence a firm's propensity to innovate (Damanpour, 1991). Organizational innovativeness studies usually have a static orientation, focusing on adoption as an event rather than its implementation and assimilation as a process (Wolfe, 1994). With exception of a few studies (Meyer and Goes, 1988; Cooper and Zmud, 1990), research on post-adoption behavior has largely been unexplored. Moreover, dynamic aspects and changes in firm adoption behavior to evolving innovations have not been studied extensively.

Prior research in strategic management and industrial organization economics has discussed the importance of viewing innovations as interrelated processes rather than single isolated events (Vickers, 1986; Dosi, 1988; Reinganum, 1989). Innovations are seen as embedded within trajectories and relational context developing along paths determined by prior experience and local search (Dosi, 1982, 1988; Nelson and Winter, 1982). Technological trajectories reflect a firm's specific knowledge accumulation from local innovation search within its technological paradigm (Dosi, 1982). Local search often causes innovations to be recombinations of existing knowledge and practice (Schumpeter, 1934). Thus innovation is a continuous process involving product and process improvements, knowledge, and experience accumulation serving as platforms for future innovation (Kim and Kogut, 1996). Despite a broad theoretical discussion, empirical research on the evolution of innovation remains sparse.

This study views innovation as evolving through subsequent levels and building on prior related innovation experience and knowledge. For instance, it regards PC banking as prior related innovation and thus a platform for transactional Internet banking. Moreover, it contrasts the effects of prior related innovation and time spent experimenting with earlier levels on more advanced innovation outcome. Finally, this study contributes to the literature by contrasting the forces that support the transition of organizations from early-level innovations to more advanced ones.

Studying patterns of industrial innovation, Abernathy and Utterback (1978) investigate changes in the character of innovations as firms mature. They propose three stages of fluid, transitional, and specific patterns that organizations pass through as their organizational characteristics and methods of coordination change with increasing product and process standardization. Standardization occurs as innovation shifts from radical to evolutionary, driven by the emergence of a dominant design. Studying technological evolution as a process of long periods of incremental change punctuated by technological breakthroughs, Tushman and Anderson (1986) build on the dominant design concept and distinguish between competence-enhancing and competence-destroying innovations. Major technological innovations are at first usually experimental (Utterback and Suarez, 1993), and rather than becoming the standard they introduce a period of intensive technical variation and selection resulting in the emergence of a dominant design (Anderson and Tushman, 1990).

Rather than assuming a strict path-dependent process in innovation development through subsequent levels, this study investigates the emergence of dynamic capabilities during a firm's move from one innovation level to the next and their relative and absolute effect on the direction of the innovation. It contrasts the influence of prior related innovations and accumulated innovation experience with the intensity of capability-building efforts during the development of an innovation's next level. Finally, the study addresses how the way in which an innovation's next level is implemented (in-house vs. outsourcing) affects an innovation's outcome relative to path-dependent effects.

Prior Research on Dynamic Capabilities and Innovation

Research on dynamic capabilities examines the way in which firms' capabilities emerge, develop, and change over time (Helfat, 2000). While extensive research has been conducted on organizational behavior and structures generating effective product innovation (Henderson and Cockburn, 1994; Helfat, 1997), few papers have studied the sources of capability development in innovation (Cockburn et al., 2000). Furthermore, dynamic capability research has widely applied case study designs to investigate the capability renewal and adaptation of single firms to environmental change (Tripsas, 1997; Holbrook et al., 2000; Rosenbloom, 2000).

Dynamic capabilities enable organizational learning, resource recombination, and routine reconfiguration necessary for organizations to adapt to changing environments (Teece and Pisano, 1994). Dynamic capability research is strongly influenced by the concept of path-dependent, evolving, and cumulative processes (Nelson and Winter, 1982) causing organizations to evolve within paths and boundaries determined by past activities and experience. For instance, breadth and similarity of past capabilities and resources have been shown to predict a firm's direction of entry into new markets (Montgomery and Hariharan, 1991; Chang, 1996; Chang and Singh, 1999) as well as its entry timing (Mitchell, 1989; Schoenecker and Cooper, 1998). Studying the entry of radio producers into the US television receiver industry, Klepper and Simons (2000) stress the importance of pre-entry experience on post-entry behavior and performance. They find pre-entry experience and timing of entry to have persisting effects on firms' future course. Thus these papers focus on path-dependent effects not accounting for the possibility of pre-entry experience and capabilities being complemented or substituted by post-entry dynamics of capability development.

A major area of dynamic capability research has been the recombination and reconfiguration of resources and capabilities leading to innovation (Schumpeter, 1934). Studying R&D intensity at the program level across firms, Henderson and Cockburn (1994) found organizational competence to explain a significant fraction of firms' variance in research productivity. For example, Henderson and Cockburn (1996) and Helfat (1997) found support for capability transfer among related areas within corporations. In this context, Helfat (1997) stresses that a firm's efforts to alter its stock of knowledge in the face of environmental changes largely depend on its existing stock of complementary know-how. Moreover, Teece (1976) notes that a firm's ability to replicate and extend capabilities depends on the breadth and variety of past activities and their replication. Finally, Helfat and Raubitschek (2000) and Kim and Kogut (1996) emphasize path dependence in capability transfer and recombination by viewing organizations' product portfolios as platforms for future product sequences. While these studies view resource recombinations as firmly embedded and emerging from within a firm's past history, dynamic capability studies on acquisitions focus on the addition of external, not yet firm-embedded resources. Acquisitions are means by which firms redeploy and reconfigure their capabilities through retention, deletion, and addition of resources (Capron et al., 1998; Karim and Mitchell, 2000). Moreover, Karim and Mitchell (2000) showed that acquisitions both deepened a firm's existing resource base and extended it into new areas requiring different resources. Singh and Mitchell (1996) note that by using external sources firms can obtain technological skills and know-how resting outside their current capabilities. Is resource addition through acquisitions the only form by which firms can potentially redirect their path or can firms also engage in redirecting internal capability-building efforts?

Studying the adoption of 'science-driven' drug discovery, Cockburn et al. (2000) distinguish between initial conditions (Stinchcombe, 1965) and differences in the rate at which firms adopt. While acknowledging path-dependent effects in the form of initial conditions that create initial resource and capability differences among firms, the Cockburn et al. (2000) study focuses on capability convergence among firms over time. Their findings show that "Stinchcombe's hypothesis is powerful, but not all encompassing" (1141) since firms that were initially behind responded most aggressively and therefore "caught up" to their more advanced competitors. This finding raises the question of to what degree dynamic capability development in innovations is embedded and determined by path-dependent effects and to what degree it is not.

This study focuses on the sources and dynamics of capability development, distinguishing between emerging capabilities directly related to and developed during the innovation's implementation process and initial capabilities rooted in accumulated past firm experience and related innovations. Studying the adoption of transactional Internet banking, this study investigates the degree to which both types of capabilities affect firm innovativeness. Furthermore, I assess the degree to which emerging capabilities are affected by path-dependent forces, trying to answer the question of whether firms can redirect their innovation path through intensified efforts in emerging capability development.

Theory

Dynamic capabilities reflect a "firm's ability to integrate, build, and reconfigure internal and external competencies to address rapidly changing environments" (Teece et al., 1997: 516). The dynamic capability view stresses a firm's ability to renew and reconfigure its competencies and to respond to shifts in its business and technological environment. It also emphasizes managers' roles in "appropriately adapting, integrating, and re-configuring internal and external organizational skills, resources, and functional competencies toward changing environments" (Teece and Pisano, 1994: 538). For organizations facing innovation and rapid technological change, such as the Internet, successful alignment of resources and capabilities with altered environmental conditions as well as integration of existing with newly developed skills is paramount.

Dynamic capabilities are comprised of a firm's processes, position, and path, with path and position shaping organizational processes (Teece et al., 1997; Teece and Pisano, 1994). A firm's *path* shapes its processes through local search, creating new initiatives from past search and experience (Nelson and Winter, 1982). For instance, firms search for new technological solutions in areas enabling them to build on their existing technological base. Thus newly developed technical capabilities end up being close to previous technological accomplishments (Teece, 1996; Stuart and Podolny, 1996). Local search, which results in inertia and path-dependent behavior of organizations, also limits a firm's future options and strategic alternatives by restricting its potential search area and preventing organizational members from considering all possible options (Cyert and March, 1963; Nelson and Winter, 1982). Therefore, the broader a firm's past set of activities and routines, the wider its current choice set of alternatives.

Additional to a firm's path, its *position*, comprising among others current endowment of technology, intellectual property, and financial assets, shapes its processes (Teece et al., 1997). The dynamic capability view adopts a resource-based perspective (Penrose, 1959; Wernerfelt, 1984) by defining a firm's position by its internal resources instead of relative to and within its industry (Porter, 1980).

Organizational processes represent routines and "patterns of current learning and practice" (Teece et al., 1997: 518) embedding organizational capabilities (Nelson and Winter, 1982). Organizational routines are repeatedly invoked patterns of activity creating continuity (96). They are made up of subroutines and build upon routines themselves creating repertoires of skills and firm behavior, thereby causing the design, implementation, and replication of routines to be a central facet of dynamic capabilities research. Replication and recombination of existing routines and capabilities as well as the integration of new with old knowledge are highly innovative activities (Schumpeter, 1934).

Similar to routines, organizational processes become more refined and reliable through repeated application, causing intensified adoption of an innovation to result in more innovation-related capability building. The degree to which a firm builds innovation-related capabilities or adopts a new technology or application indicates its *innovativeness* measured by the number of innovations adopted or speed of adoption (Iansiti, 2000; Damanpour, 1991; Wolfe, 1994). Therefore, a higher level of *innovativeness* is associated with a higher level of adoption measured by the number of total and advanced services/products offered.

Rather than representing an end state, *dynamic capabilities* are antecedents, drivers, and outcomes of innovation processes and adaptation. I distinguish between *initial* and *emerging capabilities* as two types of *dynamic capabilities*. Since dynamic capabilities evolve over time through recombination and replication of existing ones (Teece et al., 1997), firm capabilities both exist prior to the adoption of the focal innovation (*initial capabilities*) as well as develop during the adoption process (*emerging capabilities*). Similar to Makadok (2001) distinguishing between the impact of resources in the decision phase and capabilities in the implementation phase, I view *initial capabilities* as impacting a firm's decision to adopt the next level of an innovation and its future development. Hence, *initial capabilities* comprise accumulated capabilities and resources from past experience and innovation.

If not too different, accumulated knowledge and skills from prior innovation can be transferred to new technologies. *Emerging capabilities*, on the other hand, impact the implementation and deployment of innovations. They develop in direct relation to a specific innovation rather than being replicated from prior innovation experience. While *initial capabilities* reflect a firm's path dependence and the influence of past activities on firm innovativeness, *emerging capabilities* reflect a firm's efforts particularly related to the implementation of a specific innovation. Although *initial capabilities* are expected to influence the direction of emerging capabilities through path dependence, *emerging capabilities* might be a major source of innovation for firms having engaged in little prior related innovation. Hence, the difference in strength between *initial* and *emerging capabilities* on firm innovativeness is of importance.

Hypotheses

The hypotheses distinguish between the effect of initial and emerging capabilities on innovativeness. Hypotheses 1.1 and 1.2 assess the influence of accumulated past experience and related innovation capabilities on both adoption and implementation of an innovation. Hypothesis 1.3 discusses emerging capabilities operationalized as the basis (in-house building of capabilities versus outsourcing) for providing the focal innovation and their effect on implementation scope of the innovation. Hypothesis 1.4 addresses the relationship between initial and emerging capabilities and compares their effects on the dependent variable. Finally, hypothesis 1.5 tests the relationship between scope of innovation implementation and performance measures.

Initial capabilities and innovativeness

Earlier in this study I defined *initial capabilities* as the effect of path dependence on innovation implementation and capability development. Technology and capability development often evolve along a path defined by a paradigm, a pre-specified pattern of solutions to selected problems (Dosi, 1982, 1988). Technological process builds on prior firm experience, activities, and learning. Organizational learning is history dependent, routine based, and cumulative (Levitt and March, 1988). Organizational behavior is based on routines that are refined and replicated over time, thereby becoming more reliable and embedded within the organization and its memory. Over time, routines form firm-specific sets of skills and capabilities called repertoires that are applied in a reinforcing manner and gradually become resistant to change (Nelson and Winter, 1982; Cyert and March, 1963). Local search and decision rules engrained in routines cause organizations to search for new knowledge and alternatives within the realm of their prior experience, thereby keeping the organization on a history-dependent path.

A firm's stock of knowledge and capabilities is built over time through consistent patterns of activity. Especially in the case of non-tradable and firm-specific capabilities, asset stocks have to be built over time (Dierickx and Cool, 1989). Organizational capabilities reflect "a firm's capacity to deploy resources, usually in combination, using organizational processes, to affect a desired end. They are information-based, tangible or intangible processes that are firm-specific and are developed over time through complex interactions among the firm's resources" (Amit and Schoemaker, 1993: 35). Hence, capabilities evolve in cumulative learning processes by adding incremental improvements to existing capabilities over time. Therefore, learning processes are facilitated by preexisting knowledge and become easier, the more related knowledge has already been acquired (Dierickx and Cool, 1989; Cohen and Levinthal, 1989; Levitt and March, 1988). Thus, it should be easier for firms having gained innovation-related experience to adopt the next level of an innovation than for firms lacking such experience. For instance, the longer a bank has operated and maintained a website, the more familiar it has become with its underlying technology and therefore the easier it should be for the bank to upgrade to transactional Internet banking. Moreover, the more familiar a bank is with website applications, the more likely it is to comprehensively adopt transactional Internet banking and experiment with innovations online. Hence, firms already possessing related knowledge, capabilities, and routines to build on learn faster and more easily than firms lacking such prior experience and accumulated knowledge (Cohen and Levinthal, 1990).

Hypothesis 1.1a: The more innovation-related experience a firm has accumulated over time, the more likely it is to adopt the next level of an innovation.

Hypothesis 1.1b: The more innovation-related experience a firm has accumulated over time, the higher its innovativeness.

Technological advances seldom stand alone, but are built on and facilitated by prior related technological developments (Teece, 1992). The ease with which firms can advance their technological capabilities depends on the characteristics of their existing knowledge and routines upon which the focal innovation builds as well as on the relatedness of existing to new technology (Cohen and Levinthal, 1989). New product or process development is likely to lie within the technological realm of a firm's previous applications and success (Teece, 1996). Henderson and Cockburn (1996) show that a firm's research efforts in different, but similar, areas are positively related. When altering their stock of knowledge in response to environmental and technological changes firms build on established stocks of know-how (Nelson and Winter, 1982; Helfat, 1997). Hence, innovations tend to be history dependent, with firms replicating past activities, resulting in solutions through incremental change and refinement of prior search patterns and routines.

Firms with a high level of past exploration are more likely to engage in future exploration due to their more extensive search behavior and more diverse routines to build on than firms that have mainly followed a course of exploitation in the past (Cyert and March, 1963; March, 1991). Thus the broader a firm's past set of capabilities and routines, the wider its current choice set of alternatives, and the more likely it is to adopt an innovation. For example, banks that engaged in PC banking broadened their set of capabilities through the introduction of an additional remote delivery channel to call centers and ATMs. The technological innovation of PC banking caused banks to explore with the PC as a means of remote delivery, a capability on which transactional Internet banking is built. Therefore, banks that engaged in PC banking possess a wider range of local search and capabilities and therefore are also more likely to adopt transactional Internet banking, the next level of the innovation.

> Hypothesis 1.2a: Firms having engaged in a prior level of an innovation are more likely to adopt the next level of an innovation.

It is crucial for innovative firms to be able to link innovations with existing capabilities and to replicate skills in new applications (Pisano, 1994). The complexity of an innovation and speed with which it is adopted depend on the complexity and relatedness of accumulated knowledge (Cohen and Levinthal, 1989). Prior related capabilities facilitate the assimilation and application of new knowledge, helping organizations to better understand the potential of innovations. PC banking, for instance, can be seen as an early technological stage of transactional Internet banking, similar in its functions offered, but differing in its mode of account access. While PC banking technology mainly used proprietary or prepackaged software together with direct dial access to the bank's intranet, transactional Internet banking can be conducted through any computer on the World Wide Web (Frei et al., 1998). The similarity in functionality offered between both technologies, such as account inquiry, funds transfer, and bill paying, should have helped banks applying PC banking to better understand the potential of transactional Internet banking. Hence, the complexity of an advanced innovation decreases with a firm's increase in early stage innovation-related experience. Such experience and better understanding of the innovation are expected to lead to a stronger implementation of the innovation's next level.

> Hypothesis 1.2b: Firms having engaged in a prior level of an innovation are more likely to be innovative at the next level of the innovation than firms not having engaged in a prior level.

In summary, these hypotheses assess the path-dependent effect of capability building and stress the building of capability development on prior related capabilities and long-term past experience. They imply that innovations are not isolated events, but build on one another, are interrelated, and interdependent.

Emerging capabilities and innovativeness

Besides *initial capabilities, emerging capabilities* developing in direct relation to the focal innovation affect the implementation of an innovation. *Emerging capabilities* represent innovative efforts a firm undertakes at the time when adopting the focal innovation and therefore assess the effect of current capability-building efforts on innovation implementation. What is the impact of emerging relative to path-dependent capabilities on innovativeness? If emerging capabilities significantly affect the scope of innovation implementation, then path-dependent effects are reduced and capability-building decisions made in face of an innovation translate into firm innovativeness. If, however, emerging capabilities are only of minor impact on the scope of innovation implementation, then past activities entirely determine the future, and firms neither having engaged in prior related innovations nor having accumulated experience early on will be disadvantaged in the long term. Therefore, organizations can only attempt to redirect their path traveled in the past, if emerging capabilities have a strong effect on innovativeness.

Since *emerging capabilities* develop in direct relation to the focal innovation, they can be viewed as the basis for an innovation's development and realization. More specifically, is an innovation developed in-house through internal capability building or outsourced?

Frequently, particularly in service industries, innovations such as the Internet are introduced from outside the industry through licensing and outsourcing arrangements rather than designed and developed within the sector itself (Pavitt, 1984). Thus firms often rely on other firms for technological know-how necessary to compete (Hagedoorn, 1993). However, in many cases firms still face the option of developing an innovation in-house with or without third-party help, using a packaged application for in-house implementation or of outsourcing development to an external party. For example, in the financial services industry several service providers, such as Corillian, provide their customers with both in-house and outsourcing options. Hence, a firm faces the choice between in-house development of capabilities and outsourcing, a decision that affects the extent of its capability-building efforts and therefore should also impact its scope of innovation implementation.

In-house development provides a greater potential for intensive technological capability building than outsourcing does. With in-house development accumulated capabilities can be extended or redirected more easily when new opportunities emerge (Leonard-Barton, 1995), such as new online functionality. Moreover, in-house development enables firms to more fully exploit capabilities because they are already integrated within the organization. Integration of existing with new technology and capabilities is crucial in order for firms to fully leverage their potential (Dierickx and Cool, 1989; Pisano, 1994). For example, it should be easier for banks having developed their transactional Internet banking functions in-house than for banks having chosen to outsource to add additional online functionality and offer advanced Internet banking applications. Dosi (1988) notes that "substantial in-house capacity is needed to recognize, evaluate, negotiate, and finally adapt the technology potentially available from others" (1132). Furthermore, Wheelwright and Clark (1992) stress the importance of internal capability development for firms to be able to evaluate and integrate external services into internal operations, especially in situations where technology is supplied from outside the industry. While some technologies, particularly codified knowledge, can be easily copied, others can only be "learned by doing" (Arrow, 1962) and understood through experimentation (Kogut and Zander, 1992).

Due to higher resource allocation costs associated with in-house capability building rather than outsourcing, firms choosing to build capabilities in-house are expected to be more committed to an innovation than firms selecting the outsourcing option. A higher commitment to an innovation is expected to result in a broader scope of implementation of the focal innovation.

> Hypothesis 1.3: Firms developing an innovation through internal capability building at the time of adoption will adopt higher levels of innovativeness than firms pursuing external development.

It is likely that a firm's decision to develop *emerging capabilities* in-house or through outsourcing is not made in isolation, but influenced by its prior innovation-related experience. Kogut and Zander (1992) argue that a firm's decision to develop an innovation in-house versus outsource is driven by the value contribution of its past capabilities to the focal innovation as well as its ability to learn the specific capabilities needed for the innovation. If routines of prior and focal innovation are similar, replication should be easy and in-house capability development preferred. Hence, the effectiveness of in-house capability building for future innovation seems to depend on whether or not a firm has participated in earlier generations of an innovation (Teece, 1992). Due to the challenges posed by integrating innovations with routine organizational operations (Burns and Stalker, 1961; Pisano, 1994), it is easier for firms having engaged in streams of prior related innovations to adopt advanced levels of innovation. Therefore, the effect of in-house capability building on scope of innovation implementation should be positively influenced by the amount of accumulated innovation experience. For instance, a bank's accumulated website experience should positively affect its ability to develop transactional Internet banking services in-house, thereby positively influencing the degree to which its transactional website is implemented.

> Hypothesis 1.4: The more innovation-related experience a firm has accumulated over time, the higher the level of innovativeness it will achieve through internal capability building.

In summary, these hypotheses assess the absolute and relative impact of emerging capabilities during capability-building processes. They address the relative importance and ability of current innovation efforts to redirect path-dependent processes by having a stronger impact on the scope of innovation implementation. In-house capability building is expected to affect innovation outcomes positively by enabling firms to absorb and develop knowledge and skills related to the innovation in greater depth than might be possible through outsourcing.

Innovativeness and performance

This study implies that the broader the scope of a firm's innovation implementation, the more dynamic capabilities the firm has developed and therefore the higher its adaptability. Innovation activity and firm performance have been found to be positively related, especially for early adopters (Schumpeter, 1934; Geroski et al., 1993; Roberts, 1999). For instance, Banbury and Mitchell (1995) find that incumbents' early incremental product innovations led to positive changes in market share. Hence, incremental innovation increasing the scope of a firm's innovation implementation is expected to positively affect its performance. However, during early stages of an innovation such as transactional Internet banking at the end of 1998, it is oftentimes difficult to assess an innovation's performance. At the beginning of 1998 less than 2 percent of national banks offered Internet banking.

Nevertheless, it is crucial for firms to monitor the market reactions to their innovative initiatives. Firms that monitor their innovation's performance are better positioned to react in case their innovation does not prove commercially successful. Superior technological know-how does not always result in market success (McGrath et al., 1992), which emphasizes the importance of being aware of customer responses to new products (Von Hippel, 1988). For example, some banks track their number of online customers, their allocation of website expenses, as well as having implemented a formal strategy for measuring the ROI of their web initiatives. The more a firm has implemented an innovation, the more likely it is to measure the innovation's performance due to its higher commitment and resource allocation to the innovation.

Hypothesis 1.5: The more a firm has implemented an innovation, the more likely it is to measure the innovation's performance.

Data and Methods

Data

I obtained data on transactional Internet banking activities from a survey administered by a market-research firm[1] between March and October 1999. The data was collected through telephone interviews with the most senior-level manager responsible for each institution's Internet banking program. At some institutions multiple callbacks were necessary to speak to the person in charge of Internet banking. The cross-sectional survey was conducted as a structured interview with all questions and probes written beforehand and asked in the same way to all respondents. The questions were close-ended with specific response options provided, including "refuse to answer" or "do not know."

Telephone surveys are more reliable than mailed questionnaires. Nevertheless, "surveys are limited by the amount and accuracy of the information that individual respondents can capably report" (Singleton and Straits, 1999: 248) about their group or organization. Although asking respondents about the past introduces bias into the data (Golden, 1992), Miller et al. (1997) point out that focusing respondents on concrete events or facts is less likely to generate bias. The current survey's retrospective part asked managers about their institution's Internet banking and PC banking activities as of year-end 1998 and 1997. Since the questions focused on specific events and facts such as "did your institution have a website at year-end 1997?" the response bias is expected to be rather low.

A total of 430 banks and thrifts[2] from a population[3] of 9,346 US banks and thrifts were interviewed. The survey tried to sample most of the larger banks with deposits over $4 billion. The selection of banks with deposits of less than $4 billion was based on a stratified random sampling approach. A random sample of firms ordered by deposit size, geographic region, and institution type was drawn from each stratum: (1) $1–4 billion, (2) $250 million to $1 billion and (3) less than $250 million. Two hundred and fifty-two out of the 430 banks interviewed had neither a PC direct-dial program nor a website for their retail or commercial customers, which reduced the usable sample for this study to 178 banks (40.70 percent). Out of those banks 73 had adopted transactional websites, 84 information-only websites and 21 only maintained PC direct-dial programs by year-end 1998. For 3 out of those 178 banks data was incomplete, which yielded a final sample of 175 banks (Table 1.1).

Table 1.1   Survey sample

Deposit size*

US banks & thrifts	Population	Sample
< $250 million	8097	238
$250 million–$1 billion	905	72
$1–4 billion	218	68
> $4 billion	126	52
Total	9346	430

Deposit size*	Total # of banks surveyed	Total # of banks on the Internet	%	Total # of banks with transactional capabilities
< $100 million	178	16	9.0%	0
$100 million–$500 million	109	38	34.9%	9
$500 million–$1 billion	23	16	69.6%	8
$1–4 billion	68	54	79.4%	19
> $4 billion	52	51	98.1%	37
Total	430	175	40.7%	73

*The three smaller strata size categories are different from those used to select the survey sample.

To assess the survey sample size relative to total Internet banking activity in 1998 and at the time of the interviews, I compared the survey sample to aggregated data from the Office of the Comptroller of the Currency (OCC) and the Federal Deposit Insurance Corporation (FDIC). As of June 1998 the OCC reported 223 transactional banking websites[4] and 68 banks that offered both transactional Internet banking and proprietary PC banking.[5] By September 1999 the number of transactional websites had increased to 541 (OCC). As of March 1999 the FDIC reported 490 banks and thrifts with transactional websites. A comparison between the aggregated FDIC data and the survey data on transactional Internet banking, both categorized by size, indicates that the survey data seems to be a good representative sample of larger banks (> $1 billion). Therefore, the data results and findings have more prescriptive value for larger than smaller banks, the latter of which are only poorly represented in the sample of 73 banks with transactional Internet bank activity (Table 1.2). In general, larger banks are more active online. While only 17 percent

Table 1.2 Comparison between survey data and OCC and FDIC data on transactional Internet banking activity

OCC and Survey Data

	# of banks with transactional Internet banking		
Asset/Deposit* Size	Survey data 1999	OCC June 1998	%
< $100 million	0	72	0.0%
$100 million–$1 billion	17	210	8.1%
> $1 billion	56	92	60.9%
Total	73	374	19.5%

*Asset size for OCC data; deposit size for survey data.
(Missing values reduced the final sample for analysis to 67 banks with transactional website.)

FDIC and Survey Data

	# of banks with transactional Internet banking			
Deposits	Survey data 1999	%	FDIC March 1999	Assets
< $100 million	0	0.0%	197	< $150 million
$100 million–$500 million	9	5.7%	158	$150 million–$500 million
$ 500 million–$1 billion	8	20.0%	40	$500 million–$1 billion
> $1 billion	56	58.9%	95	> $1 billion
Total	73	14.9%	490	

(Missing values reduced the final sample for analysis to 67 banks with transactional website.)

of banks with assets less than $100 million had transactional websites as of July 2000, 74 and 96 percent of banks with assets of $1–10 billion and more than $10 billion had transactional websites (OCC).

Since banks are not required to report their proprietary PC banking or Internet activities, public records do not exist, making surveys and interviews the preferred way of acquiring information on banks' Internet activities. Press releases only cover selected Internet banking initiatives for larger banks (> $20 billion in assets) and annual report coverage of Internet banking activities varies widely among banks.

Measures

Dependent variables

Transactional Internet banking 1998 is measured as a dichotomous variable of 1 for banks with transactional websites at year-end 1998 and 0 otherwise. In order for a bank's website to qualify as transactional it had to at least offer account inquiry, account access, and funds transfer between existing accounts. Following the definition of the OCC, this study defines Internet banks as all banks, not just Internet-only banks, having transactional websites.

Innovativeness is operationalized as the following two dependent variables:

1 A count variable of each bank's total number of transactional online services offered at year-end 1998. The survey provided banks with a list of 14 transactional services and asked whether or not they offered each function on their transactional website at year-end 1998. The variable ranges from 2 to 12 transactional online services offered by the same bank (Table 1.3). Prior research has widely operationalized innovations as count variable (Wolfe, 1994; Damanpour, 1991; Damanpour and Evan, 1984). Studies only using a dichotomous variable to measure the adoption of an innovation ignore the fact that organizations often adopt several innovations as well as different levels of an innovation.

2 A count variable of each bank's more advanced transactional online services offered at year-end 1998. Five of the 14 transactional online services were identified as more advanced services as of year-end 1998. (1) View bill data from outside billers; (2) receive real-time approval or rejection for loan, mortgage, or credit applications; (3) purchase and trade mutual funds, stocks, bonds, and other securities; (4) file taxes; and (5) enable OFX transactions. Less than 20 percent of banks had adopted those transactional services as of year-end 1998, indicating their advanced nature. This measure was used as dependent variable to account for the fact that innovations are not homogeneous (Downs and Mohr, 1976), an assumption implicitly made when cumulating innovations. Prior research has distinguished between different types of innovation by calculating the percentage of specific relative to all innovations adopted (Damanpour and Evan, 1984) or by creating a dichotomous variable based on innovation types. Furst et al. (2000), for instance, consider websites with three or fewer online services to be basic and those with more than three to be premium (Table 1.3).

Table 1.3 Innovativeness – Service listings

Transactional Internet banking functions	# of banks with following functions	Total # of Internet functions	Total # of banks	Advanced Internet functions	Total # of banks
Download account balance into client software	50	2	5	0	27
Transfer funds between existing accounts	64	3	2	1	24
Pay bills	53	4	1	2	7
View bill data from outside billers***	11	5	10	3	9
Open new deposit accounts	32	6	15		
View credit card balance	26	7	13		
View balance on existing loan products	48	8	6		
Fill out application for a loan/mortgage/or credit product	43	9	4		
Receive real-time approval or rejection for loan, mortgage, or credit application***	7	10	5		
Purchase & trade mutual funds, stocks, bonds and other securities***	20	11	4		
E-mail a customer service inquiry	62	12	2		
File taxes***	15				
Receive targeted marketing messages	24				
Enable OFX transactions***	13				

*** transactional functions identified as "advanced services"

Independent variables

PC Banking 1997 is measured as a dichotomous variable with a value of 1 for banks that had a PC direct-dial program for their retail customers at year-end 1997 and with a value of 0 otherwise. PC banking 1997 measures the effect of early learning experience from related innovations on the adoption and scope of more advanced innovations. Although PC banking programs require customers to install proprietary or prepackaged software, e.g., Microsoft Money or Inuit's Quicken, on their PC in order to dial into the bank's intranet, thereby significantly differing from Internet banking access, PC banking functions comprise many of the transactional banking services offered online. Such services are, for example, account inquiry, transfer, and bill pay.

Number of years on the web is a continuous variable measured as the total number of years a bank had a website by year-end 1998. It assesses the effect of cumulative innovation-related experience and learning on the adoption and scope of Internet banking. Number of years has been a widely used measure for cumulative experience in strategic management (Klepper and Simons, 2000).

In-house vs. outsourcing is measured as a dichotomous variable with a value of 1 for banks whose transactional website was primarily maintained and operated in-house at the time of adoption and with a value of 0 for banks that outsourced their transactional website to an external service provider. This variable assesses the importance of in-house capabilities for implementation of an innovation.

Transactional website development is classified into two categories: While front-end represents the web interface the customer encounters and uses to navigate the site, transactional infrastructure is the application layer enabling Internet transaction and integrating front-end with back-end core processing systems. Both variables measure the degree to which a bank (1) manages the application in-house through internal development or (2) outsources at the time of adoption. The more a bank operates an application in-house using in-house staff, the more likely it is to build capabilities and skills related to specific applications.

Front-end/customer interface development is coded as two dummy variables where the first dummy variable represents a front-end application built in-house with the aid of a third-party consultant or system integrator. The second dummy variable represents a front-end application built in-house using in-house staff. Both dummy variables are compared to the outsourcing option.

Transactional infrastructure development is coded as two dummy variables where the first dummy variable is a transactional infrastructure built in-house with the aid of a third-party consultant/system integrator or based on a packaged application. The other dummy variable represents an infrastructure built in-house by in-house staff. Both dummy variables are contrasted to the outsourcing option.

Dummy 97 is a dummy variable distinguishing between banks that had adopted transactional Internet banking by year-end 1997 and banks that first adopted transactional Internet banking during 1998.

The following variables relate to the tracking of online performance, costs, and profitability at year-end 1998. They test whether scope and advanced implementation of online services is positively related to tracking online performance.

Tracking of retail customers is measured as a dichotomous variable of 1, if banks track the number of retail customers using their transactional web services, and 0 otherwise.

Tracking of allocation of costs is measured as a dichotomous variable of 1 for banks tracking the allocation of their website development and maintenance costs across categories of hardware, software, services, and telecommunications and 0 otherwise.

Strategy for measuring ROI is measured as a dichotomous variable of 1, if banks had a formal strategy for measuring the return on investment (ROI) for their web initiatives.

Control variables
All control variables are measured at year-end 1998.

Member of a multi-bank holding company is measured as a dichotomous variable of 1 for banks that were members and 0 otherwise. Furst et al. (2000) found that banks that were members of multi-bank holding companies were more likely to offer Internet banking since they could provide services to multiple subsidiaries using a single website.

Bank size is measured as a rank order variable of 1 to 5 since the survey data only provided deposit size categories. (1) < \$100 million, (2) \$100–500 million, (3) \$500 million–1 billion, (4) \$1–4 billion, and (5) > \$4 billion. Bank size has been shown to positively affect a bank's propensity to adopt Internet banking (Furst et al., 2000). It is also a widely used control variable in strategy research.

Customer focus is measured as a dichotomous variable of 1 for banks designing their transactional services for both retail and commercial customers and as a variable of 0 for banks primarily focusing on retail customers. A firm's customer focus has been shown to influence both its offerings and strategic focus (Christensen and Bower, 1996).

Statistical analysis

I estimated a logistic regression (Pindyck and Rubinfeld, 1981; Greene, 2000) to assess the path-dependent effects of prior cumulated and related innovation experience on next-level innovation adoption. Since the dependent variable is dichotomous, OLS or GLS regression models could not be used. Hence, I estimated a logit model using a maximum-likelihood estimation procedure. The model log likelihood chi-square, which is comparable to an *F*-test in the OLS regression, indicated that the logit model was significant. A two-tailed significance test of the Wald statistic was conducted for each estimate in the model.

I estimated a Poisson regression model (Greene, 2000; Barron, 1992) to analyze path-dependent and current effects on number of total and advanced online services implemented. Since the two dependent variables are count variables, GLS or OLS regression models are inappropriate. A widely used technique for modeling count data (non-negative integers) is Poisson regression. When using Poisson regression models the data has to be tested for two restrictive assumptions of the Poisson distribution: independence of events over time and equality of mean and variance of expected event counts. Overdispersion exists in the data, if the variance of event counts exceeds the mean. This might occur due to unobserved heterogeneity in events or positive contagion (Barron, 1992). Furthermore, the assumption of independence of events can be violated due to prior events affecting the probability of subsequent events (Hausman et al., 1984; Cameron and Trivedi, 1986). Therefore, I used a Lagrange Multiplier Test comparing the fit of the Poisson model to that of the negative binomial model, correcting for overdispersion (Greene, 2000; Cameron and Trivedi, 1986). In this test the Poisson regression is modeled as a parametric restriction of the negative binomial model. The Poisson model is accepted, if the limiting distribution of the Lagrange multiplier statistic is chi-squared with one degree of freedom. Since the values of the statistic range from 0.8788 to 0.9563 for the models with total number of services as dependent variable and from 0.5875 to 0.9965 for the models with number of advanced services as dependent variable, the Poisson model is appropriate and therefore used in this study.

Results

The logistic regression tests hypotheses 1.1a and 1.2a. Results for the logistic regression analysis are reported for a sample of 154 banks that had not yet adopted transactional Internet banking by year-end 1997. Out of those 154 banks, 53 adopted transactional Internet banking during 1998. Twenty-one banks out of the original sample of 175 banks had adopted transactional Internet banking by year-end 1997.

The correlation matrix shows significant correlation between the dependent variable (adoption = 1; no adoption = 0) and the independent variables of interest (Table 1.4).

Hypothesis 1.1a states that the more innovation-related experience a firm has accumulated over time, the more likely it is to adopt the next level of an innovation. Innovation-related experience was measured as the time spent experimenting with the innovation, more specifically the number of years a bank had been on the web. The parameter estimate for innovation-related experience was positive and significant in models 2 and 3 ($p < 0.01$ and $p < 0.05$ respectively). This finding indicates that the more years a bank had already spent on the web, the more likely it was to adopt transactional Internet banking, the next level of the innovation. Thus hypothesis 1.1a was supported.

Table 1.4   Descriptive statistics and correlation matrix

Variable	Mean	S.D.	1	2	3	4	5
Transactional website 1998	0.44	0.99	1				
Member of Multi-BHC	0.27	0.45	0.17**	1			
Bank Size	3.33	1.35	0.21***	0.33****	1		
Years on the web *(H1.1a)*	1.52	1.29	0.09	0.15**	0.46****	1	
PC Banking 97 *(H1.2a)*	0.27	0.44	0.16**	0.32****	0.43****	0.40****	1

$N = 154$
*$p < 0.10$; **$p < 0.05$; ***$p < 0.01$; **** $p < 0.001$.

Table 1.5   Logistic regression analysis. Adoption of transactional Internet banking

	Model 1 Control	Model 2 Path-Dependence	Model 3 Full Model
Constant	−2.62****	−1.43***	−2.66****
Member of multi-BHC	0.81**		0.82**
Bank size	0.49****		0.38**
Years on the web *(H1.1a)*		0.38***	0.27**
PC banking 97 *(H1.2a)*		0.57	0.03
−2 log likelihood	176.54	184.84	173.51
Chi-square	21.74	13.44	24.77
N	154	154	154

$*p < 0.10; **p < 0.05; ***p < 0.01; **** p < 0.001.$

Hypothesis 1.2a states that firms that engaged in a prior level of an innovation are more likely to adopt the next level of an innovation. The major prior level of innovation to transactional Internet banking is PC banking. In contrast to the other two major remote delivery channels, the telephone and ATM, PC banking, similar to transactional Internet banking, uses the PC as access medium (Frei et al., 1998). Although PC banking requires customers to install proprietary bank dial-up services or prepackaged finance software on their PC prior to dialing into the bank's intranet, many of the functions introduced on a bank's transactional website had already been offered through PC banking, thus representing a prior learning experience. PC banking is one-year lagged to ensure the causality of PC banking affecting transactional Internet banking as of year-end 1998. The parameter estimate of PC banking was not significant for both models 2 and 3, indicating that banks with PC banking operations in 1997 were not more likely to adopt transactional Internet banking by year-end 1998. Thus hypothesis 1.2a was rejected (Table 1.5). The correlation of $r = 0.4$ between PC banking and years on the web is significant ($p < 0.001$), indicating that the presence of years on the web in the model might be the cause for PC banking not being significant.

The control variable bank size was significant and positive ($p < 0.05$), a finding consistent with prior research (Furst et al., 2000) observing that size positively influenced the likelihood of innovation adoption. The control variable member of multi-holding company was also significant ($p < 0.05$).

The Poisson regression models test hypotheses 1.1b and 1.2b through 1.4. Results for the Poisson regression analysis are reported for a sub-sample of 67 banks, out of which 20 had adopted transactional Internet banking by year-end 1997 and 47 by year-end 1998. The sample of 53 banks adopting transactional Internet banking during 1998 was reduced to 47 banks due to missing values. Equally, the sample of 21 banks having adopted transactional Internet banking by year-end 1997 was reduced to 20 due to missing values. I estimated separate Poisson regression models for number of online services and number of advanced online services implemented.

The correlation matrix shows significant correlation between the dependent variables and the independent variables of interest (Table 1.6).

Table 1.6 Descriptive statistics and correlation matrix

Variable	Mean	S.D.	1	2	3	4	5	6	7	8	9	10	11	12	13	14	15	16
# of online services	6.76	2.46	1.00															
# of advanced online services	0.97	1.03	0.79****	1.00														
Member of multi-BHC	0.43	0.5	0.22*	0.29***	1.00													
Bank size	4.1	1.06	0.36***	0.48****	0.43****	1.00												
Customer focus	0.49	0.5	0.17	0.06	−0.20	−0.04	1.00											
Years on the web (H1.1b)	2.38	1.24	0.22*	0.24**	0.05	0.45****	0.01	1.00										
PC banking 97 (H1.2b)	0.42	0.5	0.39****	0.49****	0.24**	0.43****	−0.17	0.28**	1.00									
In-house vs. outsourcing (H1.3)	0.42	0.5	0.34****	0.44****	0.05	0.38****	0.14	0.38****	0.20*	1.00								
Adoption of online banking 97	0.3	0.46	0.37****	0.37****	0.02	0.37****	0.27**	0.46****	0.17	0.24**	1.00							
Front-end/customer interface with third-party help (H1.3)	0.21	0.41	0.16	0.19	−0.01	0.33***	0.23*	0.19	0.01	0.61****	0.15	1.00						
Front-end/customer interface all in-house (H1.3)	0.16	0.37	0.27**	0.33***	0.02	0.15	−0.03	0.32***	0.11	0.52****	0.24**	−0.23*	1.00					
Transactional infrastructure with third-party help (H1.3)	0.24	0.43	0.23*	0.19	0.08	0.28**	0.08	0.19	0.16	0.66****	0.02	0.57****	0.13	1.00				
Transactional infrastructure all in-house (H1.3)	0.13	0.34	0.16	0.27**	−0.08	0.13	0.14	0.27**	0.2	0.46****	0.32***	0.01	0.53****	−0.22*	1.00			
Tracking of retail customers (H1.5)	0.84	0.37	0.15	0.22*	0.01	0.06	0.01	−0.07	0.18	0.26**	0.14	0.21*	0.07	0.23*	0.04	1.00		
Tracking of allocation of costs (H1.5)	0.54	0.50	0.31***	0.24**	0.06	0.16	0.24*	0.10	0.23*	0.36***	0.17	0.29**	0.05	0.26**	0.07	0.38****	1.00	
Strategy for measuring ROI (H1.5)	0.46	0.50	0.31**	0.26**	0.07	0.32****	0.08	0.17	0.29**	0.29**	0.15	0.20	0.21	0.12	0.13	0.05	0.21*	1.00

$N = 67$.

* $p < 0.10$; ** $p < 0.05$; *** $p < 0.01$; **** $p < 0.001$.

Hypothesis 1.1b states that the more innovation-related experience a firm has accumulated over time, the higher its innovativeness. The estimate for years on the web was positive, but not significant in model 2, indicating that experience accumulated over time did not affect the number of online services a bank implemented. The estimate for the direct effect of years on the web on number of advanced online services was also not significant in model 7 (Table 1.7). Hence hypothesis 1.1b is rejected.

Hypothesis 1.2b states that firms having engaged in a prior level of innovation are more likely to be innovative than firms not having engaged in a prior level of innovation. The estimate for PC banking in 1997 was significant and positive ($p < 0.05$ and $p < 0.01$ respectively) for both models 2 and 7, showing that engagement in prior related experience positively increased the scope of a bank's online service offerings. Hence, hypothesis 1.2b was supported.

Hypothesis 1.3 states that firms developing an innovation through internal capability building at the time of adoption will adopt higher levels of innovativeness than firms pursuing external development. The estimate for in-house vs. outsourcing was significant and positive ($p = 0.05$) in model 7, but not in model 3. Hence, the effect of in-house vs. outsourcing is stronger on the number of advanced services a bank offers than on its total number of online services.

To assess the effect of in-house vs. outsourcing in greater detail, dummy variables were created distinguishing between front-end/customer interface development and transactional infrastructure development. The dummy variables front-end/customer interface development were (1) built in-house with third-party help or based on packaged application and (2) built in-house with in-house staff, (3) the constant representing outsourcing. In models 5 and 10 the dummy variable representing "built in-house with in-house staff" was significant and positive ($p < 0.05$ and $p < 0.10$ respectively), reflecting a significant positive effect on number of online services offered and an even stronger effect on advanced online services. Thus building capabilities in-house with in-house staff significantly differed in its effect from outsourcing, whereas using a third party's help or a prepackaged application for in-house implementation did not significantly differ from the outsourcing option.

I coded the dummy variables transactional infrastructure development identical to those for the front-end/customer interface development. While none of the dummy variables was significant for the dependent variable number of online services in model 4, the dummy variable representing "built in-house with in-house staff" was positive and significant ($p < 0.05$) in model 9. Thus building capabilities in-house with in-house staff significantly differed from the other two options in its effect on advanced services as well as having a stronger effect on the dependent variable. Therefore, hypothesis 1.3 was accepted. This effect might be due to the data set capturing an innovation, the Internet, in its pioneering stage. Once an innovation becomes more established and knowledge diffuses, partnering with external providers and vendors might become a more dominant solution.

Hypothesis 1.4 states that the more innovation-related experience a firm has accumulated over time, the higher the level of innovativeness it will achieve through internal capability building. The hypothesis was tested with an interaction term between in-house vs. outsourcing and number of years on the web that was positive

Table 1.7 Poisson regression analysis: innovativeness of transactional Internet banking

	Model						Model			
	Number of online services						Advanced online services			
	1	2	3	4	5	6	7	8	9	10
Constant	1.42****	1.55****	1.58****	1.55****	1.56****	2.82****	2.13***	1.94**	2.22***	2.09***
Member of multi-BHC	0.11	0.11	0.10	0.12	0.12	0.27	0.30	0.28	0.37	0.33
Bank size	0.08	0.02	0.05	0.02	0.02	0.56***	0.26	0.37*	0.29	0.26
Customer focus	0.10	0.12	0.14	0.12	0.14	0.07	0.19	0.28	0.26	0.24
Years on the web (H1.1b)		0.02	0.09	0.02	0.03		0.11	0.43**	0.13	0.12
PC banking 97 (H1.2b)		0.22**	0.23**	0.23**	0.24*		0.79***	0.78***	0.87***	0.84***
In-house vs. outsourcing (H1.3)		0.15	0.14				0.66**	0.39		
Interaction in-house vs. outsourcing * Years on the web (H1.4)			0.12					0.43*		
Dummy 97	0.18	0.18	0.15	0.15	0.15	0.42	0.49	0.36	0.45	0.39
Transactional infrastructure (H1.3) with third-party help				0.15					0.46	
Transactional Infrastructure (H1.3) all in-house				0.14					0.76**	
Front-end/customer interface (H1.3) with third-party help					0.13					0.53
Front-end/customer interface (H1.3) all in-house					0.23*					0.76**
Log likelihood	−148.54	−145.18	−144.37	−145.24	−144.75	−76.07	−69.93	−68.51	−70.30	−69.94
R-square	0.04	0.06	0.07	0.06	0.07	0.14	0.20	0.22	0.20	0.21
N	67	67	67	67	67	67	67	67	67	67

$* p < 0.10$; $** p < 0.05$; $*** p < 0.01$; $**** p < 0.001$.

and significant ($p < 0.10$) in model 8. Introducing the interaction into the full model turned the parameter for years on the web significant ($p < 0.05$) in model 8. This finding shows that the effect of internal capability building on advanced services varied as a function of the firm's innovation-related experience accumulated over time. The interaction term was not significant in model 3 testing its effect on number of online services offered. Thus hypothesis 1.4 is only partially supported.

Hypotheses 1.3 and 1.4 assessed the absolute and relative effect of initial versus emerging capabilities on innovation implementation. While the coefficient for PC banking (initial capability) was significant at $p < 0.05$ and $p < 0.01$ in models 2 and 7 respectively, the coefficient for in-house vs. outsourcing (emerging capability) was only significant in model 7 ($p < 0.05$). Hence, the path-dependent effect of PC banking was in fact stronger than the effect of emerging capabilities operationalized as in-house development vs. outsourcing. This observation might be due to the fact that transactional Internet banking was still a novel application at year-end 1998 as well as due to the fact that the effect of in-house development requires a longer time to fully materialize and surpass the effect of path dependence.

The lack of significance of the size variable across the models might result from the fact that more than 80 percent of banks in the sub-sample were larger than $1 billion and thus part of only two out of five size categories. Moreover, the lack of significance might be due to the fact that bank size was measured as a rank order variable because of lack of information on actual bank deposits.

Hypothesis 1.5 states that the more a firm has implemented an innovation, the more likely it is to measure its innovation's performance. This hypothesis was tested with correlation analysis. The correlation was significant for the variables "tracking of allocation of costs" ($r = 0.305$, $p < 0.01$, and $r = 0.242$, $p < 0.05$) and "strategy for measuring ROI" ($r = 0.301$, $p < 0.05$ and $r = 0.261$, $p < 0.05$) with both number of online services and advanced online services. The correlation between "tracking of retail customers" and number of online services was, however, not significant ($r = 0.149$, $p = 0.25$). These findings indicate that a broader scope of innovation implementation was positively correlated with a bank's use of performance measures. Therefore, hypothesis 1.5 was supported (Table 1.6).

Discussion and Conclusion

This study investigated how path-dependent effects and current capability-building efforts influenced the adoption and implementation of innovation, using survey data on transactional Internet banking activities. It distinguished between initial and emerging capabilities to assess the absolute and relative strength with which current capability-building efforts affect innovation implementation relative to initial capabilities. Is firm innovativeness predetermined by past activities and choices, or can current decisions and efforts influence and redirect it? The data set on Internet banking allowed testing for the likelihood of transactional Internet banking adoption based on past IT-related experience measured as web and PC banking experience. Differential effects of past IT-related experience (initial capabilities) and current capability-building efforts (emerging capabilities) on scope of transactional services offered were tested for a sub-sample of 67 banks.

This study found that despite a strong effect of initial capabilities on scope of innovation implementation, emerging capabilities mattered. The effect of emerging capabilities measured as in-house vs. outsourcing was stronger for advanced services than online services in general. Hence, in-house operations, which are associated with a higher level of capability building and development, resulted in a higher level of advanced service adoption. The findings show that current capability-building efforts of banks positively affect the scope of an innovation's implementation, providing support for the assumption that current decisions influence innovation processes.

Distinguishing in-house operations into "developed in-house using in-house staff" and "built in-house with the aid of a third party" showed that only the option "developed in-house using in-house staff" significantly and positively differed from outsourcing. Therefore, banks using a third party for in-house development seem not to have gained a significant advantage over outsourcing. However, in-house development with in-house staff seems to have led to a broader scope of innovation implementation. This finding was confirmed for both front-end customer interface development and transactional infrastructure development, although it was stronger for the first one and stronger for advanced than total number of services. These findings imply that the simple distinction between in-house and outsourcing is too crude and that different degrees of in-house development should be considered (Loh and Venkatraman, 1992). Kogut and Zander (1992) stress that some technologies can only be understood through experimentation and that in-house capability development leads to better knowledge integration. The lack of significant differences between "in-house development using a third party" and "outsourcing" indicates that third-party development and prepackaged applications do not provide a bank with competitive advantage relative to the outsourcing option. Hence, third-party implementation arrangements might not enable a bank to engage in learning and integration processes similar to those occurring with in-house implementation through in-house staff. Furthermore, prepackaged applications from third parties that can be easily purchased in the market and are therefore not rare, unique, or inimitable cannot provide firms with competitive advantage (Barney, 1991). However, these findings might be due to the fact that the data set covers an early period of transactional Internet adoption. Early movers adopt while innovations are still premature and evolving. Early adopters often face suppliers that are still quite inexperienced themselves with the innovation and therefore early adopters might be better off managing the development of the innovation in-house rather than outsourcing it or using third-party help. Once an innovation has further evolved and knowledge about it has diffused, third-party help in the implementation process might actually provide a superior solution to in-house implementation, a relationship future research could further investigate.

The study found that emerging capabilities built on initial capabilities, thus being influenced by path-dependent effects. The positive interaction term between related prior innovation (PC banking) and in-house operation implies that banks having engaged in PC banking are more likely to have advanced services among their transactional Internet banking applications. Comparing the strength of initial versus emerging capabilities, it becomes clear that path-dependent capabilities are dominating emerging capabilities. Prior related innovations (PC banking) positively affect innovation implementation, whereas years of accumulated prior experience (years on the web) positively affect adoption. These findings show that benefits accrued to banks that early on engaged in innovation-related learning processes as well as that experience evolved over time. However, in order to assess the longer-lasting effects of emerging capabilities and their changes in strength relative to initial capabilities, a multi-period longitudinal study is necessary.

Finally, the findings show that firms that achieved a higher level of innovation implementation were also more likely to measure their innovation's performance.

An important managerial implication of this study is that firms can "catch up" through intensive capability-building efforts in face of innovations. Thus firms that missed out on early innovation learning experience can still achieve scope and implement advanced services at the next level of the innovation through intensive current capability development. Therefore, firms can redirect paths and laggards are not doomed to stay laggards forever. Nevertheless, there is a strong path-dependent effect in innovation processes which emphasizes that learning from earlier innovations is important as well as that innovations build on innovations and therefore are not isolated, but rather interrelated processes. The findings show that PC banking, whose benefits were unclear at the time of its implementation, had a strong learning impact on later-level adoption such as transactional Internet banking. Hence, benefits that accrue to organizational adaptation from early engagement and experimentation with innovation often do not directly pay off, but yield positive returns at a much later stage in the innovation process. Therefore, investments in innovation present learning opportunities that firms should engage in even though benefits might not occur instantly.

Moreover, the finding that in-house vs. outsourcing differently affects innovation indicates that how firms organize their operations for innovation matters and has important implications. Considering the finding that in-house implementation through third parties is not significantly different from outsourcing, it is important to note that these findings are based on data from 1998 when transactional Internet banking was a "pioneering" and not yet widely adopted innovation within the banking industry. Recently, for instance, Huntington Bancshares switched from S1 to Corillian because the latter Internet service provider allowed the bank to manage its Internet platform in-house, indicating that outsourcing vs. in-house implementation through third parties represented two different options for Huntington Bancshares. Overall, however, the findings show that the more online functions a firm implements, the more important are well-integrated capabilities.

Two major limitations of this study are its cross-sectional design and small sample size. Whether in-house capability building or outsourcing is more beneficial in the long term for innovation development processes can only be assessed in a longitudinal study. While in-house development enables firms to better integrate and add additional functionality, it might also introduce lock-ins and rigidities in rapidly changing technological environments. Moreover, the effects of emerging and path-dependent capabilities on innovation might change in strength over time, causing emerging capabilities to supersede the effect of initial capabilities. Furthermore, future research could study the differences between earlier and later adopters in initial and emerging capabilities and their effect on innovation. Additionally, sequences of innovation are best studied in a longitudinal framework.

Future research could also look at how firms organize for IT innovations and build capabilities in-house. How does third-party in-house implementation differ from in-house implementation through in-house staff? Do the significant effects found in this study hold over time in a longitudinal framework? Finally, future research could study whether and how the scope to which an innovation is implemented influences firm profitability and performance.

Notes

1   I received the survey questionnaire and response data at the bank level. For confidentiality reasons, the banks' identity was not revealed to me. Thus, I was unable to collect additional data on the banks responding to the survey.
2   The survey data does not distinguish between banks and thrifts. I use the term banks for both banking and thrift institutions in this paper.
3   The population listings are based on the 1998 Third Quarter Thompson/Prospector Bank Peer Group Database and 1998 Third Quarter Thompson/Prospector S&Ls of the Nation.
4   The number of transactional websites is less than the total number of banks on the web because some bank subsidiaries provide their customers access through their parent company's website.
5   Proprietary PC banking and Internet banking are two forms of "remote" banking. Two other widely used forms of remote banking are ATMs and telephone call centers.

References

Abernathy, W. J. and Utterback, J. M. 1978: Patterns of industrial innovation. *Technology Review*, 80, 40–7.

Amit, R. and Schoemaker, P. J. H. 1993: Strategic assets and organizational rent. *Strategic Management Journal*, 14, 33–46.

Anderson, P. and Tushman, M. L. 1990: Technological discontinuities and dominant designs: a cyclical model of technological change. *Administrative Science Quarterly*, 35, 604–33.

Arrow, K. 1962: The economic implications of learning by doing. *Review of Economic Studies*, 29, 155–73.

Banbury, C. M. and Mitchell, W. 1995: The effect of introducing important incremental innovations on market share and business survival. *Strategic Management Journal*, 16, 161–82.

Barney, J. 1991: Firm resources and sustained competitive advantage. *Journal of Management*, 17, 99–120.

Barron, D. N. 1992: The analysis of count data: overdispersion and autocorrelation. In P. V. Marsden (ed.), *Sociological Methodology*. Oxford: Blackwell, 179–220.

Brynjolfsson, E. and Hitt, L. 1996: Paradox lost? Firm-level evidence on the returns to information systems spending. *Management Science*, 42, 541–58.

Burns, T. and Stalker, G. M. 1961: *The Management of Innovation*. London: Tavistock.

Cameron, A. C. and Trivedi, P. K. 1986: Econometric models based on count data: comparisons and applications of some estimators and tests. *Journal of Applied Econometrics*, 1, 29–53.

Capron, L., Dussauge, P., and Mitchell, W. 1998: Resource redeployment following horizontal acquisitions in Europe and North America, 1988–1992. *Strategic Management Journal*, 19, 631–61.

Chang, S. J. 1996: An evolutionary perspective on diversification and corporate restructuring: entry, exit, and economic performance during 1981–89. *Strategic Management Journal*, 17, 587–611.

Chang, S. J. and Singh, H. 1999: The impact of modes of entry and resource fit on modes of exit by multibusiness firms. *Strategic Management Journal*, 20, 1019–35.

Christensen, C. M. and Bower, J. L. 1996: Customer power, strategic investment, and the failure of leading firms. *Strategic Management Journal*, 17, 197–218.

Cockburn, I. M., Henderson, R. M., and Stern, S. 2000: Untangling the origins of competitive advantage. *Strategic Management Journal*, 21, 1123–45.

Cohen, W. M. and Levinthal, D. A. 1989: Innovation and learning: the two faces of R&D. *The Economic Journal*, 99, 569–96.

Cohen, W. M. and Levinthal, D. A. 1990: Absorptive capacity: a new perspective on learning and innovation. *Administrative Science Quarterly*, 35, 128–52.

Cooper, R. B. and Zmud, R. W. 1990: Information technology implementation research: a technological diffusion approach. *Management Science*, 36, 123–39.

Cyert, R. M. and March, J. G. 1963: *A Behavioral Theory of the Firm*. Englewood Cliffs, NJ: Prentice-Hall.

Damanpour, F. 1991: Organizational innovation: a meta-analysis of effects of determinants and moderators. *Academy of Management Journal*, 34, 555–90.

Damanpour, F. and Evan, W. M. 1984: Organizational innovation and performance: the problem of organizational lag. *Administrative Science Quarterly*, 29, 392–409.

Dierickx, I. and Cool, K. 1989: Asset stock accumulation and sustainability of competitive advantage. *Management Science*, 35, 1504–11.

Dosi, G. 1982: Technological paradigms and technological trajectories. *Research Policy*, 11, 147–62.

Dosi, G. 1988: Sources, procedures, and microeconomic effects of innovation. *Journal of Economic Literature*, 26, 1120–71.

Downs, G. W. and Mohr, L. B. 1976: Toward a theory of innovation. *Administrative Science Quarterly*, 21, 700–14.

Frei, F. X., Harker, P. T., and Hunter, L. W. 1998: Innovation in retail banking. *Working Paper, Financial Institutions Center, The Wharton School, University of Pennsylvania*, 97-48-B.

Furst, K., Lang, W. W., and Nolle, D. E. 2000: Who offers Internet banking? *Quarterly Journal, Special Studies on Technology and Banking, OCC*, 19, 29–48.

Geroski, P., Machin, S., and Van Reenen, J. 1993: The profitability of innovating firms. *Rand Journal of Economics*, 24, 198–211.

Golden, B. R. 1992: The past is the past – or is it? The use of retrospective accounts as indicators of past strategy. *Academy of Management Journal*, 35, 848–60.

Greene, W. H. 2000: *Econometric Analysis*. Upper Saddle River, NJ: Prentice Hall.

Hagedoorn, J. 1993: Understanding the rationale of strategic technological partnering: interorganizational modes of cooperation and sectoral differences. *Strategic Management Journal*, 14, 371–85.

Hausman, J., Hall, B. H., and Griliches, Z. 1984: Econometric models for count data with an application to the patents–R&D relationship. *Econometrica*, 52, 909–38.

Helfat, C. E. 1997: Know-how and asset complementarity and dynamic capability accumulation: the case of R&D. *Strategic Management Journal*, 18, 339–60.

Helfat, C. E. 2000: Guest editor's introduction to the special issue: the evolution of firm capabilities. *Strategic Management Journal*, 21, 955–9.

Helfat, C. E. and Raubitschek, R. S. 2000: Product sequencing: co-evolution of knowledge, capabilities and products. *Strategic Management Journal*, 21, 961–79.

Henderson, R. and Cockburn, I. 1994: Measuring competence? Exploring firm effects in pharmaceutical research. *Strategic Management Journal*, 15, 63–84.

Henderson, R. and Cockburn, I. 1996: Scale, scope and spillovers: the determinants of research productivity in ethical drug discovery. *RAND Journal of Economics*, 27, 32–59.

Holbrook, D., Cohen, W. M., Hounshell, D. A., and Klepper, S. 2000: The nature, sources, and consequences of firm differences in the early history of the semiconductor industry. *Strategic Management Journal*, 21, 1017–41.

Iansiti, M. 2000: How the incumbent can win: managing technological transitions in the semiconductor industry. *Management Science*, 46, 169–85.

Karim, S. and Mitchell, W. 2000: Path-dependent and path-breaking change: reconfiguring business resources following acquisitions in the US medical sector, 1978–1995. *Strategic Management Journal*, 21, 1061–81.

Kim, D. and Kogut, B. 1996: Technological platforms and diversification. *Organization Science*, 7, 283–301.

Klepper, S. and Simons, K. L. 2000: Dominance by birthright: entry of prior radio producers and competitive ramifications in the US television receiver industry. *Strategic Management Journal*, 21, 997–1016.

Kogut, B. and Zander, U. 1992: Knowledge of the firm, combinative capabilities, and the replication of technology. *Organization Science*, 3, 383–97.

Leonard-Barton, D. 1995: *Wellsprings of Knowledge: Building and Sustaining the Sources of Innovation*. Boston: Harvard Business School Press.

Levitt, B. and March, J. G. 1988: Organizational learning. *Annual Review of Sociology*, 14, 319–40.

Loh, L. and Venkatraman, N. 1992: Determinants of information technology outsourcing. *Journal of Management Information Systems*, 9, 7–24.

Makadok, R. 2001: Toward a synthesis of the resource-based and dynamic-capability views of rent creation. *Strategic Management Journal*, 22, 387–401.

March, J. G. 1991: Exploration and exploitation in organizational learning. *Organization Science*, 2, 71–87.

McGrath, R., MacMillan, I., and Tushman, M. 1992: The role of executive team actions in shaping dominant designs: towards the strategic shaping of technological progress. *Strategic Management Journal*, 13, 137–61.

Meyer, A. D. and Goes, J. B. 1988: Organizational assimilation of innovations: a multilevel contextual analysis. *Academy of Management Journal*, 31, 897–923.

Miller, C., Cardinal, L., and Glick, W. 1997: Retrospective reports in organizational research: A reexamination of recent evidence. *Academy of Management Journal*, 40, 189–204.

Mitchell, W. 1989: Whether and when? Probability and timing of incumbents' entry into emerging industrial subfields. *Administrative Science Quarterly*, 34, 208–30.

Montgomery, C. A. and Hariharan, S. 1991: Diversified expansion by large established firms. *Journal of Economic Behavior and Organization*, 15, 71–89.

Nelson, R. R. and Winter, S. G. 1982: *An Evolutionary Theory of Economic Change*. Cambridge, MA: Harvard University Press.

Pavitt, K. 1984: Sectoral patterns of technical change: towards a taxonomy and a theory. *Research Policy*, 13, 343–73.

Penrose, E. T. 1959: *The Theory of the Growth of the Firm*. White Plains, NY: M. E. Sharpe, Inc.

Pindyck, R. S. and Rubinfeld, D. L. 1981: *Econometric Models and Economic Forecasts*. New York: McGraw-Hill.

Pisano, G. P. 1994: Knowledge, integration, and the locus of learning: an empirical analysis of process development. *Strategic Management Journal*, 15, 85–100.

Porter, M. E. 1980: *Competitive Strategy: Techniques for Analyzing Industries and Competitors*. New York: Free Press.

Reinganum, J. F. 1989: The timing of innovation: research, development, and diffusion. In R. Schmalensee and R. D. Willig (eds.), *Handbook of Industrial Organization*. Amsterdam, North-Holland: Elsevier Science Publishers, 850–908.

Roberts, P. W. 1999: Product innovation, product-market competition and persistent profitability in the US pharmaceutical industry. *Strategic Management Journal*, 20, 655–70.

Rogers, E. M. 1995: *Diffusion of Innovations*. New York: The Free Press.

Rosenbloom, R. S. 2000: Leadership, capabilities, and technological change: the transformation of NCR in the electronic era. *Strategic Management Journal*, 21, 1083–103.

Schoenecker, T. S. and Cooper, A. C. 1998: The role of firm resources and organizational attributes in determining entry timing: a cross-industry study. *Strategic Management Journal*, 19, 1127–43.

Schumpeter, J. A. 1934: *The Theory of Economic Development*. Cambridge, MA.

Singh, K. and Mitchell, W. 1996: Precarious collaboration: business survival after partners shut down or form new partnerships. *Strategic Management Journal*, 17, 95–115.

Singleton, Jr., R. A. and Straits, B. C. 1999: *Approaches to Social Research*. New York: Oxford University Press.

Stinchcombe, A. L. 1965: Social structure and organizations. In J. G. March (ed.), *Handbook of Organizations*. Chicago: Rand-McNally, 142–93.

Stuart, T. E. and Podolny, J. M. 1996: Local search and the evolution of technological capabilities. *Strategic Management Journal*, 17, 21–38.

Teece, D. J. 1976: *The Multinational Corporation and the Resource Cost of International Technology Transfer*. Cambridge, MA: Ballinger.

Teece, D. J. 1992: Competition, cooperation, and innovation: organizational arrangements for regimes of rapid technological progress. *Journal of Economic Behavior and Organization*, 18, 1–25.

Teece, D. J. 1996: Firm organization, industrial structure, and technological innovation. *Journal of Economic Behavior and Organization* 31, 193–224.

Teece, D. J. and Pisano, G. 1994: The dynamic capabilities of firms: an introduction. *Industrial and Corporate Change*, 3, 537–56.

Teece, D. J., Pisano, G., and Shuen, A. 1997: Dynamic capabilities and strategic management. *Strategic Management Journal*, 18, 509–33.

Tripsas, M. 1997: Unraveling the process of creative destruction: complementary assets and incumbent survival in the typesetter industry. *Strategic Management Journal*, 18, 119–42.

Tushman, M. L. and Anderson, P. 1986: Technological discontinuities and organizational environments. *Administrative Science Quarterly*, 31, 439–65.

Utterback, J. M. and Suarez, F. F. 1993: Innovation, competition, and industry structure. *Research Policy* 22, 1–21.

Vickers, J. 1986: The evolution of market structure when there is a sequence of innovations. *Journal of Industrial Economics*, 35, 1–12.

Von Hippel, E. 1988: *Sources of Innovation*. New York: Oxford University Press.

Wernerfelt, B. 1984: A resource-based view of the firm. *Strategic Management Journal*, 5, 171–80.

Wheelwright, S. C. and Clark, K. B. 1992: *Revolutionizing Product Development*. New York: Free Press.

Wolfe, R. A. 1994: Organizational innovation: review, critique and suggested research directions. *Journal of Management Studies*, 31 (3), 405–31.

CHAPTER 2

Competitive Advantage, Knowledge Assets and Group-Level Effects: An Empirical Study of Global Investment Banking

Boris Durisin and Georg von Krogh

Keywords: Strategic groups, competitive advantage, knowledge-based view, knowledge assets.

Abstract

Using data from the investment banking industry (1991–1999), this paper investigates the question whether strategic group-level effects on knowledge assets exist on firms, using data from the investment banking industry. The existence of such group-level effects is affirmed. We discuss implications of the study for theory and research on strategic groups as well as for the knowledge-based view of the firm.

Introduction

During the 1990s, established frameworks of competitive advantage (Porter, 1980, 1985, 1990, 1991) were increasingly questioned by scholars who believed that sustainable competitive advantage resulted from unique assets, such as resources (e.g., Barney, 1986a, 1986b; 1991; Rumelt, 1991, Wernerfelt, 1984), dynamic capabilities (e.g., Teece et al., 1997), and knowledge (e.g., Nonaka et al., 2000). An important body of literature sought to compare and contrast divergent premises between these two perspectives (e.g., Teece et al., 1997; Conner, 1991; Peteraf, 1993). Another stream of research sought to decompose performance variation amongst firms empirically, and thus to examine the relative impact of an industry's vs. firms' resources and capabilities (e.g., Anand and Singh, 1997; Brush and Bromiley,

1997; Chang and Singh, 2000; Hansen and Wernerfelt, 1989; Jacobson, 1988; Mauri and Michaelis, 1998; McGahan and Porter, 1997, Powell, 1996; Schmalensee, 1985; Spanos and Lioukas, 2001; Roquebert et al., 1996; Rumelt, 1991). A third group of scholars recognized that both perspectives were complementary in explaining a firm's performance (e.g., Amit and Schoemaker, 1993; Mahoney and Pandian, 1992; McGahan and Porter, 1997; Conner, 1991). According to Wernerfelt (1984), they constitute the two sides of the same coin. They share the view that consistently above-normal returns to the firm are possible, and that to this end an attractive positioning is of crucial importance (Conner, 1991).

In this paper, we share this compound view and argue that a firm's attributes are not the sole determinant of its performance: Even among a firm's most idiosyncratic resources, such as its knowledge assets, group-level effects may exist. Strategic group membership may therefore affect firms' performance. In a longitudinal study of the global investment banking industry, we explore whether group-level effects exist among knowledge assets. We find that strategic groups have an effect on the conditions of rivalry. Firms are members of a strategic group if their competitive positions are affected by group-level effects which, in turn, differ among strategic groups. In the case of resource stocks, or knowledge assets, "rivalry effects" are group-level effects that change the behavior of members of the group. This perspective is key for our study. We do not attempt to establish a direct link between strategic group membership and intra-industry performance differences, as most studies of strategic group research do. Group configuration adds explanatory power, because for given asset stock levels a firm will generate different levels of performance depending on the degree of rivalry it faces, which in turn depends on the location in the form of attribute space of other firms (Cool et al., 1994: 223). Hence, if conditions of rivalry among knowledge assets exist, firms exposed to group-level effects. We find empirical support for this assertion.

We begin with a brief review of the knowledge-based view of the firm, defining the concept of "knowledge assets," a firm's most idiosyncratic resources. Next, we present our research framework and variables, the industry setting of global investment banking activities, our analysis, and results. Finally, we discuss the implications of our findings for strategic management research.

Knowledge Assets and Strategic Groups

Edith Penrose noted that most economists broaching the subject of knowledge have found it "too slippery to handle," though they have "always recognized the dominant role that increasing knowledge plays in economic processes" (1959: 77). As Marshall noted, "knowledge is our most powerful engine of production" (1969: 115). Consequently, it has been claimed that the knowledge-based view of the firm "promises to be one of the most profound changes in management thinking since the scientific revolution of the early decades of this century" (Grant, 1997: 454).

An increasing number of scholars have investigated the nature of knowledge in firms on the assumption that knowledge is a source of competitive advantage (e.g., Blackler, 1995; Hedlund, 1994; von Krogh and Roos, 1995, 1996; von Krogh et al., 1994; Nonaka, 1994). Knowledge may be "the key inimitable resource managers need to appreciate, if not understand" (Schendel, 1996: 3), and firms within an industry may differ in their ability to create and apply knowledge (Conner and Prahalad, 1996). Basing their work on Polanyi (1958, 1967, 1969), Nelson and Winter (1982), Nonaka (1991, 1994), Ichijo et al. (1998), Kogut and Zander (1992) and others have discussed how knowledge embodied in individuals and embedded in organizational practices cannot be readily articulated and shared. Knowledge created by skilled individuals (e.g., Simon, 1991, 1993) is generally tacit (Nonaka, 1991, 1994).

Knowledge is also embedded in organizational processes (Nelson and Winter, 1982; Winter, 1982, 1988) and relationships (Aoki, 1990, 45). Dosi and Marengo stress that "knowledge is neither presupposed nor derived from the available information but rather emerges as a property of the learning system and is shaped by the interaction among the various learning-processes that constitute organization" (1994: 162). As Teece and Pisano (1994, 544) elaborate: "Learning-processes are intrinsically social and collective and occur not only through the imitation and emulation of individuals." Thus, knowledge is embedded in social structure and not immediately transparent (Hodgson, 1998: 185). The firm becomes a repository of tacit knowledge (e.g., Fransman, 1994). A knowledge-based view indicates that different organizational settings, routines in the firm, and processes must be established to maximize the effectiveness with which different types of knowledge are created and shared (Nonaka and Takeuchi, 1995; Kotha, 1995; von Krogh et al., 2000).

Although we can assert that "tacit knowledge" is "of critical strategic importance because, unlike explicit knowledge, it is both inimitable and appropriable" (Spender and Grant, 1996: 8), so far there is limited understanding of how firms can gain competitive advantage from knowledge in general (von Krogh and Nonaka, 1999). We address this issue in clarifying the notion of knowledge or, rather, "knowledge assets."

Knowledge assets and competitive advantage

Competitive advantage is sustainable only if it endures after efforts to duplicate that advantage have ceased (Barney, 1991: 102; Lippman and Rumelt, 1982), and in competitive markets the ease of imitation determines the sustainability of competitive advantage (Teece et al., 1997: 526). Among the factors that make a resource inimitable are tacitness (Reed and DeFillippi, 1990), causal ambiguity (Lippman and Rumelt, 1982), time compression diseconomies and interconnectedness (Dierickx and Cool, 1989), path dependence and social complexity (Barney, 1991; Reed and DeFillippi, 1990). All of these are characteristics that apply to (tacit) knowledge in firms. Thus, differences in firm performance may represent differences in their stocks of knowledge, or *knowledge assets*.

Teece suggested (1998: 72): "The asset structure of a firm is perhaps the most relevant aspect of its positioning when the commercialization of knowledge in tangible products and processes is at issue." The essence of knowledge assets is that they cannot be readily assembled through markets (Teece, 1982, 1986b; Zander and Kogut, 1995). Advantages must be found in the "rare, imperfectly imitable, and non-substitutable resources already controlled by firms" (Barney, 1991: 117).

McGrath et al. point out that "the most potent of such assets are posited to be intangible or tacit" (1995: 252). Miller and Shamsie remark that there appear to be two fundamentally different bases of inimitability. Assets cannot be imitated either because they are protected by property rights or because they are protected by "knowledge barriers" (1996: 521). Itami (1987: 12) gives an indication of what might constitute such barriers: He defines invisible [knowledge] assets as information-based resources and gives examples such as consumer trust, corporate culture, and management skills. According to Boisot, knowledge assets are "stocks of knowledge from which services are expected to flow" (1998: 3). Knowledge assets are different from physical assets; there are no strict time limits to what services are expected to flow from them, and there are no clear one-to-one relationships between investments to build up these assets, and the services that are expected to flow from them (Boisot, 1998). They constitute both a major source of competitive advantage and a major challenge to managers in understanding them (Griffiths et al., 1999).

A key decision for companies to make is how broad or narrow their knowledge assets should be. With limited resources, it is usually best to focus on specific domains of knowledge. However, with a broader knowledge base the firm will be in a better position to combine related knowledge in/between various domains in a more complex manner. A broader knowledge base should result in increased strategic flexibility, and adaptability to environmental change (Volberda, 1996). In studies of the pharmaceutical industry, Henderson and Cockburn (1994) and Pisano (1994) showed that a broader knowledge base is essential to the successful integration of different knowledge areas and provides a source of advantage. Here, one should note that increased sustainability of competitive advantage can stem from combining different knowledge assets that generate causal ambiguity (Reed and DeFillippi, 1990). Also, if a firm's knowledge assets are pitched too narrowly, they may constitute core rigidities and thus prevent effective responses to technological advances in different but related fields (Leonard-Barton, 1992).

Knowledge assets and isolating mechanisms

The term "isolating mechanism" denotes mobility barriers that protect a firm's competitive advantage. The concept of mobility barriers is inextricably linked to strategic groups and performance differences; or, as McGee and Thomas (1986: 153) note, "mobility barriers are a corollary to the existence of strategic groups." Isolating mechanisms are features of assets that prevent other firms from obtaining or replicating them (Mahoney and Pandian, 1992; Rumelt, 1984). Rumelt (1984: 566f.) introduced the term "isolating mechanism" and explicitly linked it to the concept of mobility barriers. He provided a list of isolating mechanisms, including causal ambiguity, specialized assets, switching and search costs, consumer and producer learning, team-embodied learning, special information, unique resources, patents and trademarks, reputation and image, etc. (1984: 568). Isolating mechanisms explain a stable stream of rents and provide a rationale for intra-industry differences among firms. Mahoney and Pandian suggest that ". . . isolating mechanisms are the result of the rich connections between uniqueness and causal ambiguity" (1992: 373) and reject the opinion that "invisible (intangible) assets are the most likely candidates for resources that are unique and causally ambiguous" (1992: 373). Mahoney stresses that "the accumulation and deployment of these invisible (intangible) resources and capabilities are the primary source of sustainable competitive advantage. The heart of invisible resources involves tacit understanding and articulable information both as a stock and as a flow" (1995: 94). Hence, an analysis of strategic groups based on knowledge assets should focus on the capacity of knowledge assets to erect mobility barriers in order to protect a firm's competitive advantage.[1]

Competitive advantage and strategic groups

The concept of strategic group has spawned a "plethora of research" (e.g., McGee and Thomas, 1986; Thomas and Venkatraman, 1988, Ketchen et al., 1993), and testing for the existence and performance implications of strategic groups in an industry has become one of the dominant areas of empirical research in strategic management literature (Barney and Hoskisson, 1990; Peteraf and Shanley, 1997). However, the empirical findings on the ability of strategic groups to explain intra-industry performance differences are conflicting (e.g., Porter, 1979; Caves and Pugel, 1980; Oster, 1982; Frazier and Howell, 1983; Dess and Davis, 1984; McGee and Thomas, 1986; Cool and Schendel, 1987, 1988; Fiegenbaum and Thomas, 1990; Ketchen et al., 1993; Reger and Huff, 1993). Critics have also commented on inadequate theory building (e.g., McGee and Thomas, 1986; Hatten and Hatten, 1987; Cool and Schendel, 1987; Thomas and Venkatraman, 1988; Barney and Hoskisson, 1990). Barney and Hoskisson (1990) pointed out two untested assertions in strategic group theory: the existence of strategic groups, and the dependence of a firm's performance, at least to some extent, on the strategic group within which it finds itself (1990: 188).

Fiegenbaum and Thomas (1993: 72) argue that "there are strong theoretical reasons for the existence of strategic groups." For instance, spatial competition is promoting the principle of locating in (resource) space (Tang and Thomas, 1992). Further support for the existence of strategic groups is found in studies of managerial cognition. Using such concepts as strategic group identity (Peteraf and Shanley, 1997), reference groups (e.g., Fiegenbaum and Thomas, 1993, 1995; Fiegenbaum et al., 1996), and cognitive categorization theory (e.g., Porac, Thomas and Baden-Fuller, 1989), authors have suggested that managers tend to view their industries in terms of groups of firms (e.g., Gripsrud and Gronhaug, 1985; Porac et al., 1987; Porac and Thomas, 1990, 1994; Porac et al., 1995; Reger and Huff, 1993; Lant and Baum, 1995). At the same time, other fields such as organization theory (e.g., DiMaggio and Powell, 1983; Hannan and Freeman, 1977) and oligopoly theory (e.g., Stigler, 1964) also point to consensus-building and coordinating mechanisms among firms within industries. One could even take this convergence among perspectives to advance the empirical support of the existence of strategic groups (Nath and Gruca, 1997).

The lack of robust, convincing evidence and theory on strategic group–performance relationship may have some fundamental reasons: improper conceptualization of the strategic group concept (see, e.g., Cool and Schendel, 1987: 1103); inadequate definition of the concept and conceptualization of variables (see, e.g.,

Cool and Schendel, 1987: 1103); improper conceptualization of the performance concept (see, e.g., Thomas and Venkatraman, 1988: 548); and an indirect relationship between strategic group structure and firm profitability may exist to the extent that rivalry within groups has a different effect on firm profitability than that of rivalry between groups (Cool and Dierickx, 1993: 49). In their critique, Barney and Hoskisson claim that "(it) may be necessary to abandon [the strategic group] concept and develop models where the strategically relevant attributes to firms are those that are idiosyncratic" (1990, 188). Cool et al. (1994: 225) remark that Barney and Hoskisson imply that asset positions are the unique determining factor in explaining performance differences. Thus, if assets alone were to drive product market positions, a change in in-group membership should not have explanatory power beyond the effect of changes in the asset stock. The issue boils down to whether firm attributes are the unique determinant, or whether firm performance is also a function of the location of firms in an attribute space of a group (Cool et al., 1994: 232). We will address those issues in our research framework.

As we have reasoned so far, knowledge assets are key to firm performance. They may constitute mobility barriers, so that a group of firms may share attributes of assets. Our study, therefore, explores whether group-level effects among knowledge assets exist.

> Hypothesis 2.1: Group-level effects among knowledge assets exist.

A confirmation of this hypothesis would add to the research on the contribution of strategic groups to explaining firm performance, as well as to the empirical grounding of a knowledge-based theory of the firm.

Research Framework

A robust theory of strategic groups articulates what constitutes a group-level effect. Group-level effects originate in group-level processes resulting in market power, efficiency, or differentiation effects, and they can be static or dynamic (Dranove et al., 1998: 1032f.). These processes take the form of interaction among group members that alters the orientations, decisions, and actions of the individual group members. Group-level effects change the behavior of members as compared to how they would behave in the absence of the group. Group-level effects can be applied to the analysis of the resource flows (e.g., Peteraf, 1993) or resource stocks (e.g., Cool et al., 1994).

Knowledge assets can originate in group-level processes and result in group-level effects on market power, efficiency, or differentiation.[2] Moreover, as knowledge assets constitute mobility barriers, and as mobility barriers are necessary to ensure that the profits generated by group-level effects are not competed away (Dranove et al., 1998: 1037), knowledge assets can constitute strategic groups and have a persistent effect on profits. The relative contribution of group-level effects and mobility barriers to the explanation of a firm's profits in a given industry context is an empirical matter (Dranove et al., 1998: 1033).

For the purpose of this study, a single industry, the global investment banking activities of financial companies, has been selected. Peteraf and Shanley "advocate industry studies rather than inter-industry settings" (1997: 183) because group-level effects can only be studied in single industries (see also Ketchen et al., 1993: 1290; Hatten and Hatten, 1987: 330; Dranove et al., 1998: 1039). The global investment banking activities have been chosen for three reasons. First, this represents a service industry (unlike the greater proportion of previous strategic group studies). In this industry, lack of physical assets, such as manufacturing technologies (ref. "techno-logical relatedness") create "fuzzy" competitive boundaries overlapping into various other domains within the sector of financial services. Secondly, over the past 20 years investment banking has exhibited turbulence in terms of stronger competitive and regulatory dynamics. This is well illustrated by the development of an increasing range of innovative competitive products. Thirdly, global investment banking is also a knowledge-intensive industry (International Monetary Fund, 1998).

Industry context

Investment banking activities engage issuers and investors in capital markets, and are in demand because they lead to lower funding costs, new funding sources, risk reallocation, reduction in transaction costs, and reduced tax costs. Firms with investment banking activities perform financial intermediation. Their core activities are underwriting of debt and equity, and advisory functions such as mergers and acquisitions (M&A).

In 1998, The International Monetary Fund (IMF) noted that the structural changes that have occurred in national and international finance during the past two decades could best be described as the globalization of finance and financial risk. The key elements of this ongoing transformation have been (1) an increase in the tech-nical capabilities for engaging in precision finance, that is, unbundling, repackaging, pricing, and redistributing financial risks; (2) the integration of financial markets, investor bases, and borrowers into a global financial market place; (3) the blurring of distinctions between financial institutions and the activities and markets they engage in; and (4) the emergence of the global bank and the international financial conglomerate, each providing a mix of financial products and services in a broad range of markets and countries. These changes have altered investors' and borrowers' perceptions of financial risks and rewards around the world, and their behavior across national and international financial markets (International Monetary Fund, 1998: 180).

Globalization of financial markets implies accessing the complete array of financial contracts, suppliers, and users of finance worldwide in order to optimize the object-ive functions of financial end-users, and it covers intermediaries' services in structuring financial contracts, underwriting and distribution of securities, and corporate finance and M&A advisory work (Walter, 1998: 98–123, 4f.). Ingo Walter noted (1999a: 159; 1999b: 459) that investment banking is "basically a globalized industry." Similarly, Smith (1997: 29) remarked in the mid-1990s: "[Global] market integra-tion at the wholesale level is now virtually a done deal." Today's globalization of finance leads to a new financial landscape or, as an OECD report put it, to "a new brave world of finance" (OECD, 2000, *Financial Market Trends*, 75: 188). As the

OECD report concluded: "The globalization of finance has resulted in an unprecedented number of financial institutions that are active internationally, highly leveraged and, in many cases, using new financial technologies and instruments" (OECD, 2000, *Financial Market Trends*, 75: 177).

Various secular trends shaped investment banking and its globalization and further contributed to its growth. The most important are commoditization and derivatization, technological developments, institutionalization, deregulation, privatization, securitization, and corporate governance and corporate restructuring. Their effects were enhanced by events such as the introduction of the European Monetary Union. As a consequence, the 1990s saw the emergence of a relatively small group of global, full-service investment banking. In addition to the secular trends, the importance of frequent financiers, the economics of the compensation method in investment banking, alterations in syndication practices, falling margins and the secular trends contributed to the emergence of global, full-service firms (Durisin, 2001). In the globalizing business world, only a limited number of firms are able to do the deals of the size and complexity frequent financiers require. These frequent financiers are the customers for which firms with global investment banking activities compete fiercely to sign. The complexity and size of such deals led to the emergence of *global, full-service investment-banking firms*. As David Komansky, then president and chief operating officer at Merrill Lynch, characterized it: "Just a handful of firms will be able to say to a client wishing to raise capital through debt or equity: 'We can do that for you in whatever world capital market is cheapest, because we research, trade, sell and originate in every major capital market in the world'."(*Euromoney*, 1995, September, Smith runs with the herd, p. 75).

Variable selection

In this study we use no balance sheets data to measure knowledge assets.[3] As Teece et al. (1997) stressed: "We point out that the assets that matter for competitive advantage are rarely reflected in the balance sheet, while those that do not are." Vicari et al. (1999), referring to a study by former Arthur Andersen, notes that "Cost is not relevant for determining the value of an intangible asset, which is derived from future economic benefits. There is no direct correlation between expenditures on an asset and its value." Srivastava et al. suggest that 'efforts to replicate intangible assets often necessitate extensive investments in marketing, sales, service, and human resources development with little, if any, guarantees of success" (1998: 7), and they stress that "the value of any assets ultimately is realized, directly or indirectly, in the marketplace" (1998: 4). Thus in our study we focus on the outward manifestations of the knowledge asset in the marketplace. The common feature of the select variables is that they result from customer appraisals, preferences, or choices.

The required substantive contribution of strategic group studies is the identification of variables or classes of variables that systematically affect profitability (McGee and Thomas, 1986: 149). For this study we identified reputation, market orientation, and complementary assets as variables. These are sources of dissimilarity between firms and exhibit the properties of knowledge assets and isolating mechanisms, as will be demonstrated.

Reputation

A firm's *reputation* is an asset that can generate future rents (Fombrun, 1996; Wilson, 1985). A variety of academic and practitioner literatures explored different facets of the construct (Fombrun and van Riel, 1998). Reputation-building behavior is strategically important in incomplete information settings and affects strategic choice by generating future rents (Weigelt and Camerer, 1988: 443, 452). Reputations summarize a good deal of information about firms and shape the responses of customers, suppliers, and competitors (Teece et al., 1997: 521). Amit and Schoemaker conceptualize it as a strategic asset (1993: 37) and as a unique property of the firm from which rents ensue (1993: 42). Hall (1993: 613) identifies reputation as an asset based on a unique and defendable position. Hall also remarks that reputation is one of the perceived assets that make the most important contribution to business success (1993: 143). A firm's reputation is a private good that is difficult to accumulate, imitate, substitute, or transfer (Rao, 1994: 39). Reputation can be seen as a form of collective knowledge (between a firm and its constituents), and is thus as Spender (1996: 52) argues "the most secure and strategically significant kind of organizational knowledge."

Fombrun notes that reputational rankings create barriers to competition (1996: 11). According to Rindova and Fombrun, "reputational rankings are another manifestation of constituents' differential perceptions of firms that affect competitive advantage. . . . reputational rankings reflect an ordering, a status hierarchy with implications about the superiority and inferiority of its members," and they stress that "reputational rankings incorporate the demands of resource-holders" (1999: 700). In investment banking, institutional investors are the key financial capital holders. We use them in our study for our reputational rankings. As Fombrun expresses it: "Perceptual ratings establish the relative value of one firm's intangible assets against another's" (1996: 181) and, when discussing investment banks, he notes that "there are significant barriers that derive from perceptions of the bank by clients and from self-perceptions within banks."[4] He adds: "Favorable rankings in reputational surveys . . . create intangible barriers to competition that lesser rivals find difficult to overcome" (1996: 329).

A high reputation is one of the most important knowledge assets a bank can enjoy. As Eccles and Crane point out, "the production process of investment banking makes differentiation through reputation especially important" for both investment banks and customers (Eccles and Crane, 1988: 109f.). It is difficult to distinguish oneself from others "in a deal-based service business, particularly when the characteristics of a deal are determined as much by the customer as by the investment bank, and when products are easily imitated. Thus investment banks attempt to create positive reputations to distinguish themselves from their competitors" (Eccles and Crane, 1988: 92). Well-known investor Warren Buffet made a remark to a group of Salomon Brothers managers in the aftermath of a trading scandal that shook the bank in 1991: "If you lose dollars for the firm by bad decisions, I will be very understanding. If you lose reputation for the firm, I will be ruthless" (Fombrun, 1996: 85).

Euromoney, a journal of the financial community, polled treasurers and financial officers at corporations, financial institutions, state agencies, and supranational organizations. Respondents were asked to nominate firms providing the best service in raising capital. The opinions of customers about which firm they think of first when they want to raise money, and which serves their needs best, have been ranked for underwriting in different areas and regions. As reputation is a multidimensional construct, we need to identify the reputational capital of the firms for the various, distinct investment banking activities that comprise a firm's activity scope. The following categories have been included to analyze reputation as a knowledge asset: borrowers' vote in capital raising, peers' vote in bond underwriting, and peers' vote in equity underwriting. They are weighted equally. These rankings, their methodology, and their use in our study are outlined in Appendix A.

Market orientation

A firm's *market orientation* is also essential to success (e.g., Dougherty, 1990; Leonard-Barton, 1995; Quinn, 1985). Research suggests that market orientation is an aspect of organizational culture (Day, 1994; Desphandé et al., 1993; Slater and Narver, 1995, 1998), a set of specific behaviors and activities related to customer understanding (Kohli and Jaworski, 1990), a basis for decision making (Shapiro, 1988), and an asset (Hunt and Morgan, 1995). Market orientation is considered to be distinct from marketing orientation in the sense that the former addresses organization-wide concerns, whereas the latter reflects a functional focus of marketing (Shapiro, 1988). Researchers have investigated a direct causal link between market orientation and performance (Narver and Slater, 1990; Ruekert, 1992), and others have found this to be moderated by innovation (Diamontopoulos and Hart, 1993; Greenley, 1995; Jaworski and Kohli, 1993; Slater and Narver, 1994b). Yet others have studied the roles of market orientation's antecedents (Jaworski and Kohli, 1993). Studies have shown that market orientation has a significant influence on relative returns on assets (Narver and Slater, 1990), sales growth (Slater and Narver, 1994a; Pelham and Wilson, 1996), new-product success (Slater and Narver, 1994a; Pelham and Wilson, 1996), product quality (Pelham and Wilson, 1996), and "overall performance" (Jaworski and Kohli, 1993; Pelham and Wilson, 1996). The ability of a market-oriented firm to outperform less market-oriented competitors is based on the premise that the former can create long-term superior value for the firm's customers in comparison with the latter (Pelham and Wilson, 1996; Reed and DeFillippi, 1990). The market orientation of a firm is a private good that is difficult to accumulate, imitate, substitute, or transfer, and can be seen as a form of collective knowledge which is "the most secure and strategically significant kind of organizational knowledge" (Spender, 1996, 52).

Desphandé et al. (1993: 27) stress that "the evaluation of how [market] oriented an organization is should come from its customers rather than merely from the company itself. This point is a critical one." Aaker (1991) notes that many firms pay "lip service" to being customer oriented. And Shapiro (1988: 122) identifies as a problem with the concept the fact that "just about every company thinks of itself as market oriented." Kohli and Jaworski (1990: 16) noted: "We were apprised of several instances in which members of an organization felt they were very customer oriented, but in fact were hardly so." However, Eccles and Crane (1988: 178) stress that "in investment banking, strategy is ultimately implemented at the level of the individual customer. For this reason, it is necessary to have individuals who are concerned with what strategy means for specific issuers and investors. These relationship managers are an important integrating force at the customer interface."

In this industry, since products are information and therefore easily imitable (products constitute a temporary advantage), investment bankers must differentiate through the way they provide solutions for the customer's needs and maintain the relationships with their customers. This market orientation would help to build market share. Eccles and Crane note that "the greater a firm's market share in a particular activity, the more extensive its information about both sides of the market interface concerning potential deals" (1988: 42) and "the more information it has on the market. This improves the firm's ability to price deals and lowers the risk of loss" (1988: 159). Investment bankers refer to it as "being in the deal stream." This contributes to the creation of a knowledge cycle, in the sense that deals generate knowledge, which in turn generates more deals, which in turn generate knowledge, and so on (Eccles and Crane, 1998: 57).

Our study removes the bias of managerial perceptions of market orientation. Moorman and Slotegraaf (1999: 251) investigated a firm's ability to develop and maintain relationships with customers, using market share as the variable. Mitchell (1992) used market share to measure the market-related capabilities of a firm. Accordingly, we also will employ in our study market share as the outward manifestation of a market orientation. Slater and Narver (1994b: 25) depicted market share as the outcome of a market orientation in achieving competitive advantage. Varadarajan and Jayachandran (1999: 123, 140) have also shown market share as the outcome of competitive strategy. A high and stable market share is one of the indicators of a company's ability to acquire and maintain customer preferences, and as Slater and Narver (1994b: 22) note, "The heart of a market orientation is its customer focus." Market share is one of the most important indicators for assessing a company's market orientation effectiveness.

Market orientation is measured as the relative market share a firm has in the total market of the respective segment. The data has been lagged by one year, e.g., 1997 data indicates the stock of market orientation a firm could access in 1998. Thus the market activities, experience, and relationships built up in a single year indicate the stock of knowledge on which a firm can build its activities of the following year. A firm's current activities are thus built up on the accumulated market orientation of the previous year.

As market orientation is a multidimensional construct, we need to identify the firm's propensity to have a market orientation for the various, distinct investment-banking activities that comprise its activity scope. The public market for securities underwriting, the private placement of securities, and the market for mergers and acquisitions can be distinguished. The Appendix gives an overview of the segments incorporated in our study of the respective markets. The various segments have been weighted according to their relative volume. Data was provided by *Thomson Financial Securities Data*, included the market shares of the top 25 or 30 firms respectively. This data can be used as a good approximation, since the top 25 firms normally claim about 99 percent of the total market volume. Moreover, the firm ranked as 25th usually has a market share of about 0.1 percent of the total market.[5]

Complementary assets

The value of a firm's core competences can be enhanced by combination with the appropriate *complementary assets* (Teece et al., 1997: 516). Various authors have pointed out the role of complements (Amit and Schoemaker, 1993; Dierickx and Cool, 1989; Grant, 1991; Lippman and Rumelt, 1982; Reed and DeFilippi, 1990; Teece et al., 1997). Teece points out that "the successful commercialization of an innovation requires the know-how in question to be utilized in conjunction with other capabilities or assets" (1986a: 288). If innovation is not tightly protected, and, once "out," is easy to imitate, then securing control of complementary capacities is likely to be the key success factor, particularly if those capacities are in fixed supply, – so called "bottlenecks" (1986a: 292), and he stresses that "with weak intellectual property protection, it is quite clear that the innovator will often lose out to imitators and/or asset-holders, even when the innovator is pursuing the appropriate strategy" (1986a, 297). Empirical support for this argument is found in the medical diagnostic imaging industry (Mitchell, 1989, 1992). Teece et al. (1997: 523) note "in Mitchell's study, firms already controlling the relevant complementary assets could in theory start last and finish first." Other studies support this view (Henderson and Clark, 1990; Mitchell and Singh, 1992; Nagarajan and Mitchell, 1998; Rosenbloom and Christensen, 1994; Tripsas, 1997).

As the most important complementary assets one can identify the breadth and depth of research, execution and distribution in equities, secondary market support for bonds, and risk management.

1 *Breadth and depth of research.* The stronger a bank's research capability in a given market, the more knowledge is available to the front-line investment bankers. Analysts working in the research department are the most important source of internally generated knowledge. Their activity exceeds traditional equity and bond analysis for investors, providing knowledge to the employees at the issuer interface, i.e., the investment bankers, and at the investor interface, i.e., traders and salespeople. John S. Chalsty, then CEO of Donaldson, Lufkin & Jenrette, remarked: "I have always thought that the analysts should be the best source of transactions. They know the industry, its directions and who can benefit" (Eccles and Crane, 1998: 173f.).

Research teams are ranked to measure the research strength of a firm. The firms with the most comprehensive bundle of valuable knowledge assets in North America, Europe, Latin America, Asia, and Japan (see Appendix A for more details) were identified and ranked. The categories are weighted according to the stock-market capitalization of the region relative to the other regions. The *Institutional Investor*, a journal of the financial community, publishes annual ratings of research. In our study, the number of nominations in the *Institutional Investor* rankings in relation to the investment specialties considered in the respective categories are used as a variable of research strength.

2 *Execution/distribution in equities.* Relationships based on expertise in equity execution can lead to close relationships with corporate clients. In choosing a firm for a public offering, the secondary market in equities is an important decision factor. Such relationships can also lead to fee-producing advisory roles. Equity execution is

no longer regarded as the inevitable by-product of highly rated research. Rather, fund managers view it as a separate capability.

The firms with the most comprehensive bundle of valuable knowledge assets in US broking, European and Asian equity execution, were identified and ranked. The categories are weighted according to the value traded in equities of the region relative to the other regions. In broking, "the more people you have on the ground in more places making phone calls, the more business you are going to bring in," remarked Brad England, head of equity capital markets for Europe, the Middle East and Africa at Merrill Lynch; "and the more you do, the better idea you have of who has and wants which positions in various stocks" (*Euromoney*, 1998, September, Selling Europe to the US). Thus, the number of Registered Representatives (RRs) a firm has can be taken as an indicator of the knowledge stock a company can access in acquiring business from customers.[6] The top 50 firms by number of retail and institutional representatives are ranked by the *Securities Industry Association*. A registered representative is defined as a full-time employee of a member firm of the Securities Industry Association primarily engaged (more than 50 percent) in selling securities on a full-time work-week basis (not less than 35 hours) (e.g., Securities Industry Association, 2000: 73). In our study, the number of RRs in relation to the total number of RRs of the top 50 firms is used. Retail and institutional RRs were considered separately and weighted equally.

Over 300 European equity fund managers active in Europe were polled by *Euromoney* and asked to nominate their three preferred houses in providing execution in individual countries.[7] The results produced a rank order for each category. The top five firms in each category were used in our study.[8] The categories "overall best in Europe" (A) and "best in individual countries" (B) (see Appendix A for more information) were considered for our study.[9] The category A was weighted a quarter, as the overall strength is a good indicator for the ability to assemble the bundles of locally dispersed knowledge assets in a concentrated effort satisfying the customers' needs. Within B the measures were weighted according to the value traded in equities of the country concerned in relation to the European value traded.

3   *Secondary market support for bonds.* The activities of underwriting securities and secondary trading have been identified as complementary in a study of research project of the City of London (Dermine, 1996). Expertise in secondary market support is key to obtaining primary business. Shaun Wyles, then head of Eurobond trading at Paribas, noted in 1998: "The leverage you gain from primary on secondary business is one of the biggest sea changes in the nineties" (*Euromoney*, 1998, May, Tussling for the bulge bracket). The ability to offer full services from origination to secondary market support sets the leading firms apart from the rest.

The firms with the most comprehensive bundle of valuable knowledge assets in secondary market support for Eurobonds and government bonds were identified and ranked. Both categories are weighted equally. For Eurobonds, the five most reputed firms for trading in the major currencies were identified and ranked (see Appendix A for more details). The currencies were weighted according to the relative volume of outstanding amounts of currency issues in international bonds and notes. As government bonds are used as a hedging vehicle in the corporate bonds markets, and as having a strong position in the secondary government bond markets gives a firm a high profile with clients, the variable was included in our study. The five most highly reputed firms for trading in the major currencies were identified and ranked. The currencies were weighted according to the relative volume of outstanding amount by both (a) nationality of issuer of international debt securities by government agencies, plus (b) public domestic debt securities.

4 *Risk management.* Risk management is at the core of investment banking activity, due to market changes. The last years have seen an increase in the pace of financial innovations, a rapid expansion of cross-border financial transactions, the faster pace of transmitting shocks or mistakes throughout the international financial systems, and greater sensitivity on the part of financial market prices in preferences. These and other factors contribute to the increased risk in the new financial landscape (OECD, 2000, *Financial Market Trends*, 75: 174).

As a complementary asset, risk management enables the firm to assist a client's risk management on liabilities and to win underwriting business. The firms with the most comprehensive bundle of valuable knowledge assets were identified. The five most highly reputed firms in overall risk management, in providing risk management advice, and in providing interest-rate and currency swaps in various currencies were identified and ranked (see Appendix for more details). The categories were weighted equally.

In any single year, our study measures the stock of knowledge in global investment banking activities for each company with 49 measures for reputation, 65 measures for market orientation, 155 measures for depth and breadth of research, 33 measures for execution/distribution in equities, 40 measures for secondary market support for bonds, and 16 for risk management. Table 2.1 gives as a summary an overview of the variables used in our study.

Sample selection

The sample consists of firms with investment banking activities in any of the markets on a global scale. The first sample of firms that appeared in our measures was intentionally kept very broad for two reasons. First, it should be noted that we were not identifying investment banks per se, but firms with investment banking activities. This allowed us to investigate firms with valuable knowledge assets for acquiring business from customers rather than to try to identify and mirror an industry structure defined by "insiders" (in this case "investment banks") and "outsiders," as is done in studies relying on classification of industry structures such as, for example, the SIC code. Secondly, starting with such a large sample we should be able to address problems concerning statistical power.[10] Based on these two considerations we included about 360 firms in our study. These were not investment banks per se, but firms engaged in investment banking activities, i.e., firms with capital market activities in securities underwriting, private placement of securities, and merger and acquisitions transactions in our sample. Further firms with valuable complementary assets were also included.[11] This sample was then used to identify the firms with *global investment banking activities.*

We performed a cluster analysis using both our weighted measures and the factor-scored measures (see below for a detailed discussion of the measures and clustering

Table 2.1 Operationalization of knowledge assets: overview

Construct	Variables	Categories/ measures	Source
Reputation	Borrowers' vote in capital raising	3/10	*Euromoney,*
	Peers' vote in bond underwriting	3/19	Global Financing Guide
	Peers' vote in equity underwriting	5/20	(1991–99)
		Categories (segment)	
Market orientation	Public securities underwriting	30	Thomson Financial
	Private placement of securities	11	Securities Data
	Merger and acquisitions	24	(1990–98)
		Categories/ measures	
Complementary	All-America Research Team	4/73–90	*Institutional Investor*
assets: depth and	All-Europe Research Team	4/30–76	(1991–99)
breadth of	All-British Research Team (91)		
research	All-Latin America Research Team (93–99)	4/19–26	
	All-Asia Research Team (94–99)	4/13–24	
	All-Japan Research Team (94–99)	4/20–32	
Complementary	US broker rating	2/2	*SIA yearbook* (90/91–99/2000)
assets: execution/			
distribution	European equity execution	2/17	*Euromoney* (90/91–99)
in equities			
	Asian equity execution	1/14	*Euromoney* (90/91–99)
Complementary	Secondary market support for	2/22	*Euromoney,*
assets: secondary	Eurobonds		Bond trading poll
market support	Secondary market support for	2/18	(1991–99)
for bonds	government bonds (93–99)		
Complementary	Borrowers' vote in liability	4/16	*Euromoney,*
assets: risk	management		Global Financing Guide
management			(1991–99)

methodology used) in order to identify the groups composed of firms with global investment banking activities. The number of clusters was identified by using dendograms, content analysis, and, in addition, agglomeration coefficients for some years. A clear picture emerged: Very few firms can claim to be part of such a group, because customers regard the vast majority of firms as being able to provide valuable services only within a focused and locally embedded sphere. Based on these groupings we identified our sample. In a first step we eliminated all the firms that appeared only in the <rest> group cluster. These are virtually always firms that provide services only to a limited local market, or in a very specific segment. These firms neither intend to provide, nor are capable of providing, global investment banking services. None of these was identified as a firm that should be included in our study in order to avoid distortions in the sample. This means there are far fewer than 360 investment banks, and even fewer firms performing global investment banking activities. We then excluded some of the remaining firms in order to avoid distortions when performing the clustering procedures. Table 2.2 gives an overview of the roughly 50 firms[12] with global investment banking activities that were selected for the purpose of our study.

Table 2.2 Firms active in global investment banking

North America	The Netherlands
United States	ABN Amro
Bear Stearns	IING Group
Chase Manhattan	
Donaldson, Lufkin & Jenrette (DLJ)	*Switzerland*
Goldman Sachs	CSFB [Credit Suisse Group]
J.P. Morgan	SBC [SBC] (1991–94)
Lehman Brothers	SBC Warburg [SBC] (1995–96)
Merrill Lynch	SBC Warburg Dillon Read [SBC] (1997)
Nationsbank	Warburg Dillon Read [UBS] (1998–1999)
PaineWebber	UBS (Schweizerische Bankgesellschaft) (1991–98)
Prudential	
Wasserstein Perella	*UK*
	Barclays
Bankers Trust (1991–98)	HSBC
Morgan Stanley Dean Witter (1997–99)	NatWest
Morgan Stanley (1991–96)	Robert Fleming (incl. Jardine Fleming)
Dean Witter Reynolds (1991–96)	Schroders
Salomon Smith Barney (1997–99)	Baring Brothers (1991–1994)
Salomon Brothers (1991–96)	Kleinwort Benson (1991–94)
Smith Barney (1991–96)	Smith New Court (1991–94)
	S. G. Warburg (1991–94)
Citicorp (1991–98)	
Canada	**Japan**
CIBC	Bank of Tokyo-Mitsubishi
	Daiwa
Europe	Industrial Bank of Japan
Germany	Nikko
Commerzbank	Nomura
Deutsche Bank	Yamaichi
Dresdner Bank	
	US/UK/France
France	Lazard Houses
Banque Paribas	Rothschild Group
Banque Indosuez	
SociétéGénérale	

Identification of strategic groups

Most studies in strategic group research use cluster analysis. However, its application has come under frequent attack (e.g., Barney and Hoskisson, 1990; Meyer,
1991; Thomas and Venkatraman, 1988). In order to ensure the validity of findings,
the technique must be applied prudently (Ketchen and Shook, 1996). The critical
issues when using cluster analysis are: selecting clustering variables; addressing
multicollinearity among variables; selecting appropriate clustering algorithms; determining the number of clusters; and validating clusters.

In our study, the variables were chosen with a solid theoretical foundation (see
above). But as it is an exploratory study investigating the role of knowledge assets in
acting as mobility barriers, we do not have theoretical foundation on the expected
number and nature of groups in our cluster solution. Our study used a two-pronged
approach. First, we examined a cluster analysis based on our weighted measures.
Because in this approach the distance between the measures is already meaningful,
we did not use standardization, as it might distort results. In a second step we
subjected our measures to factor analysis and used the resulting uncorrelated factor
scores for each construct as the basis for clustering. In this second step, controlling
for issues of multicollinearity, we used standardization as recommended.[13,14]

Selecting the appropriate clustering algorithms, we used a two-stage procedure. In
a first step, applying a hierarchical algorithm, we determined the number of clusters
using two approaches: we clustered groups by using our weighted measures, and
by using the factor-scored measures. As Ward's method matches the features of the
underlying structure of the data in our study best, we apply this method. In both
approaches we used dendogram observations and for some years also changes in
agglomeration coefficients to determine the number of clusters. Using both approaches, we identified the same number of clusters in every single year. This can be
regarded as a confirmation of identifying a number of clusters that reflect the data
set. In addition, in all cases the group membership of the vast majority of the firms
is found identical in both approaches; this adds to confidence in the reliability of our
results.

In a second step, we identify the groups clustering with non-hierarchical K-means,
using both our weighted measures and the factor-scored measures. Distances between
final cluster centers and the creation of a dendogram are used for determining the
ranking order of the clusters. In both approaches, group membership of the vast
majority of the firms is found to be identical in both approaches for every single
year; again, this adds to the reliability of our results. Reliability is further increased
through a comparison with the grouping obtained through hierarchical clustering and the creation of a dendogram based on non-hierarchical clustering.

Determining the time period

We chose 1991 to 1999 for our analysis because this period covered the globalization
process that transformed the investment banking business. A second motivation
was data availability. This was a period of great instability in every single region
and especially on a global scale, characterized by a new regulatory environment in

Europe and the United States, the institutionalization of financial markets, various crises such as those in 1994 and 1998, the introduction of the euro, etc. The challenge for firms was to cultivate the timely creation of valuable knowledge assets under these conditions. Our focus on a time period of strategic instability distinguishes our study from recent research in which stable strategic time periods (SSTPs) have been chosen to study strategic group structures (e.g., Bogner et al., 1996; Cool and Schendel, 1987, 1988; Fiegenbaum and Thomas, 1990, 1993, 1995; Fiegenbaum et al., 1987, 1990; Mascarenhas, 1989; Mascarenhas and Aaker, 1989; Sudharshan et al., 1991). Once the SSTPs had been identified, these studies analyzed changes in group structures by comparing group compositions of adjacent stable time periods. As Bresser et al. (1994: 189) note: "the unstable time periods, where the strategic reorientation actually takes place, are treated as black box." Peteraf and Shanley (1997) argue that groups are more likely to be important for firm performance during periods of industry instability. Also Dranove et al. (1998, 1037) criticize the restriction of strategic group studies to stable time periods. Acknowledging the importance of the strategic instability for globalization of investment banking activities in the 1990s, we will follow Bresser et al. (1994) by not separating strategic stable time periods from periods of discontinuity and focusing solely on those SSTPs.

Strategic groups are identified for every single year, as this represents a business cycle.[15] Furthermore, the *Financial Times* (1999, January, 29, i) remarked in a review of the global investment banking activities in 1999, "one year on, those [. . .] days seem a generation ago," so in this industry a yearly comparison of the investment effects and strategic efforts on changes in stocks might be appropriate. This approach allows us to investigate the stability of groupings over time in the face of strategic instability, an important condition for the usefulness of a strategic group analysis (e.g., Oster, 1982).

Rivalry measures

As we wanted to measure rivalry in order to investigate "configurations in 'resource space'" (Cool et al., 1994: 233), an adjusted Herfindahl index is an appropriate measure. This was calculated for each firm in the market segment in which it participates, excluding the focal firm's own market share. In order to assess the extent to which rivalry cut into a given firm's profits, one must exclude that firm's individual market share from the traditional concentration measures (Shepherd, 1972; Kwoka, 1979; Clarke et al., 1984). In the case of the Herfindahl index, a measure of rivalry faced by a given firm is obtained by excluding its own share from the overall industry Herfindahl index (Cool and Dierickx, 1993: 50), and hence, the effects of a firm's own market share are separated from the effects of intensity of rivalry from its competitors (Cool et al., 1989). A further benefit of this measure is that it reduces the interdependence between market power and rivalry variables (Cool et al., 1999: 7).

For rivalry to exist, firms must be interdependent; that is to say, firms must compete directly with one another for the same customer (Caves and Porter, 1977). Our definition and operationalization of global investment banking fits this definition, as each firm is capable of contesting for customers with each and every other firm (Baumol, 1982). If markets are segmented, market share and rivalry will have to be evaluated at the segment level (Cool and Dierickx, 1993: 50).

One should be concerned about the interdependencies of the various measures, since they all draw on the same data. A change in market share of one firm affects the market share of the others and thus the aggregate share of actual rivalry. However, the measures of market share and actual rivalry do not necessarily create systematic interdependencies because they could move in opposite directions, the same direction, or one could change but not the other. This is so because the indicator for actual rivalry is based on the adjusted Herfindahl measure. For instance, if the market share of one firm increases, actual rivalry could increase or decrease, depending on whether the remaining firms' market shares are becoming more or less symmetric (Cool et al., 1999: 8).

Results

In this paper, our concern is to examine whether group-level effects among knowledge assets exist. We investigated whether group membership is a significant explanatory variable for the share of within-group rivalry with respect to total rivalry for each single firm. Our study supports this at a significant level.

Table 2.3 shows the various groupings we obtained by performing the different cluster algorithms. From the 1990s on, the clusters are dominated by US firms and, in specific, by the traditional bulge bracket firms.[16] This reflects the weight of the American markets. In 1991, the American stock market accounted for 38.8 percent of world market capitalization, and 24.8 percent of the market value of bonds was traded in the US. It is clear that the leading firms in this market dominate the

Table 2.3 Groupings *K*-means weighted and factors scored, secondary firms identified

1991	Group 1	Group 2	Group 3	Group 4
Groupings (*K*-means method, weighted variables)	CSFB Goldman Sachs Merrill Lynch Morgan Stanley	Lazard Lehman Brothers Salomon Brothers	Deutsche J.P. Morgan *Nomura*	\<Rest\>
Groupings (*K*-means method, using factor scores)	Goldman Sachs Lazard Lehman Brothers Merrill Lynch Morgan Stanley Salomon Brothers	CSFB	Bankers Trust Citi Deutsche Dresdner J.P. Morgan	\<Rest\>

The factor scores-based clustering attributed to 39 out of 47 firms the same cluster membership. One firm of cluster 1, CSFB, fell back to cluster 2 when using factor scores. Cluster 2 of the *k*-means method using weighted variables was folded into cluster 1 when using factor scores.

Table 2.3 *(cont'd)*

1992	Group 1	Group 2	Group 3	Group 4
Groupings (*K*-means method, weighted variables)	*CSFB* Goldman Sachs Merrill Lynch	Lehman Brothers Morgan Stanley Salomon Brothers	*Deutsche* J.P. Morgan *Nomura* UBS *S.G. Warburg*	<Rest>
Groupings (*K*-means method, using factor scores)	Goldman Sachs Lehman Brothers Merrill Lynch Morgan Stanley Salomon Brothers	CSFB	Bankers Trust Barclays J.P. Morgan UBS	<Rest>

The factor scores-based clustering attributed to 38 out of 47 firms the same cluster. Cluster 2 of the *k*-means method using weighted variables was folded into cluster 1 when using factor scores.

1993	Group 1	Group 2	Group 3	Group 4
Groupings (*K*-means method, weighted variables)	CSFB Goldman Sachs Merrill Lynch	*Deutsche* *J.P. Morgan* Morgan Stanley	Kidder Peabody Lehman Brothers Salomon Brothers	<Rest>
Groupings (*K*-means method, using factor scores)	Goldman Sachs Merrill Lynch	CSFB Kidder Peabody Lehman Brothers Morgan Stanley Salomon Brothers	Bankers Trust Barclays Citi J.P. Morgan SBC	<Rest>

The clustering according to *k*-means method using factor scores attributed to 37 out of 47 firms the same cluster membership as the clustering according to *k*-means method using weighted variables. One firm of cluster 1, CSFB, fell back to cluster 2 when using factor scores. Cluster 3 of the *k*-means method using weighted variables was folded into cluster 2 when using factor scores. Overall, both methods produced similar clusterings.

1994	Group 1	Group 2	Group 3	Group 4
Groupings (*K*-means method, weighted variables)	Goldman Sachs Merrill Lynch Morgan Stanley	CSFB Lehman Brothers *J.P. Morgan*	Barclays/Citi/Daiwa/ Deutsche/HSBC/ Kidder Peabody/ NatWest/Nomura/ Salomon Brothers/ SBC/S.G. Warburg/ UBS	<Rest>

Table 2.3 (*cont'd*)

Groupings (*K*-means method, using factor scores)	Goldman Sachs Merrill Lynch	Kidder Peabody Lehman Brothers Morgan Stanley Salomon Brothers	Bankers Trust/ Barclays/Citi/CSFB/ J.P. Morgan	\<Rest\>

The clustering according to *k*-means method using factor scores attributed to 41 out of 46 firms the same cluster membership as the clustering according to *k*-means method using weighted variables.

1995	Group 1 (1+2)	Group 2	Group 3	Group 4
Groupings (*K*-means method, weighted variables)	Merrill Lynch (cluster 1) Goldman Sachs Morgan Stanley	SBC Warburg J.P. Morgan	CSFB Lehman Brothers Salomon Brothers	\<Rest\>
Groupings (*K*-means method, using factor scores)	Goldman Sachs Merrill Lynch (cluster 1) CSFB Morgan Stanley	SBC Warburg J.P. Morgan	Lazard Lehman Brothers Salomon Brothers	\<Rest\>

The factor scores-based clustering attributed to 39 out of 41 firms the same cluster membership.

1996	Group 1 (1+2)	Group 2	Group 3	Group 4
Groupings (*K*-means method, weighted variables)	Merrill Lynch (cluster 1) Goldman Sachs Morgan Stanley	*SBC Warburg* J.P. Morgan	CSFB Lehman Brothers Salomon Brothers *UBS*	\<Rest\>
Groupings (*K*-means method, using factor scores)	Merrill Lynch (cluster 1) CSFB Goldman Sachs Morgan Stanley Lehman Brothers Salomon Brothers	J.P. Morgan	*Chase* Citi SBC Warburg UBS	\<Rest\>

The factor scores-based clustering attributed to 34 out of 41 firms the same cluster. Cluster 4 of the *k*-means method using weighted variables was folded into cluster 2 when using factor scores.

1997	Group 1 (1+2)	Group 2	Group 3	Group 4
Groupings (*K*-means method, weighted variables)	Merrill Lynch (cluster 1) Goldman Sachs MSDW	J.P. Morgan SBC Warburg Dillon Read	CSFB Lehman Brothers Salomon Smith Barney	\<Rest\>

Table 2.3 (*cont'd*)

| Groupings (*K*-means method, using factor scores) | Merrill Lynch (cluster 1) Goldman Sachs MSDW J.P. Morgan SBC Warburg Dillon Read | *Chase* *UBS* | Citi CSFB Deutsche DLJ Lehman Brothers PaineWebber Salomon Smith Barney | <Rest> |

The factor scores-based clustering attributed to 31 out of 39 firms the same cluster membership. Cluster 3 of the *k*-means method using weighted variables was folded into cluster 2 when using factor scores.

1998	Group 1 (1+2)	Group 2	Group 3	Group 4
Groupings (*K*-means method, weighted variables)	Merrill Lynch (cluster 1) Goldman Sachs MSDW	*Deutsche* J.P. Morgan Warburg Dillon Read	*ANB Amro* CSFB Lehman Brothers Salomon Smith Barney	<Rest>
Groupings (*K*-means method, using factor scores)	Merrill Lynch (cluster 1) Goldman Sachs MSDW	J.P. Morgan Warburg Dillon Read	CSFB Deutsche Lazard Lehman Brothers Salomon Smith Barney	<Rest>

The factor scores-based clustering attributed to 36 out of 39 firms the same cluster membership.

1999	Group 1 (1+2)	Group 2	Group 3	Group 4
Groupings (*K*-means method, weighted variables)	Merrill Lynch MSDW (cluster 1) Goldman Sachs Salomon Smith Barney	Deutsche J.P. Morgan *Warburg Dillon* *Read*	CSFB Lazard Lehman Brothers	<Rest>
Groupings (*K*-means method, using factor scores)	Goldman Sachs Merrill Lynch MSDW (cluster 1) Deutsche J.P. Morgan Salomon Smith Barney	*ABN Amro* *Chase* Warburg Dillon Read	CSFB Lazard Lehman Brothers	<Rest>

The factor scores-based clustering attributed to 31 out of 36 firms the same cluster membership. Cluster 3 of the *k*-means method using weighted variables was folded into cluster 2 when using factor scores.

Group 1 (1+2): clusters 1 and 2 were merged to form group 1, as cluster 1 comprised often just one firm.

rankings. Firms in other regions of the world face fragmentation, with a comparatively very low capitalization in their dispersed markets. The 1991 groupings indicate the head start of the strong US houses in global investment banking. Through their strong home presence they already dominated the rankings. Most firms were still very much domestically oriented. This was influenced by their clients, too. Both issuers and investors, though possibly internationally present, would still seek their financing and investing in their home country. The globalization of finance had not yet taken place. At the beginning of the 1990s, global investment banking was a concept rather than reality. By the end of the decade, this picture would be drastically altered, and a significant position in the major markets would be a precondition to establish a global, full-service investment bank.

The k-means and the Ward's method attributed to most firms the same or similar cluster membership. The results obtained varied from being identical (e.g., 1991) to differences, such as one (1993) to four (1998). The most substantial difference was obtained in 1996, with six firms being attributed to different clusters. A tentative explanation might be the wave of acquisitions that took place in 1995, due to regulatory changes in the US and financial problems of UK merchant banks. Overall, both methods produced similar clusterings.

Table 2.3 indicates the groupings according to the k-means method using factor scores. Further reported is the overlap of the groupings according to k-means method using weighted variables and the groupings according to k-means method using factor scores. A high overlap indicates that "meaningful" strategic groups are identified. It should be noted that a complete overlap of the groupings using the weighted variables and using factor scores could not be hoped for; this would mean that using weighted variables did not contain any additional information. Here we used basically only replicated factor scores. We introduced the weighted variables to capture information that would be lost when applying factor scores only. Thus, neither a complete overlap nor a completely disparate picture is hoped for. The former would imply that using different methods does not yield any advantage. The latter would call into question the significance of the results altogether. A closer look at the groupings reveals that in using weighted variables we were able to capture more fully the global presence of firms and their positioning in the different markets worldwide. By applying factor scores only, we would have been even more dominated by the weight of the US markets and would have missed out on the crucial (re)positioning of firms in Europe and the emerging markets in the 1990s. With our method of weighted variables we were able to distinguish strategic groups of firms with a global, full-service investment banking presence from strategic groups of firms with a US, full-service investment banking presence. Taking this into consideration, we therefore think that overlap generated when applying the different methods indicates that in our study meaningful strategic groupings were identified.

Recognizing the similarity of groupings over time, the issue might be raised whether an analysis of stable strategic time periods (SSTPs) rather than business cycles might not have been more fruitful, capturing the essential information. Over time the cluster centroids became more distanced. The identified strategic groups became more and more distinctive. Considerable strategic reorientation took place in global investment banking throughout the 1990s. Our approach was able to capture this information, which would have been lost if an analysis based on SSTPs had been performed.

In a next step, we identified whether the strategic groups identified were significantly distinctive from each other. We put this to a formal test when checking whether group membership is a significant explanatory variable for the share of within-group rivalry with respect to total rivalry for each single firm. Our data support this at a significant level. In some years, one could raise the issue whether, in our sample, the result of significance in membership is driven solely by the first group. This is put to a more formal test whereby we check whether group membership remains significant for the remaining groups 2 to 4. We find support for this too. Table 2.4 shows for the respective years the averaged within-and between-rivalries of the different groups, and the share of within-rivalry to total rivalry. Assuming normality,

Table 2.4 Group-level (rivalry) effects among groupings

1991	Group 1	Group 2	Group 3	Group 4
Groupings (K-means method)	CSFB Goldman Sachs Merrill Lynch Morgan Stanley	Lazard Lehman Brothers Salomon Brothers	Deutsche J.P. Morgan Nomura	<Rest>
Rivalry between Rivalry within Within to total	0.0469 0.0352 0.4287 F-statistic 4.10363	0.0663 0.0171 0.2401 p-value 0.013	0.0602 0.0021 0.0021 F*-statistic 6.86569	0.0590 0.0240 0.2947 p*-value 0.002
1992	Group 1	Group 2	Group 3	Group 4
Groupings (K-means method)	CSFB Goldman Sachs Merrill Lynch	Lehman Brothers Morgan Stanley Salomon Brothers	Deutsche J.P. Morgan Nomura UBS S.G. Warburg	<Rest>
Rivalry between Rivalry within Within to total	0.0497 0.0292 0.3692 F-statistic 2.71539	0.0734 0.0137 0.1567 p-value 0.058	0.0713 0.0059 0.0787 F*-statistic 4.70407	0.0600 0.0161 0.2547 p*-value 0.015
1993	Group 1	Group 2	Group 3	Group 4
Groupings (K-means method)	CSFB Goldman Sachs Merrill Lynch	Deutsche J.P. Morgan Morgan Stanley	Kidder Peabody Lehman Brothers Salomon Brothers	<Rest>

Table 2.4 (*cont'd*)

1993	Group 1	Group 2	Group 3	Group 4
Rivalry between	0.0473	0.0670	0.0800	0.0455
Rivalry within	0.0300	0.0046	0.0077	0.0237
Within to total	0.3877	0.0685	0.0877	0.3602
	F-statistic	p-value	F^*-statistic	p^*-value
	4.96333	0.005	9.44931	0.000

1994	Group 1	Group 2	Group 3	Group 4
Groupings (*K*-means method)	Goldman Sachs Merrill Lynch Morgan Stanley	CSFB Lehman Brothers J.P. Morgan	Barclays/Citi/Daiwa/ Deutsche/HSBC/ Kidder Peabody/ NatWest/Nomura/ Salomon Brothers/ SBC/S.G. Warburg/ UBS	<Rest>
Rivalry between	0.0415	0.0675	0.0417	0.0608
Rivalry within	0.0256	0.0085	0.0193	0.0089
Within to total	0.3796	0.1106	0.3184	0.1313
	F-statistic	p-value	F^*-statistic	p^*-value
	20.9334	0.000	28.5696	0.000

1995	Group 1 (1+2)	Group 2	Group 3	Group 4
Groupings (*K*-means method)	Merrill Lynch (cluster 1) Goldman Sachs Morgan Stanley	SBC Warburg J.P. Morgan	CSFB Lehman Brothers Salomon Brothers	<Rest>
Rivalry between	0.0423	0.0585	0.0591	0.0409
Rivalry within	0.0253	0.0008	0.0162	0.0143
Within to total	0.3704	0.0191	0.2149	0.3065
	F-statistic	p-value	F^*-statistic	p^*-value
	1.72024	0.181	4.15890	0.024

1996	Group 1 (1+2)	Group 2	Group 3	Group 4
Groupings (*K*-means method)	Merrill Lynch (cluster 1) Goldman Sachs Morgan Stanley	SBC Warburg J.P. Morgan	CSFB Lehman Brothers Salomon Brothers UBS	<Rest>
Rivalry between	0.0440	0.0637	0.0605	0.0430
Rivalry within	0.0263	0.0030	0.0135	0.0161
Within to total	0.3726	0.0498	0.1797	0.3255
	F-statistic	p-value	F^*-statistic	p^*-value
	3.23987	0.034	6.5308	0.004

Table 2.4   *(cont'd)*

1997	Group 1 (1+2)	Group 2	Group 3	Group 4
Groupings (*K*-means method)	Merrill Lynch (cluster 1) Goldman Sachs MSDW	J.P. Morgan SBC Warburg Dillon Read	CSFB Lehman Brothers Salomon Smith Barney	<Rest>
Rivalry between Rivalry within Within to total	0.364 0.0371 0.5019 *F*-statistic 2.87977	0.0661 0.0028 0.0497 *p*-value 0.051	0.0748 0.0105 0.1223 *F**-statistic 5.08570	0.0466 0.0178 0.3457 *p**-value 0.012
1998	**Group 1 (1+2)**	**Group 2**	**Group 3**	**Group 4**
Groupings (*K*-means method)	Merrill Lynch (cluster 1) Goldman Sachs MSDW	Deutsche J.P. Morgan Warburg Dillon Read	ANB Amro CSFB Lehman Brothers Salomon Smith Barney	<Rest>
Rivalry between Rivalry within Within to total	0.0438 0.0297 0.4042 *F*-statistic 22.80450	0.0696 0.0067 0.0883 *p*-value 0.000	0.0738 0.0103 0.1219 *F**-statistic 4.66992	0.0597 0.0109 0.1582 *p**-value 0.017
1999	**Group 1 (1+2)**	**Group 2**	**Group 3**	**Group 4**
Groupings (*K*-means method)	Merrill Lynch MSDW (cluster 1) Goldman Sachs Salomon Smith Barney (cluster 2)	Deutsche J.P. Morgan Warburg Dillon Read	CSFB Lazard Lehman Brothers	<Rest>
Rivalry between Rivalry within Within to total	0.0325 0.0459 0.5851 *F*-statistic 38.93189	0.0703 0.0055 0.0785 *p*-value 0.000	0.0727 0.0111 0.1301 *F**-statistic 3.67059	0.0575 0.0083 0.1508 *p**-value 0.037

Rivalry between: Rivalry between groups.
Rivalry within: Rivalry within groups.
Within to total: Share of within-group rivalry to total group rivalry.
F*-statistic/p*-value: statistics for groups 2 to 4.
Group 1 (1+2): clusters 1 and 2 were merged to form group 1, as cluster 1 comprised often just one firm.

the resulting test statistic is F-distributed, and, as the figures show, all of them are significant or highly significant. This indicates that for our groupings group-level effects exist. There is one single non-significance of our test statistic for the overall share of within to total rivalry in 1995; we speculate that this result can be explained by the turmoil in the bond markets in 1994 and the repercussion this had for investment banking firms. 1994 was a disastrous year. Volume in many markets halved as the Fed raised interest rates. American firms were much less likely to refinance; this led to a shake-up of many underwriters, who, moreover, had in many cases a large inventory on many positions that were now much less valuable. Many firms reported substantial losses, some, e.g., Kidder Peabody, even went bankrupt. Overall, the results show that our groupings exhibit group-level effects and, thus, are "true" strategic groups. Group-level effects among collective knowledge assets exist.

Discussion

By the end of the 1990s, a clear picture of three dominant groups emerged in global investment banking. Firms with "regional or special focus" complemented them. Table 2.5 gives an overview. The firms in the leading group 1 can be characterized as *global, full-service investment banks*. Members of this group are the leading firms on Wall Street. In the course of the 1990s, they succeeded in establishing a firm presence among the top banks in every major investment banking center in the world. By the end of the 1990s, they were characterized by "seamless integrated"

Table 2.5 Strategic groups in global investment banking in the 1990s (with cluster membership)

Groups	Member firms	Representation
Group 1 cluster 1: 1991–94 cluster 1+2: 1995–99	Merrill Lynch Goldman Sachs Morgan Stanley/MSDW	global, full-service firms
Group 2 cluster 3: 1991–92 cluster 2: 1993–94 cluster 3: 1995–99	J.P. Morgan Deutsche SBC Warburg/SBC Warburg Dillon Read/Warburg Dillon Read [1995–1999]	global contenders
Group 3 cluster 2: 1991, 92, 94 cluster 3: 1993 cluster 4: 1995–99	CSFB Lehman Brothers Salomon Brothers/Salomon Smith Barney	globalizing, full-service contenders
Group 4 Group "regional or special focus" cluster 4: 1991–94 cluster 5 (4): 1995–99	<Rest>	"regional or special focus" firms

Author's analysis.

global, full-services. They were able to build up valuable knowledge assets and were as regards reputation, market orientation and complementary assets among the leading firms with the highest score.

The second group that emerged can be described as *global contenders*, comprising mainly cluster 2 of the groupings in the second half of the 1990s. Members of this group are the "new entrants" in investment banking in the 1990s. At the end of the 1980s, firms belonging to this group were mainly commercial banks whose management decided that due to secular trends their firms' future lay in investment banking. In the 1990s, they made a forceful and systematic effort to build up their investment banking capabilities. They all started by establishing a leading position in Eurobonds and from there expanded their businesses into new business lines and into new geographic regions. These firms successfully established an investment banking presence in many parts of the world. Nevertheless, they had gaps in some geographical and business areas and were not able to offer seamless integrated full-service investment banking on the same level of expertise as the first group. They cannot be described as competitors of a "regional or special focus," but they were, in fact, global rivals to the firms of group 1. But they still had to complement their knowledge assets to achieve both a stronger presence in some parts of the world and acquire stronger full-service capabilities. On some dimensions, they had a gap in knowledge assets, compared to the members of group 1.

The third group that dominated the groupings can be described as *globalizing, full-service contenders*, comprising mainly cluster 3 of the groupings in the second half of the 1990s. Members of this group are the rivals on Wall Street of the firms on group 1. The group contained a set of American firms that were part of the US "bulge bracket firms" that offered full-service investment banking in the US. At the beginning of the 1990s, they could rival the firms of group 1 on almost every respect. Then, all the firms of group 3 had to go through a period of turmoil in the first half of the 1990s. This absorbed management attention, and the managers of these firms were unwilling or unable to build up their firms' global investment banking in the same way as the firms of group 1. In the first half of the 1990s, some of the members of group 2 had valuable activities even on a global scale, but their globally dispersed units acted as autonomous entities and opposed the creation of integrated global businesses. This distinguishes the firms of group 1. Whereas the firms of group 1 were able to build up integrated, global investment banking activities, the firms of group 3 struggled to do so. Only in the second half of the 1990s, when the existence of valuable knowledge assets in global, full-service investment banking became imperative, did the firms of group 3 overcome their internal struggles and start to globalize their US-centered full-service activities. Some of them had a strong presence in selected areas in Europe or other parts of the world, but they were not able to offer the same kind of global, full-service investment banking as the firms of group 1. They would still have to complement their gaps and build up valuable knowledge assets on a global scale.

Finally, all other firms occupied cluster 4 (respectively 5 in the second half of the 1990s). This group covered the firms with *"regional or special focus,"* i.e., firms that are either strong in a specific market, traditionally their domestic market, or strong in specific product categories, sometimes on a worldwide level; nevertheless, they cannot claim to dominate global, full-service investment banking.

The objective of this paper is not to show that firms in different strategic groups differ in terms of their endowment of idiosyncratic resources. Rather, our objective is to show that group-level effects among knowledge assets, the firm's most idiosyncratic resources, do in fact exist. In this sense, we have argued that if firms are affected by rivalry effects in their competitive positioning, a discussion of knowledge assets may not be satisfactory if it does not include group-level effects in its investigation of performance differences. We find empirical support for the existence of group-level effects among knowledge assets in global investment banking in the 1990s. Hence, an empirical study investigating performance differences based on knowledge assets/resources/capabilities/competences might need to include strategic group membership considerations. We have shown that firm performance is a function of both the stock of knowledge assets and the group membership in knowledge-asset space, if group-level effects among the knowledge assets in the businesses we researched do take place.

The contributions of the paper are as follows. First, by its design the current study is one of the few studies that control for group-level effects (e.g., Cool and Dierickx, 1993; Peteraf, 1993) and thereby lends additional empirical support for a concept of strategic groups as still relevant to the strategic management field. This would hold true at least concerning future research in the investment banking industry. Secondly, our study also distinguishes between core and secondary members of a group as proposed by Reger and Huff (1993). It explicitly identifies and lists issues to be considered in an empirical strategic group study. Also, we use weighted variables as advocated by Ketchen and Shook (1996: 453). Thirdly, our study uses a two-stage procedure for identifying strategic groups, with hierarchical and non-hierarchical methods applied in tandem as advocated by Ketchen and Shook (1996: 452). Prior to 1996 (the year of Ketchen and Shook's review article), only six studies applying cluster analysis had used this dual technique. Fourthly, prior studies of investment banking in strategic management research investigated the dynamics of the industry in the 1980s in the US (e.g., Eccles and Crane, 1988; Hayes et al., 1983; Chung et al., 2000). The dynamics of the business altered substantially in the 1990s and were of a different quality on a global level. Our study sheds light on these issues.

Fifthly, the most important contribution of our study is to strategic group theory and research made by developing and empirically exploring a strategic group concept based on knowledge assets. Our study does not identify the firm sample based on a predefined industry structure, but based on knowledge assets. Thus, we do not investigate "investment banks" (as, for example, identified by the SIC-code or similar classification), but "firms with investment banking activities," as we selected the firms that have knowledge assets in global investment banking. It is one of the few studies that identify strategic groups based on mobility barriers (e.g., Mascarenhas and Aaker, 1989) and not based on firm strategies. In so doing, this study seeks to find the root causes of competitive advantage and group level effects: our focus is on the knowledge assets that allow firms to implement value-creating strategies, such as global full service, rather than on the dimensions of the strategies themselves.

Because knowledge assets are path dependent and take time to build up (see Arthur, 1996), and as they can indeed be collective in the manner outlined, future studies of mobility barriers should be well served by focusing on knowledge assets. The study does not base its identification of variables based on balance-sheet activities. As Teece et al. (1997, 518) stress: "We point out that the assets that matter for competitive advantage are rarely reflected in the balance sheet, while those that do not, are." We have contributed by operationalizing the construct of knowledge assets by reputation, market orientation, and complementary assets. Because knowledge assets are collective in the asset space and constitute mobility barriers, we elected to use data and reports external to the firm, and not those reflected in the balance-sheet activities.

Sixthly, the operationalization of collective knowledge assets found here should be of relevance to the knowledge-based view of the firm, and so should the performance implications of the identified collective knowledge assets. Firm performance is a function of both the stock of knowledge assets and the group membership in knowledge-asset space, if group-level effects among the knowledge assets in the researched businesses do indeed take place. This raises the issue as to why group-level effects among knowledge assets exist. We speculate that among firms in the same strategic group *knowledge spillover* effects exist. Knowledge spillover could result from a history of interactions among strategic group members, from learning through benchmarking among strategic group members, and from inter-firm mobility of individual employees and whole teams of experts. Strategic group members can draw upon a history of common interactions, events, and experiences. These interactions help the firms to understand their competitors better and chart their position in knowledge-asset space. Such understanding and information are crucial for prudent strategizing in firms, and hence future studies should uncover the extent to which, and by what means, knowledge spillover occurs in strategic groups.

Because group-level effects are present, benefits to group members may ensue from comprehensive spillover of knowledge within the group, and group members might therefore initiate, embrace, or allow activities to build up collective knowledge assets. Especially in knowledge-intensive industries, learning through benchmarking can produce valuable insights. Firms will have to manage this systematically by removing barriers (see, e.g., Szulanski, 1995, 1996). Furthermore, the inter-firm mobility of employees is a major issue in investment banking and also in other knowledge-intensive industries. Virtually no week passes without reports of senior managers defecting to rival firms. This not only causes a seemingly never-ending increase in salaries, it also leads to substantial knowledge-transfer among firms, especially if people move laterally among in-group firms. A study of the patterns of inter-firm mobility of employees within and across strategic group members, as well as of the firms' responses and level of commitment to such mobility, could shed light on knowledge spillover among group members.

Clearly, our research is only a first step in the field of knowledge-based strategic groups, and further work is much needed. Such research should not only investigate the existence of group-level effects among knowledge assets, but should also direct attention to the extent of group-level effects and their consequences for the explanation of intra-industry performance differences, and examine the relations that give rise to the existence of group-level effects.

Conclusion

The purpose of the research underlying this paper was to draw up and test the hypothesis that strategic group-level effects among knowledge assets exist. Knowledge assets are collective and exist in the strategic group's asset space. Using data from the investment banking industry (1991–99), we found support for this hypothesis. Our findings have implications for theory and research on strategic groups as well as for the knowledge-based view of the firm. It is our hope that this study will inspire more work on various types of interaction among firms, including the question of how knowledge spillovers shape mobility barriers.

Acknowledgements

This research project received support from the Swiss National Science Foundation. Thanks are due to Thomson Financial Securities Data for permitting access to their databases.

Notes

1   Knowledge assets are mobility barriers on a firm rather than on a group level. Yet if we are able to identify even group-level effects among knowledge assets, we can argue that strategic group considerations should be included in empirical studies investigating resources/capabilities/knowledge assets and their ability to influence a firm's competitive advantage. This should be a major contribution to our current understanding of strategic groups.

2   Take the example of reputation. For group-level *market power effects* to occur, managers of member firms must recognize their mutual interdependence – that each member's actions affect the outcomes of other members. As a result, they take into account the activities of other firms to improve their own results. This mutual modeling and its resulting coordinated interactions at the group level produce market power effects on profitability (Dranove et al., 1998: 1032). Clearly, this is the case for reputation; the value of a firm's reputation is affected by the actions of other firms. Tirole's (1996) work on collective reputations finds that individual reputations are determined by collective reputations, and vice versa. In investment banking the firms coordinate their actions in an attempt to increase their reputational capital (Chung et al., 2000).

   Efficiency effects are a second type of group-level effects. A strategic alliance among group members would be an example of a static efficiency effect due to interactions among group members (Dranove et al., 1998: 1033). Chung et al. (2000) show that in investment banking strategic alliances are created with the intent to access each other's complementary assets and enhance through collaboration their reputational capital Dynamic efficiency effects may also result from group-level effects (Dranove et al., 1998: 1033). Increased levels of interaction within a group may intensify competitive activity, which may induce greater numbers of new product introductions, higher-quality products, and faster competitive response (Smith et al., 1996). To be perceived by customers as a premium bank, membership of the "(global) bulge bracket" in investment banking inspires lots of effort put into excelling at the superiority of its services (Eccles and Crane, 1988).

A third type of group-level effects on profitability occurs through the effect of group interactions on *differentiation* (Dranove et al., 1998: 1033). Firms may interact to create reputational capital; just as the professional interactions of the "Big Four" accounting firms help to differentiate them from smaller competitors (Dranove et al., 1998: 1033), so do the interactions among the "bulge bracket" firms differentiate them from the rest and create reputational capital.

Moreover, Baden-Fuller et al. (2000: 624) indicate group-level effects in the case of reputation when they remark that reputation resources may flow between one organization and another because of perceived or actual linkages.

3   Note that, for example, Cool et al. (1994) used balance-sheet data for identifying the stocks of firms in their study. For a conceptualization of asset stocks without reliance on balance-sheet data, see also DeCarolis and Deeds (1999).

4   See Rao (1994: 32) for a discussion of the interaction of certification contest and path-dependent processes, extrinsic criteria of fitness, and the "Matthew effect."

5   One might raise the issue whether the market orientation variables and the reputation variables are not correlated. First, they measure distinct properties. Whereas the market orientation variables identify how a firm is embedded in the "deal flow," the reputation variables detect customers' evaluation of a firm's activities. Second, a look at the ranking shows that often firms, which do not have high market share, end up scoring highly in the reputation rankings.

6   Consequently, though customers did not directly evaluate this variable, it represents a valuable knowledge asset, too. It should be noted that this is the sole variable that does not directly use customers' evaluations for our analysis. Still, as indicated by Brad England of Merrill Lynch, the number of RRs contributes to the stock of knowledge a firm has in this area.

7   1990 data was used for substituting missing 1991 data (equivalent to the Asian equity execution variable).

8   In the category of "Overall best in Europe" the top 8 firms were identified and ranked in 1999; the top 10 firms in 1998, 1997; the top 7 in 1996; the top 4 in 1995; and the top 3 in 1994, 1993, 1992.

9   If any category was not completed in any single year, due to reasons of distortion the category was dropped in the respective year for our study – rather than considering the results of the previous year.

10 If, for example, a researcher expects a small effect and compares four groups of observations (e.g., four strategic groups), and adopts $\alpha = 0.05$, then a sample size of 274 is needed.

11 See above for the various segments and complementary assets identified.

12 Since strategic group analysis is concerned with activity within an industry, the term "firm" stands for "division" or "strategic business unit" in the case of diversified corporations.

13 In order to be able to perform our cluster analysis, we had to transform our rank-ordered measures into meaningful distance measures. As the variables are industry specifically defined, we applied a transformation procedure which reflects the nature of the industry-specific distribution: 1st: 0.25, 2nd: 0.2, 3rd: 0.2, 4th: 0.1, and 5th: 0.1. And for those few cases where we had more than the first five firms in a ranking order, we applied, based on the same considerations, the following transformation procedure for the other ranks: 6th: 0.05, 7th: 0.02, 8th: 0.02, 9th: 0.01, 10th: 0.01, and 11th to 25th: 0.002.

14 However, this technique is controversial, because researchers often omit all factors with low eigenvalues (statistic representing the amount of variance explained by a factor). The excluded factors may represent unique, important information (Dillon, Mulani, and

Frederick, 1989), meaning that a less-than-optimal set of clusters may result. Thus, similarly to standardization, any remedy for multicollinearity has potential pitfalls (Ketchen and Shook, 1996: 444). In our study, in the weighted measures approach, no factors are omitted: we avoid these issues in using both approaches concurrently. If in using both approaches we find the same results, we are able to be confident of our results.

15 This implicitly assumes that a calendar year encompasses a stable time period for strategic group membership. The reason for this is the nature of the financial reporting and its impact on companies' needs for financial advice, and it represents a business cycle.

16 Eccles and Crane (1988) identified First Boston (CSFB), Goldman Sachs, Lehman Brothers, Morgan Stanley, Merrill Lynch, and Salomon Brothers as bulge bracket, resp. special bracket firms. They are distinguished from second-tier firms and niche firms.

17 In the category of "overall best in capital raising through the international markets" the top 10 firms were identified and ranked.

18 All volume data, except for international equity issuance from Thomson Financial Securities Data. Data on international equity issuance from OECD for 1990–1995 and BIS for 1996–1998; OECD, 1996, *Financial Market Trends*, 63: 74; OECD, 1994, *Financial Market Trends*, 57: 113; BIS, *International Banking and Financial Markets Developments*, August 1999: 78. US and Euro-MTN arrangers and dealers are weighted equally (see, e.g., *Euromoney*, 1999, *International Capital Markets Review*), volume data for US and Euro-MTN programs are taken from Thomson Financial Securities Data.

19 IFC, *Emerging Stock Markets Factbook 1999*, 16f. for stock market capitalization. FIBV for the market value of bonds listed (see publications on www.fibv.org). Data as available from FIBV has been considered: For the US the markets of Amex, Chicago, NASDAQ, and NYSE are considered; For Japan the markets of Osaka and Tokyo are considered.

20 In the category of "overall best house in lead managing Eurobonds" the top 10 firms were identified and ranked from 1992 to 1998.

21 Australian and New Zealand dollar issues as the sum of issues in Australian dollars and New Zealand dollars.

22 In the category of "overall best house in lead managing Eurobonds" the top 10 firms were identified and ranked in 1998; top 8 in 1997; top 5 in 1999, 1996, 1994, and 1993; top 4 in 1995.

23 "Global issuers" indicates issues by non-US and non-European issuers.

24 Traditional p.p. excludes CDs, deposit notes, acquisition-related, lease-related, high-yield, placements with registration rights, mortgage-backed, asset-backed, MTNs, ESOP-related and 144A deals.

25 Yankee placements include foreign issuers (and US issuers with a foreign parent) issuing in the US.

26 Securitized private placements include asset-backed and mortgage-backed securities.

27 Plain equity p.p. excludes CDs, deposit notes, acquisition-related, lease-related, high-yield, placements with registration rights, mortgage-backed, asset-backed, MTNs and ESOP-related deals.

28 "Global transactions" defined as non-US and non-European transactions.

29 i.e., not the above mentioned. It should be noted that very few transactions fall into this category.

30 Countries incorporated as considered in the 1999 editions of the *Institutional Investor* research team surveys.

31 If any category was not completed in any single year, due to reasons of distortion the category was dropped in the respective year for our study – rather than considering the results of the previous year.

32 If any category was not completed in any single year, due to reasons of distortion the category was dropped in the respective year for our study – rather than considering the results of the previous year.

33 In the category of "voted as overall best house in trading Eurobonds" the top 25 firms were identified and ranked in 1999, 1998; the top 20 firms in 1997, 1996, 1995; the top 15 in 1994, 1993, 1992, and 1991. In the category of "voted as best in trading fixed bonds in US dollars" the top 10 firms were identified and ranked from 1991 to 1999. In the category of "voted as best in trading floating rates notes in US dollars" the top 10 firms were identified and ranked from 1991 to 1999.

34 In the category of "voted as overall best house in trading government bonds" the top 25 firms were identified and ranked in 1999, 1998; the top 20 firms in 1997; the top 22 firms in 1996; the top 15 in 1995; the top 13 in 1994; and the top 14 in 1993. For the categories of B the top 10 firms were identified and ranked in 1999. In the category of "voted as best in trading government bonds in US dollars" the top 6 firms were identified and ranked in 1994 and 1995.

35 If any measure was not completed in any single year, due to reasons of distortion the measure was dropped in the respective year for our study – rather than considering the results of the previous year.

36 In the category of "overall best in risk management" the top 10 firms were identified and ranked in 1999, 1998, 1996, 1995, 1994, 1993, 1992; the top 15 in 1997; and the top 20 in 1991. In the category of "best in risk management advice" the top 10 firms were identified and ranked in 1999, 1998, 1997, 1996, 1995, and 1994.

37 If any measure was not completed in any single year, due to reasons of distortion the measure was dropped in the respective year for our study – rather than considering the results of the previous year.

References

Aaker, D. A. 1991: *Managing Brand Equity: Capitalizing on the Value of a Brand Name.* New York: The Free Press.

Amit, R. and Schoemaker, P. J. H. 1993: Strategic assets and organizational rent. *Strategic Management Journal*, 14 (1), 33–46.

Anand, J. and Singh, H. 1997: Asset redeployment, acquisitions and corporate strategies in declining industries. *Strategic Management Journal*, 18 (Summer Special Issue), 99–118.

Aoki, M. 1990: The participatory generation of information rents and the theory of the firm. In M. Aoki, B. Gustafsson, and O. E. Williamson (eds.), *The Firm as Nexus of Treaties.* London: Sage, 26–51.

Arthur, W. B. 1996: Increasing returns and the New World of Business. *Harvard Business Review*, 74, July–August, 100–9.

Baden-Fuller, C., Ravazzolo, F., and Schweizer, T. 2000: Making and measuring reputations: the research ranking of European Business Schools. *Long Range Planning*, 33, 621–50.

Barney, J. B. 1986a: Organizational culture: can it be a source of sustained competitive advantage? *Academy of Management Review*, 11 (3), 656–65.

Barney, J. B. 1986b: Strategic factor markets: expectations, luck and business strategy. *Management Science*, 32 (10), October, 1231–41.

Barney, J. B. 1991: Firm resources and sustained competitive advantage. *Journal of Management*, 17 (1), 99–120.

Barney, J. B. and Hoskisson, R. E. 1990: Strategic groups: untested assertions and research proposals. *Managerial and Decision Economics*, 11 (3), 187–98.

Baumol, W. J. 1982: Contestable markets: an uprising in the theory of industry structure. *American Economic Review*, 72 (1), March, 1–15.

Blackler, F. 1995: Knowledge, knowledge work and organizations: an overview and interpretation. *Organization Studies*, 16 (6), 1021–46.

Bogner, W. C., Thomas, H., and McGee, J. 1996: A longitudinal study of the competitive positions and entry paths of European firms in the US pharmaceutical market. *Strategic Management Journal*, 17, February.

Boisot, M. H. 1998: *Knowledge Assets: Securing Competitive Advantage in the Information Economy.* Oxford: Oxford University Press.

Bresser, R., Dunbar, R., and Jithendranathan, T. 1994: Competitive and collective strategies: an empirical examination of strategic groups. In P. J. Shrivastava, A. S. Huff, and J. E. Dutton (eds.), *Advances in Strategic Management*, Vol. 10B. Greenwich, CT: JAI Press, 187–211.

Brush, T. H. and Bromiley, P. 1997: What does a small corporate effect mean? A variance-components simulation of corporate and business effects. *Strategic Management Journal*, 18, 825–35.

Caves, R. E. and Porter, M. E. 1977: From entry barriers to mobility barriers: conjectural decisions and contrived deterrence to new competition. *Quarterly Journal of Economics*, 91, May, 241–62.

Caves, R. E. and Pugel, T. 1980: *Intra-industry differences in conduct and performance: Viable strategies in US manufacturing industries.* Monograph, New York University, Salomon Brothers Center for the Study of Financial Institutions, New York.

Chang, S.-J. and Singh, H. 2000: Corporate and industry effects on business unit competitive position. *Strategic Management Journal*, 21, 739–52.

Chung, S., Singh, H., and Lee, K. 2000: Complementarity, status similarity and social capital as drivers of alliance formation. *Strategic Management Journal*, 21, 1–22.

Clarke, R., Davis, S., and Waterson, M. 1984: The profitability–concentration relationship: market power or efficiency. *Journal of Industrial Economics*, June, 435–50.

Conner, K. R. 1991: A historical comparison of resource-based theory and five schools of thought within industrial organization economics: do we have a new theory of the firm? *Journal of Management*, 17, 121–54.

Conner, K. R. and Prahalad, C. K. 1996: A resource-based theory of the firm: knowledge versus opportunism. *Organization Science*, 7 (5), September–October, 477–501.

Cool, K. and Dierickx, I. 1993: Rivalry, strategic groups and firm profitability. *Strategic Management Journal*, 14, 47–59.

Cool, K., Dierickx, I., and Jemison, D. 1989: Business strategy, market structure and risk–return relationships: a structural approach. *Strategic Management Journal*, 10 (6), 507–22.

Cool, K., Dierickx, I., and Martens, R. 1994: Asset stocks, strategic groups and rivalry. In H. Daems and H. Thomas (eds.), *Strategic Groups, Strategic Moves and Performance.* Oxford: Pergamon Press, 219–34.

Cool, K., Röller, L.-H., and LeLoux, B. 1999: The relative impact of actual and potential rivalry on firm profitability in the pharmaceutical industry. *Strategic Management Journal*, 20, 1–14.

Cool, K. and Schendel, D. E. 1987: Strategic group formation and performance: the case of the US pharmaceutical industry, 1963–1982. *Management Science*, 33 (9), 1102–24.

Cool, K. and Schendel, D. E. 1988: Performance differences among strategic group members. *Strategic Management Journal*, 9 (3), 207–23.

Day, G. S. 1994: The capabilities of market-driven organizations. *Journal of Marketing*, 58, October, 37–52.

DeCarolis, D. M. and Deeds, D. L. 1999: The impact of stocks and flows of organizational knowledge on firm performance: an empirical investigation of the biotechnology industry. *Strategic Management Journal*, 20, 953–68.

Dermine, J. 1996: European banking with a single currency. *Financial Markets, Institutions & Instruments*, 5 (5), 62–101.

Desphandé, R., Farley, A. U., and Webster, F. E., Jr. 1993: Corporate culture, customer orientation, and innovativeness in Japanese firms: a quadrad analysis. *Journal of Marketing*, 57, January, 23–7.

Dess, G. G. and Davis, P. S. 1984: Porter's (1980) generic strategies as determinants of strategic group membership and organizational performance. *Academy of Management Journal*, 27, 467–88.

Diamontoplous, A. and Hart, S. 1993: Linking market orientation and company performance: preliminary work on Kohli and Jaworski's framework. *Journal of Strategic Marketing*, 1, 93–122.

Dierickx, I. and Cool, K. 1989: Asset stock accumulation and the sustainability of competitive advantage. *Management Science*, 35, 12, December, 1504–11.

Dillon, W. R., Mulani, N., and Frederick, D. G. 1989: On the use of component scores in the presence of group structure. *Journal of Consumer Research*, 16, 106–12.

DiMaggio, P. J. and Powell, W. W. 1983: The iron cage revisited: institutional isomorphism and collective rationality in organizational fields. *American Sociological Review*, 48, April, 147–60.

Dosi, G. and Marengo, L. 1994: Some elements of an evolutionary theory of organizational competencies. In R. W. England (ed.), *Evolutionary Concepts in Contemporary Economics*. Ann Arbor, MI: University of Michigan Press, 157–78.

Dougherty, D. 1990: Understanding new markets for new products. *Strategic Management Journal*, 11 (Summer Special Issue), 59–78.

Dranove, D., Peteraf, M. A., and Shanley, M. 1998: Do strategic groups exist? An economic framework for analysis. *Strategic Management Journal*, 19, 1029–44.

Durisin, B. 2001: *Global investment banking: Competing on knowledge assets in the quest for competitive advantage*. Unpublished dissertation, mimeo, University of St. Gallen, St. Gallen.

Eccles, R. G. and Crane, D. B. 1988: *Doing Deals: Investment Banks at Work*. Boston, MA: Harvard University Press.

Euromoney 1995: Smith runs with the herd, September.

Euromoney 1998: Tussling for the bulge bracket, May.

Euromoney 1998: Selling Europe to the US, September.

Fiegenbaum, A., Hart, S., and Schendel, D. E. 1996: Strategic reference point theory. *Strategic Management Journal*, 17 (3), 219–35.

Fiegenbaum, A., Sudharshan, D., and Thomas, H. 1987: The concept of stable strategic time periods in strategic groups research. *Managerial and Decision Economics*, 8, 139–48.

Fiegenbaum, A., Sudharshan, D., and Thomas, H. 1990: Strategic time periods and strategic group research: concepts and an empirical example, *Journal of Management Studies*, 27, 133–48.

Fiegenbaum, A. and Thomas, H. 1990: Strategic groups and performance: the US insurance industry, 1970–1984. *Strategic Management Journal*, 11, 197–215.

Fiegenbaum, A. and Thomas, H. 1993: Industry and strategic group dynamics: competitive strategy in the insurance industry, 1970–1984. *Journal of Management Studies*, 30, 69–105.

Fiegenbaum, A. and Thomas, H. 1995: Strategic groups as reference groups: theory, modeling and empirical examination of industry and competitive strategy. *Strategic Management Journal*, 16, 461–76.

Financial Times 1999: Good, but it could be much better. Separate section on global investment banking, January 29.

Fombrun, C. J. 1996: *Reputation: Realizing Value from the Corporate Image.* Boston, MA: Harvard Business School Press.

Fombrun, C. J. and van Riel, C. 1998: The reputational landscape. *Corporate Reputation Review*, 1 (1), 5–13.

Fransman, M. 1994: Information, knowledge, vision and theories of the firm. *Industrial and Corporate Change*, 3 (3), 713–57.

Frazier, G. L. and Howell, R. D. 1983: Business definition and performance. *Journal of Marketing*, 47, Spring, 59–67.

Grant, R. M. 1991: The resource-based theory of competitive advantage. *California Management Review*, 33, 114–34.

Grant, R. M. 1997: A knowledge-based view of the firm: implications for management practice. *Long Range Planning*, 30 (3), 450–4.

Greenley, G. E. 1995: Market orientation and company performance: empirical evidence from UK companies. *British Journal of Management*, 6, 1–13.

Griffiths, D., Boisot, M., and Mole, V. 1999: *Strategies for managing knowledge assets: A tale of two companies.* Working paper 162, ESADE, Barcelona.

Gripsrud, G. and Gronhaug, K. 1985: Strategy and structure in grocery retailing: a sociometric approach, *Journal of Industrial Economics*, 33, 339–47.

Hall, R. 1993: A framework linking intangible resources and capabilities to sustainable competitive advantage. *Strategic Management Journal*, 14 (8), 607–18.

Hannan, M. and Freeman, J. 1977: The population ecology of organizations. *American Journal of Sociology*, 82, 929–64.

Hansen, G. S. and Wernerfelt, B. 1989: Determinants of firm performance: the relative importance of economic and organizational factors. *Strategic Management Journal*, 10 (5), 399–411.

Hatten, K. J. and Hatten, M. L. 1987: Strategic groups, asymmetrical mobility barriers, and contestability. *Strategic Management Journal*, 9, 329–42.

Hayes, S. L., III, Spence, M. A., and Marks, D. V. P. 1983: *Competition in the Investment Banking Industry.* Cambridge, MA: Harvard University Press.

Hedlund, G. 1994: A model of knowledge management and the N-form corporation. *Strategic Management Journal*, 15, 73–90.

Henderson, R. and Cockburn, I. 1994: Measuring competence? Exploring firm effects in pharmaceutical research. *Strategic Management Journal*, 15 (Winter Special Issue), 63–84.

Henderson, R. M. and Clark, K. B. 1990: Architectural innovation: the reconfigurations of existing product technologies and the failure of established firms. *Academy Science Quarterly*, 35 (1), 9–30.

Hodgson, G. M. 1998: Competence and contract in the theory of the firm. *Journal of Economic Behavior and Organization*, 35, 179–201.

Hunt, S. D. and Morgan, R. M. 1995: The comparative advantage theory of competition. *Journal of Marketing*, 59, April, 1–15.

Ichijo, K., von Krogh, G., and Nonaka, I. 1998: Knowledge enablers. In G. von Krogh, J. Roos, and D. Kleine (eds.), *Knowing in Firms: Understanding, Managing and Measuring.* London: Sage.

International Monetary Fund 1998: Globalization of finance and financial risks. *International Capital Markets: Developments, Prospects, and Key Policy Issues.* Washington, DC: IMF, September.

Itami, H. 1987: *Mobilizing Invisible Assets.* Cambridge, MA: Harvard University Press.

Jacobson, R. 1988: Distinguishing among competing theories of the market share effect. *Journal of Marketing*, 52, October, 68–80.

Jaworski, B. J. and Kohli, A. K. 1993: Market orientation: antecedents and consequences. *Journal of Marketing*, 57, July, 53–70.

Ketchen, D. J., Jr., and Shook, C. L. 1996: The application of cluster analysis in strategic management research: a comparison of theoretical approaches. *Strategic Management Journal*, 17, 441–58.

Ketchen, D. J., Jr., Thomas, J. B., and Snow, C. C. 1993: Organizational configurations and performance: a comparison of theoretical approaches. *Academy of Management Journal*, 36 (6), 1278–313.

Kogut, B. and Zander, U. 1992: Knowledge of the firm, combinative capabilities, and the replication of technology. *Organization Science*, 3, 383–97.

Kohli, A. K. and Jaworski, B. J. 1990: Market orientation: the construct, research propositions, and managerial implications. *Journal of Marketing*, 54, April, 1–18.

Kotha, S. 1995: Mass customization: implementing the emerging paradigm for competitive advantage. *Strategic Management Journal*, 16 (Summer Special Issue), 21–42.

von Krogh, G., Ichijo, K., and Nonaka, I. 2000: *Enabling Knowledge Creation: How to Unlock the Mystery of Tacit Knowledge and Release the Power of Innovation.* New York: Oxford University Press.

von Krogh, G. and Nonaka, I. 1999: *Knowledge assets, knowledge conversion and competitive advantage.* Working paper, University of St. Gallen, St. Gallen.

von Krogh, G. and Roos, J. 1995: *Organizational Epistemology.* London: Macmillan and St. Martin's Press.

von Krogh, G. and Roos, J. 1996: *Managing Knowledge: Perspectives on Cooperation and Competition.* London: Sage Publications.

von Krogh, G., Roos, J. and Slocum, K. 1994: An essay on corporate epistemology. *Strategic Management Journal*, 15 (Summer Special Issue), 53–71.

Kwoka, J. 1979: The effect of market share distribution on industry performance. *American Economic Review*, February, 351–63.

Lant, T. K. and Baum, J. A. C. 1995: Cognitive sources of socially constructed competitive groups: examples from the Manhattan hotel industry. In R. Scott and S. Christensen (eds.), *The Institutional Construction of Organizations: International and Longitudinal Studies.* Thousand Oaks, CA: Sage, 15–38.

Leonard-Barton, D. 1992: Core capabilities and core rigidities: a paradox in managing new product development. *Strategic Management Journal*, 13 (Summer Special Issue), 111–25.

Leonard-Barton, D. 1995: *Wellsprings of Knowledge: Building and Sustaining the Source of Innovation.* Boston, MA: Harvard Business School Press.

Lippman, S. A. and Rumelt, R. P. 1982: Uncertain imitability: an analysis of interfirm differences in efficiency under competition. *Bell Journal of Economics*, 13, Autumn, 418–38.

Mahoney, J. T. 1995: The management of resources and the resource of management. *Journal of Business Research*, 33 (Special Issue), 91–101.

Mahoney, J. T. and Pandian, J. R. 1992: The resource-based view within the conversation of strategic management. *Strategic Management Journal*, 13 (5), 363–80.

Marshall, A. 1969: *Principles of Economics.* London: Macmillan.

Mascarenhas, B. 1989: Strategic group dynamics. *Academy of Management Journal*, 32, 333–52.

Mascarenhas, B. and Aaker, D. A. 1989: Mobility barriers and strategic groups. *Strategic Management Journal*, 10 (5), 475–85.

Mauri, A. J. and Michaelis, M. P. 1998: Firm and industry effects within strategic management: an empirical examination. *Strategic Management Journal*, 19, 211–19.

McGahan, A. M. and Porter, M. E. 1997: How much does industry matter, really? *Strategic Management Journal*, 18, Summer Special Issue, 15–30.

McGee, J. and Thomas, H. 1986: Strategic groups: theory, research and taxonomy. *Strategic Management Journal*, 10, 141–60.

McGrath, R., MacMillan, I. C., and Venkataraman, S. 1995: Defining and developing competence: a strategic process paradigm. *Strategic Management Journal*, 16, 251–75.

Meyer, H. 1991: A solution to the performance feedback enigma. *Academy of Management Executive*, 5 (1), 68–76.

Miller, D. and Shamsie, J. 1996: The resource-based view of the firm in two environments: the Hollywood studios from 1936 to 1965. *Academy of Management Journal*, 39 (3), 519–43.

Mitchell, W. 1989: Whether and when? Probability and timing of incumbents' entry into emerging industrial subfields. *Administrative Science Quarterly*, 34, 208–30.

Mitchell, W. 1992: Are more good things better, or will technical and market capabilities conflict when a firm expands? *Industrial and Corporate Change*, 1 (2), 327–46.

Mitchell, W. and Singh, H. 1992: 'Incumbents' use of preentry alliances before expansion into new technological subfields in an industry. *Journal of Economics, Behavior and Organization*, 18 (3), 347–72.

Moorman, C. and Slotegraaf, R. J. 1999: The contingency value of complementary capabilities in product development. *Journal of Marketing Research*, XXXVI, May, 239–57.

Nagarajan, A. and Mitchell, W. 1998: Evolutionary diffusion: internal and external methods used to acquire encompassing, complementary, and incremental technological changes in the lithography industry. *Strategic Management Journal*, 19, 1063–77.

Narver, J. C. and Slater, S. F. 1990: The effect of market orientation on business profitability. *Journal of Marketing*, 54, October, 20–35.

Nath, D. and Gruca, T. S. 1997: Convergence across alternative methods for forming strategic groups. *Strategic Management Journal*, 18, 745–60.

Nelson, R. R. and Winter, S. G. 1982: *An Evolutionary Theory of Economic Change*. Cambridge, MA: Belknap Press of Harvard University Press.

Nonaka, I. 1991: The knowledge-creating company. *Harvard Business Review*, 69 (6), 96–104.

Nonaka, I. 1994: A dynamic theory of organizational knowledge creation. *Organization Science*, 5 (1), February, 14–37.

Nonaka, I. and Takeuchi, H. 1995: *The Knowledge-Creating Company: How Japanese Companies Create the Dynamics of Innovation*. New York: Oxford University Press.

Nonaka, I., Toyama, R., and Nageta, A. 2000: A firm as a knowledge-creating entity: a new perspective on the theory of the firm. *Industrial and Corporate Change*, 9 (1), 1–20.

OECD 2000: Cross-border trade in financial services: economics and regulation. *Financial Market Trends*, 75, July, 23–60.

Oster, S. 1982: Intra-industry structure and the ease of strategic change. *Review of Economics and Statistics*, 64, August, 376–83.

Pelham, A. M. and Wilson, D. T. 1996: A longitudinal study of the impact of market structure, firm structure, strategy, and market orientation culture on dimensions of small-firm performance. *Journal of the Academy of Marketing Science*, 24 (1), 27–43.

Penrose, E. T. 1959: *The Theory of the Growth of the Firm*. New York: John Wiley.

Peteraf, M. A. 1993: The cornerstones of competitive advantage: a resource-based view. *Strategic Management Journal*, 14 (3), 179–91.

Peteraf, M. A. and Shanley, M. 1997: Getting to know you: a theory of strategic group identity. *Strategic Management Journal*, 18 (Summer Special Issue), 165–86.

Pisano, G. P. 1994: Knowledge integration and the locus of learning: an empirical analysis of process development. *Strategic Management Journal*, 15 (Winter Special Issue), 85–100.

Polanyi, M. 1958: *Personal Knowledge: Towards a Post-critical Philosophy*. Chicago, IL: University of Chicago Press.

Polanyi, M. 1967: *The Tacit Dimension*. Garden City, NY: Anchor Books.

Polanyi, M. 1969: *Knowing and Being*. Chicago, IL: University of Chicago Press.

Porac, J. F. and Thomas, H. 1990: Taxonomic mental models in competitor definition. *Academy of Management Review*, 15 (2), 224–40.

Porac, J. F. and Thomas, H. 1994: Cognitive categorization and subjective rivalry among retailers in a small city. *Journal of Applied Psychology*, 79 (1), 54–66.

Porac, J. F., Thomas, H., and Baden-Fuller, C. 1989: Competitive groups as cognitive communities: the case of Scottish knitwear manufacturers. *Journal of Management Studies*, 26 (4), July, 397–416.

Porac, J. F., Thomas, H., and Emme, B. 1987: Understanding strategists' mental model of competition. In G. N. Johnson (ed.), *Business Strategy and Retailing*. New York: Wiley, 59–79.

Porac, J. F., Thomas, H., Wilson, F., Paton, D., and Kanfer, A. 1995: Rivalry and the industry model of Scottish knitwear producers. *Administrative Science Quarterly*, 40, 203–27.

Porter, M. E. 1979: The structure within industries and companies performance. *Review of Economics and Statistics*, 61, 214–27.

Porter, M. E. 1980: *Competitive Strategy: Creating and Sustaining Superior Performance*. New York: The Free Press.

Porter, M. E. 1985: *Competitive Advantage*. New York: The Free Press.

Porter, M. E. 1990: *The Competitive Advantage of Nations*. New York: Macmillan.

Porter, M. E. 1991: Toward a dynamic theory of strategy. *Strategic Management Journal*, 12 (Winter Special Issue), 95–117.

Powell, T. C. 1996: How much does industry matter? An alternative empirical test. *Strategic Management Journal*, 17, 323–34.

Quinn, J. B. 1985: Managing innovation: controlled chaos. *Harvard Business Review*, 63 (3), May–June, 73–84.

Rao, H. 1994: The social construction of reputation: certification contests, legitimation, and the survival of organizations in the American automobile industry: 1895–1912. *Strategic Management Journal*, 15 (Winter Special Issue), 29–44.

Reger, R. K. and Huff, A. S. 1993: Strategic groups: a cognitive perspective. *Strategic Management Journal*, 14, 103–24.

Reed, R. and DeFillippi, R. J. 1990: Causal ambiguity, barriers to imitation, and sustainable competitive advantage. *Academy of Management Review*, 15 (1), 88–102.

Rindova, V. P. and Fombrun, C. J. 1999: Constructing competitive advantage: the role of firm-constituent interactions. *Strategic Management Journal*, 20, 691–710.

Roquebert, J., Phillips, R., and Westfall, P. 1996: Market versus management: what drives profitability. *Strategic Management Journal*, 17 (8), 653–64.

Rosenbloom, R. S. and Christensen, C. M. 1994: Technological discontinuities, organizational capabilities, and strategic commitments. *Industrial and Corporate Change*, 3 (3), 655–85.

Ruekert, R. W. 1992: Developing a market orientation: an organizational strategy perspective. *International Journal of Research in Marketing*, 9, August, 225–45.

Rumelt, R. P. 1984: Towards a strategic theory of the firm. In L. B. Lamb (ed.), *Competitive Strategic Management*. Englewood Cliffs, NJ: Prentice-Hall, 556–70.

Rumelt, R. P. 1991: How much does industry matter? *Strategic Management Journal*, 12, 167–85.

Schmalensee, R. 1985: How much do markets matter? *American Economic Review*, 75 (3), 341–51.

Schendel, D. E. 1996: Knowledge and the firm: editor's introduction. *Strategic Management Journal*, 17 (Winter Special Issue), 1–4.

Securities Industry Association 2000: *Securities Industry Factbook 1999*. New York: SIA.

Shapiro, B. P. 1988: What the hell is 'market oriented'? *Harvard Business Review*, 66, November–December, 119–25.

Shepherd, W. G. 1972: The elements of market structure. *Review of Economics and Statistics*, 54, 25–37.

Simon, H. A. 1991: Bounded rationality and organizational learning. *Organization Science*, 2, 125–34.

Simon, H. A. 1993: Strategy and organizational evolution. *Strategic Management Journal*, 14, 131–42.

Slater, S. F. and Narver, J. C. 1994a: Does competitive environment moderate the market orientation–performance relationship? *Journal of Marketing*, 58, January, 46–55.

Slater, S. F. and Narver, J. C. 1994b: Market orientation, customer value, and superior performance. *Business Horizons*, March–April, 22–8.

Slater, S. F. and Narver, J. C. 1995: Market orientation and the learning organization. *Journal of Marketing*, 59, July, 63–74.

Slater, S. F. and Narver, J. C. 1998: Customer-led and market-oriented: let's not confuse the two. *Strategic Management Journal*, 19, 1001–6.

Slater, S. F. and Narver, J. C. 1999: Market-oriented is more than customer-led. *Strategic Management Journal*, 20, 1165–8.

Smith, K. G., Grimm, C. M., and Wally, S. 1996: Strategic groups and rivalrous firm behaviour: towards a reconciliation. *Strategic Management Journal*, 18 (2), 149–57.

Smith, R. C. 1997: World financial integration. In F. D. S. Choi (ed.), *International Accounting and Finance Handbook*. New York: Wiley & Sons.

Spanos, Y. E. and Lioukas, S. 2001: An examination into the causal logic of rent generation: contrasting Porter's competitive strategy framework and the resource-based perspective. *Strategic Management Journal*, 22, 907–34.

Spender, J.-C. 1996: Making knowledge the basis of a dynamic theory of the firm. *Strategic Management Journal*, 17 (Winter Special Issue), 45–62.

Spender, J.-C. and Grant, R. M. 1996: Knowledge and the firm: overview. *Strategic Management Journal*, 17 (Winter Special Issue), 5–9.

Srivastava, R. K., Shervani, R. A., and Fahey, L. 1998: Market-based assets and shareholder value: a framework for analysis. *Journal of Marketing*, 62, January, 2–18.

Stigler, G. J. 1964: A theory of oligopoly. *Journal of Political Economy*, 72, 55–9.

Sudharshan, D., Thomas, H., and Fiegenbaum, A. 1991: Assessing mobility barriers in dynamic strategic group analysis. *Journal of Management Studies*, 28, 429–38.

Szulanski, G. 1995: Unpacking stickiness: an empirical investigation of the barriers to transfer best practice inside the firm. *Academy of Management Journal*, Best Papers Proceedings, 437–41.

Szulanski, G. 1996: Exploring internal stickiness: impediments to the transfer of best practice within the firm. *Strategic Management Journal*, 17 (Winter Special Issue), 27–44.

Tang, M.-J. and Thomas, H. 1992: The concept of strategic groups: theoretical construct or analytical convenience? *Managerial and Decision Economics*, 13, 323–9.

Teece, D. J. 1982: Towards an economic theory of the multiproduct firm. *Journal of Economic Behavior and Organization*, 3, 39–63.

Teece, D. J. 1986a: Profiting from technological innovation: implications for integration, collaboration, licensing and public policy. *Research Policy*, 15, 285–305.

Teece, D. J. 1986b: Transaction cost economics and the multinational enterprise. *Journal of Economic Behaviour and Organization*, 7, 21–45.

Teece, D. J. 1998: Capturing value from knowledge assets: the new economy, markets for know-how and intangible assets. *California Management Review*, 40 (3), Special Issue, 55–79.

Teece, D. J. and Pisano, G. 1994: The dynamic capabilities of firms: an introduction. *Industrial and Corporate Change*, 3 (3), 537–56.

Teece, D. J., Pisano, G., and Shuen, A. 1997: Dynamic capabilities and strategic management. *Strategic Management Journal*, 18 (7), 509–33.

Thomas, H. and Venkatraman, N. 1988: Research on strategic groups: progress and prognosis. *Journal of Management Studies*, 25, 537–55.

Tirole, J. 1996: A theory of collective reputations (with applications to the persistence of corruption and to firm quality). *The Review of Economic Studies*, 63, 1–22.

Tripsas, M. 1997: Unraveling the process of creative destruction: complementary assets and incumbent survival in the typesetter industry. *Strategic Management Journal*, 18 (Summer Special Issue 0, 119–42.

Varadarajan, P. R. and Jayachandran, S. 1999: Marketing strategy: an assessment of the state of the field and outlook. *Journal of the Academy of Marketing Science*, 27 (2), 120–43.

Vicari, S., Bertoli, G., and Busacca, B. 1999: *Valutazione dei beni immateriali nella prospettiva della fiducia*. Working paper 52, SDA Boccni, Milano.

Volberda, H. W. 1996: Toward the flexible form: how to remain vital in hypercompetitive environments. *Organization Science*, 7 (4), 359–74.

Walter, I. 1998: Globalization of markets and financial center competition. *Salomon Center at the New York University Working Paper Series*, S-98-23.

Walter, I. 1999a: Financial services strategies in the euro-zone. *European Investment Bank Papers*, 4 (1), 145–64.

Walter, I. 1999b: Financial services strategies in the euro-zone, *European Management Journal*, 17 (5), 447–65.

Weigelt, K. and Camerer, C. 1988: Reputation and corporate strategy: a review of recent theory and applications. *Strategic Management Journal*, 9, 443–54.

Wernerfelt, B. 1984: A resource-based view of the firm. *Strategic Management Journal*, 5 (2), 171–80.

Wilson, R. 1985: Reputations in games and markets. In A. E. Roth (ed.), *Game-theoretic Models of Bargaining*. Cambridge, MA: Cambridge University Press, 27–62.

Winter, S. G. 1982: An essay on the theory of production. In S. H. Hymans (ed.), *Economics and the World Around It*. Ann Arbor, MI: University of Michigan Press, 55–91.

Winter, S. G. 1988: On Coase, competence and the corporation. *Journal of Law, Economics, and Organization*, 4 (1), Spring, 163–80.

Zander, U. and Kogut, B. 1995: Knowledge and the speed of the transfer and imitation of organizational capabilities: an empirical test. *Organization Science*, 6 (1), 76–92.

Appendix A

Reputation

Borrowers' vote in capital raising

Capital raising for clients is the defining feature of investment banking activities and should thus be included in our study. The 5 most reputable firms in following categories were identified and ranked.[1]

A. overall best in capital raising through the international markets

B. Eurobonds; Private placement; International equity; US MTN program arrangers; US MTN program dealers; Euro-MTN program arrangers; Euro-MTN program dealers

C. US public domestic markets; Japanese public domestic markets

The categories A, B, and C are weighted equally. Within B, the measures were weighted according to their relative volume. Data was taken from Thomson Financial Securities Data, *Financial Market Trends* of the OECD and *International Banking and Financial Markets Developments* of the Bank for International Settlements (BIS).[2] Within C, the measures were weighted according to the relative market capitalization of their equity and bond markets. Data was taken from the *Emerging Stock Markets Factbook 1999* of the International Financial Corporation (IFC) and the publications of the FIBV.[3] *Euromoney* polled treasurers and financial officers at corporates, financial institutions, state agencies, and supranational organizations. Respondents were asked to nominate banks providing the best service in capital raising.

Peers' vote in bond underwriting

Investment banking raises capital through bond and equity markets; both should thus be included in our study. The 5 most reputable firms in following categories were identifed and ranked.[4]

A. overall best in lead managing Eurobonds

B. Eurobonds: bringing most innovative issues; best in: origination of new issues/pricing of new issues/ syndication of new issues/retail placement of new issues/institutional placement of new issues/serving as co-manager/best at lead managing new issues in convertibles

C. best at lead-managing new issues of Eurobonds in: US dollars; Japanese Yen; ECU; Deutschmark; pound sterling; French francs; Italian lira; Swiss francs; Canadian dollars; Australian and New Zealand dollars

The categories A, B, and C are weighted equally. Within B, the measures were weighted equally. Within C, the measures were weighted according to their relative volume. Data was taken from Thomson Financial Securities Data.[5] *Euromoney* polled heads of bond syndication. Respondents were asked to nominate banks providing the best service in bond and equity raising.

Peers' vote in equity underwriting

For the investment banking activity of raising capital through equity markets, the 5 most reputable firms in following categories were identifed and ranked.[6]

1. best as global coordinator

2. best as regional lead manager in the US; Western Europe; Eastern Europe; Japan; Asia (ex Japan)

3. having best institutional penetration in the US; Western Europe; Japan; in Asia (ex Japan); Latin America

4. having best retail penetration in the US; in Western Europe; Japan; Asia (ex Japan); Latin America

5. best ad advising on privatizations; best at managing IPOs; secondary offerings

The categories A, B, C, D and E are weighted equally. Within E, the measures are weighted equally. Within B, C, D, and E the measures are weighted according to the relative stock market capitalization of the region to the other regions. *Euromoney* polled heads of equity syndicate chiefs. Respondents were asked to nominate banks providing the best service in bond and equity raising. Participants were asked not to nominate their own institutions. The following box gives an overview of the countries selected as considered

US: USA; Japan: Japan

Western Europe: Belgium, Denmark, Finland, France, Germany, Ireland, Italy, The Netherlands, Norway, Portugal, Spain, Sweden, Switzerland, and the U.K.

Eastern Europe: The Czech Republic, Hungary, Poland, Russia, Slovakia

Asia: Australia, China, Hong Kong, India, Indonesia, Korea, Malaysia, New Zealand, The Philippines, Taiwan, Thailand

Latin America: Argentina, Brazil, Chile, Mexico, Colombia, Peru, Venezuela

Market orientation

Public securities underwriting

US domestic investment grade debt/Eurobonds/Yankee bonds/Eurobonds by US issuers

Eurobonds by European issuers

High-yield debt by US/European/Global issuers

Convertible debt by US/European/Global issuers[7]

Asset-backed securities by US/European/Global issuers

Mortgage-backed securities by US/European/Global issuers

Common stock by US/European/Global issuers

Yankee common stock: non US issuers selling equity in the US.

Preferred stock by US/European/Global issuers

IPOs by US/European/Global issuers

Public medium-term notes by US/European/Global issuers

Private placement of securities

Traditional issues[8] by US issuers

Traditional issues by European issuers

Yankee issues[9]

High-yield debt by US/European issuers

Securitized issues[10] by US/European issuers

Plain equity[11] by US/European issuers

Rule 144A issues by US/European issuers

Merger and acquisitions

Top financial advisors in the financials sector on: US/European/Global transactions[12]/Global Cross-border transactions

Top financial advisors in the services sector on: US/European/Global transactions/Global Cross-border transactions

Top financial advisors in the manufacturing sector on: US/European/Global transactions/Global Cross-border transactions

Top financial advisors in the trade sector on: US/European/Global transactions/Global Cross-border transactions

Top financial advisors in the natural resources sector on: US/European/Global transactions/Global Cross-border transactions

Top financial advisors in the other[13] sectors on: US/European/Global transactions/Global Cross-border transactions

Complementary assets

Depth and breath of research

First-teamers/Second-teamers/Third-teamers/Runners-up:

A. All-America Research Team
B. All-Europe Research Team
C. All-Latin America Research Team
D. All-Asia Research Team
E. All-Japan Research Team

The categories A, B, C, D, and E are weighted according to the relative stock market capitalization of the region to the other regions. The following box gives an overview of the countries selected as considered.[14]

America: USA., Canada; Japan: Japan
Europe: Belgium, The Czech Republic, Denmark, Finland, France, Germany, Greece, Hungary, Ireland, Israel, Italy, The Netherlands, Norway, Poland, Portugal, Russia, South Africa, Spain, Sweden, Switzerland, Turkey and the UK
Latin America: Argentina, Brazil, Chile, Mexico, Colombia, Peru, Venezuela
Asia: Australia, China, Hong Kong, India, Indonesia, Korea, Malaysia, New Zealand, The Philippines, Taiwan, Thailand

Execution/distribution in equities

For European equity execution, the following categories were considered for our study;[15] overall best in Europe: The overall results are not a compilation of the results in the individual categories, but responses to a separate question; best in individual countries: Austria, Belgium, Denmark, Finland, France, Germany, Greece, Ireland, Italy, The Netherlands, Norway, Portugal, Spain, Sweden, Switzerland, and the UK.
For Asian equity execution, the following categories were considered for our study;[16] Best in individual countries: Australia, China, Hong Kong, India, Indonesia, Japan, Korea, Malaysia, New Zealand, Pakistan, The Philippines, Singapore, Taiwan, Thailand

Secondary market support for bonds

For Eurobonds, the 5 most reputable firms in following categories were identified and ranked:[17]

The overall measure was weighted by turnover as this reflects support activities. Within B the measures were weighted according to the relative volume of outstanding amounts of currency issues in intern. bonds and notes.

A. overall best in trading Eurobonds (The category A was weighted a quarter.)
B. best in trading: in US dollars: fixed bonds/floating rates notes/convertibles/asset-backed /illiquid/high-yield /brady bonds; bonds issued in Australian and New Zealand dollars/Canadian dollars/Deutschmark/Dutch guilders/benchmark ECUs/illiquid ECUs/French francs/Italian lira/Scandinavian currencies/Spanish pesetas/South African Rand/pound sterling/Swiss francs/Japanese yen

For government bonds, the 5 most reputable firms in following categories were identified and ranked:[18]

Within B, it was according to the relative volume of outstanding amount by both (a) nationality of issuer of international debt securities by government agencies, plus (b) public domestic debt securities.[19]

A. overall best in trading government bonds (The category A was weighted a quarter.)
B. best in trading government bonds issues in:
Canadian dollars/US dollars/Belgium francs/Danish krona/Finnish markka/French francs/Deutschmark/Irish pounds/Italian lira/Dutch guilders/Norwegian krona/Portuguese escudo/Spanish pesetas/Swedish krona/pound sterling/Swiss francs/Japanese yen

Risk management

The five most reputable firms in following categories were identified and ranked:[20]

The categories A, B, C, and D are weighted equally. Within C the measures are weighted according to the relative volume of outstanding amounts of interest-rate swaps. Within D, according to the relative volume of outstanding amounts of currency swaps.[21]

A. Overall best in risk management
B. Best in risk management advice
C. Best at providing OTC interest-rate swaps in: S dollars; Japanese Yen; Deutschmark; pound sterling; French francs; Swiss francs; Italian lira
D. Best at providing OTC currency swaps in: US dollars; Japanese Yen; Deutschmark; pound sterling; French francs; Swiss francs; Italian lira

1 In the category of "overall best in capital raising through the international markets" the top 10 firms were identified and ranked.

2 All volume data, except for international equity issuance from OECD for 1990–95 and BIS for 1996–98; OECD, 1996, *Financial Market Trends*, 63, 74; OECD, 1994, *Financial Market Trends*, 57, 113; BIS, *International Banking and Financial Markets Development*, August 1999, 78.

3 IFC, *Emerging Stock Markets Factbook 1999*, 16f. for stock market capitalization. FIBV for the market value of bonds listed (see publications on www.fibv.org). Data as available from FIBV has been considered: For the US the markets of Amex, Chicago, NASDAQ, and NYSE are considered; For Japan the markets of Osaka and Tokyo are considered.

4 In the category of "overall best house in lead managing Eurobonds" the top 10 firms were identified and ranked from 1992 to 1998.

5 Australian and New Zealand dollar issues as the sum of issues in Australian dollars and New Zealand dollars.

6 In the category of "overall best house in lead managing Eurobonds" the top 10 firms were identified and ranked in 1998; top 8 in 1997; top 5 in 1999, 1996, 1994, and 1993; top 4 in 1995.

7 "Global issuers" indicates issues by non-US and non-European issuers

8 Traditional p.p. excludes CDs, deposit notes, acquisition-related, lease-related, high-yield, placements with registration rights, mortgage-backed, asset-backed, MTNs, ESOP-related and 144A deals.

9 Yankee placements include foreign issuers (and US issuers with a foreign parent) issuing in the US.

10 Securitized private placements include asset-backed and mortgage-backed securities.

11 Plain equity p.p. excludes CDs, deposit notes, acquisition-related, lease-related, high-yield, placements with registration rights, mortgage-backed, asset-backed, MTNs and ESOP-related deals.

12 "Global transactions" defined as non-US and non-European transactions.

13 i.e., not the above mentioned. It should be noted that very few transactions fall into this category.

14 Countries incorporated as considered in the 1999 editions of the *Institutional Investor* research team surveys.

15 If any category was not completed in any single year, due to reasons of distortion the category was dropped in the respective year for our study – rather than considering the results of the previous year.

16 If any category was not completed in any single year, due to reasons of distortion the category was dropped in the respective year for our study – rather than considering the results of the previous year.

17 In the category of "voted as overall best house in trading Eurobonds" the top 25 firms were identified and ranked in 1999, 1998; the top 20 firms in 1997, 1996, 1995; the top 15 in 1994, 1993, 1992, and 1991. In the category of "voted as best in trading fixed bonds in US dollars" the top 10 firms were identified and ranked from 1991 to 1999. In the category of "voted as best in trading floating rates notes in US dollars" the top 10 firms were identified and ranked from 1991 to 1999.

18 In the category of "voted as overall best house in trading government bonds" the top 25 firms were identified and ranked in 1999, 1998; the top 20 firms in 1997; the top 22 firms in 1996; the top 15 in 1995; the top 13 in 1994; and the top 14 in 1993. For the categories of B the top 10 firms were identified and ranked in 1999. In the category of "voted as best in trading government bonds in US dollars" the top 6 firms were identified and ranked in 1994 and 1995.

19 If any measure was not completed in any single year, due to reasons of distortion the measure was dropped in the respective year for our study – rather than considering the results of the previous year.

20 In the category of "best in risk management" the top 10 firms were identified and ranked in 1999, 1998, 1996, 1995, 1994, 1993, 1992; the top 15 in 1997; and the top 20 in 1991. In the category of "best in risk management advice" the top 10 firms were identified and ranked in 1999, 1998, 1997, 1996, 1995, and 1994.

21 If any measure was not completed in any single year, due to reasons of distortion the measure was dropped in the respective year for our study – rather than considering the results of the previous year.

CHAPTER 3

Happy Kids and Mature Losers: Differentiating the Dominant Logics of Successful and Unsuccessful Firms in Emerging Markets

Krzysztof Obłój and Michael G. Pratt

Dominant logics, sensemaking, decision making and choice, managerial action and cognition, learning, entrepreneurs

Abstract

With the onset of the cognitive revolution managerial sciences, there has been increasing interest in how top managers come to think and act strategically in response to ever-changing environmental opportunities and constraints (Walsh, 1995). With this interest has come a variety of concepts, tools, and lenses for conceptualizing the strategy of a firm – such as marketplace position, strategic configurations, core resources and capabilities, and rule sets (e.g., Porter, 1980; Hamel and Prahalad, 1994; Porter, 1996; Miler, 1999; Eisenhardt and Martin, 2000). Common to most of these conceptualizations is the assumption that firms develop shared and consistent cognitive frameworks and behavioral routines – just as individuals develop schemas, heuristics, and other mental models (Cyert and March, 1963) – in order to deal with environmental complexity. Such frameworks, referred to as "strategic frames" (Huff, 1982), "dominant themes" or "configurations" (Miler, 1999), "interconnected choices" (Siggelkow, 2000), and most often "dominant logic" (Prahalad and Bettis, 1986), are effective in cognitive terms because they help prevent managers from being overwhelmed by a flood of information and decision options.

Research has shown that over time firms develop such logics, and that such logics become a central force in shaping an organization's actions. What is less clear, however, is whether (or how) these logics are effective in enhancing a firm's performance. For example, dominant logics may enhance performance by facilitating strategic fit and alignment (Miles and Snow, 1978; Powell, 1992; Jennings and Seaman, 1994; Zajac et al., 2000). By contrast, these same logics may create perceptual "blind spots" and rigidities in practice that overly constrain or inhibit (via inertia) an organization's ability to respond to environmental demands (Tushman et al., 1986; Kelly and Amburgey, 1991). Similarly, it is not yet clear whether the development of some types of dominant logics or mindsets is associated with effective companies, or whether there exists some level of equifinality (Cavaleri and Obloj, 1993) such that different dominant logics can result in the same performance – or that similar dominant logics can result in dissimilar performance – depending upon the external set of opportunities and threats that firms face in the industry. In short, we do not know whether certain types of dominant logics differentiate winners and losers in the dynamic marketplace.

In this chapter, we address these performance-related questions at a fine-grained level by studying the strategic frameworks (i.e., dominant logic) of both successful and less successful firms in a turbulent environment: emerging markets in Poland during the decade of 1989–99.

Developing Dominant Logics: An Overview

In order to more effectively examine the development and impact of dominant logics on organizational success and failure in turbulent markets, we examined several literatures that addressed the issue of how managers and organizations come to develop cognitive schemas and strategic frameworks. The purpose of this review was to identify some basic characteristics of these structures that would help focus our case study observations.

Our review revealed several elements central to most conceptualizations of managerial and organizational cognitive structures: (a) perception and sense-making; (b) actions, such as choices; (c) learning; and (d) codification of learning via routines. At its most basic level, strategic frameworks are believed to filter managers' perception about the environment: that is, they serve as sense-making devices (Starbuck and Miliken, 1988; Weick, 1995). Hence, Prahalad and Bettis (1986) refer to dominant logics as a "mindset."

These strategic frameworks or mind sets are created through a combination of action, learning, and codification of learning via routines. The action–cognition link is often inherent in the very descriptions of strategic frameworks. To illustrate, Prahald and Bettis (1986: 491, emphasis ours) define dominant logic as a "*mind set . . .* or conceptualization of the business and administrative tools to *accomplish goals and make decisions in that business.*" This definition is similar to Schein's (1985: 9, emphasis ours) conceptualization of culture as "a *pattern of basic assumptions* – invented, discovered, or invented by a given group as it learns to *cope with its problems of external adaptation and internal integration . . .*" In these two concepts, action is represented by decision making and problem solving, and cognition by mind sets and patterns of basic assumptions. Gioia and Poole (1984: 450, emphasis theirs) more clearly delineate this linkage by describing the utility of "scripts" as those knowledge structures that "enable *understanding* of situations . . . and they provide a *guide to behavior* appropriate to those situations."

Delving deeper into this relationship between thinking and doing, we see that action is central to cognitive structures, such as dominant logics, in two ways. To begin, dominant logics are believed to guide strategic action, such as decision making. That is, they serve as justifications for initiating certain activities and not others. These activities may be conceptualized broadly as *exploration* – the search for new opportunities, knowledge, and solutions, or *exploitation* – the mobilization and use of resources and knowledge (March, 1994, 1996). Alternatively, dominant logics may be viewed as guiding more specific sets of strategic actions, such as: allocating resources, formulating business strategies, and setting and monitoring performance targets (Grant, 1988).

Alternatively, action can facilitate the creation (and possibly the refinement) of dominant logics. This linkage often requires learning. For example, following rules of operant conditioning, Prahald and Bettis (1986) suggest that activities, such as strategic choices (Child, 1997), that are rewarded will be learned and repeated. March (1996) further argues that learning in firms may also be vicarious as members combine "learning from an organization's own experience, learning from others, and selection stemming from differential organizational growth and survival." This learning from action ultimately becomes codified in organizations via various cognitive structures. As noted, these cognitive structures have gone by a variety of names in the strategy literature, such as scripts, recipes, knowledge structures, strategic frames, dominant themes or configurations, rules, and routines (Huff, 1982; March, 1996; Miler, 1999; Nelson and Winter, 1982; Starbuck, 1983; Walsh, 1995).

Taken together, the elements of acting/choosing, learning, codifying, and sensemaking should be seen as mutually reinforcing in the creation of knowledge structures such as strategic frameworks [see also Giddens' (1984) structuration theory]. One knowledge structure that captures these dynamics is "identity." In psychology (Markus and Nurius, 1986) and in some areas of organizational behavior (Pratt, 2003), repetition of action is said to lead to schemas and identities, which then, in turn, serve as sense-making devices that influence future behaviors. In the area of strategic management, these elements are captured in Bettis and Prahalad (1995) and Bettis (2000) as revised conceptualization of dominant logic. Here, they view dominant logics as involving the interplay of perceptions (e.g., of environment and organization) and actions (e.g., operational plans and routines). Because of the strategic focus of our paper, we chose to concentrate on "dominant logics" – and the elements of dominant logics (sensemaking, learning, action/choices, and codification) – as a general guide when determining what processes to examine when collecting data.

More specifically, we were interested in how dominant logics formed, and the utility of those logics on firm behavior. That is, were certain types of dominant logics common to successful firms versus unsuccessful ones? Unfortunately, conceptual research in dominant logics far outstrips empirical work in this area. As such, relatively few studies have actually examined how organizations form and utilize dominant logics (see Cote et al., 1999; Lampel and Shamsie, 2000; Siggelkow, 2000 as exceptions). Hence, we saw an opportunity to further research by taking advantage of a rare opportunity (large-scale development of a market) for the purpose of building theory about the dominant logics of market "winners" and "losers."

Methods

As noted, the aforementioned review became a springboard for examining dominant logic and its influence on firm performance. We operationalized performance as the position that each firm achieved in an industry, i.e., whether it was a dominant or peripheral player. Below, we describe the context of our study – new ventures in Poland – as well as data-gathering and analysis techniques.

Context and sampling

Since 1989, Poland has made great strides in converting from a state-controlled economy under communism to a free market economy. During this decade of rapid but uneven growth,[1] Poland privatized most of its 7,000+ state-owned firms, changed its industry structure by moving from a focus on heavy industry to manufacturing and services, and reoriented its business relations from the East to the West. One of the key success factors responsible for the growth of Polish economy has been a surge of entrepreneurship, driven by unparalleled opportunity as well as by necessity. Today, close to 2 million individuals are registered as independent entrepreneurs. The *Total Entrepreneurial Activity Index* – which indicates the number per every 100 adults of individuals who are trying to start a new firm or who own an active business less than 42 months old (Reynolds et al., 2001) – of Poland is 10, which places it between the European average of about 9 and the North American average of 12. The systematic transformation of this economy offered a unique chance to study the evolution of dominant logics in new ventures – an opportunity comparable only to the study of emerging high-tech industries (Zyglidopoulos, 1999). Many markets did not exist under the communist regime; and if they existed, they were artificially designed and managed by central planners. Therefore, entrepreneurs and venture capitalists often created new markets and industries.

Within this context, we examined "extreme cases" of high and low performers so that the effects of dominant logics on the firm's market position would be "transparently observable" (Eisenhardt, 1989; Pettigrew, 1990). Moreover, to minimize the influence of different economic, legal, and other forces, we concentrated on companies that were in similar industries (and thus had a similar set of opportunities and constraints), and that had similar "launch" dates. Specifically, we focused on companies in industries that – in 1989–90 – met the following conditions: (1) were relatively new, (2) had low entry barriers, (3) initially lacked economies of scale, and (4) were of limited strategic significance to multinational companies (MNCs). This set of conditions was agreed upon by a panel of eight entrepreneurs, managers, and consultants as being the most fertile for creating new ventures, in a relatively level playing field (i.e., did not favor particular companies), during the early 1990s.

Our rationale for each of these conditions is as follows. Relative newness meant that such in industry did not exist under the communism regime or was marginally developed because of its lack of strategic importance in a planned economy. Hence, neither the market structure nor regulations were limiting entry for entrepreneurs. Low barriers to entry were considered a necessary condition for new, undercapitalized entrepreneurs from former communist Poland to start their business in a particular industry. Third, initial lack of economies of scale was considered important to level the chances of most entrepreneurs and prevent quick creation of barriers to entry. Finally, limited significance of market for MNCs was an important factor in allowing new ventures ample time to develop their strategies and attain significant market shares.

After identifying industries or industry segments that met all or most of these conditions we looked to see if we could identify a clear *industry leader* and its *alter ego* – a peripheral firm that should have had the same chances to attain a leadership position because of similar starting conditions and resources. In most of the cases we examined, companies did not meet the aforementioned conditions for several reasons (e.g., the leader was acquired by an MNC in an early stage of the game, a peripheral firm that had the same chances as the leader was impossible to identify, etc.) However, after screening all of the cases, we were able to focus on five pairs of such "twin companies" (leader and alter ego) in the following industries: cosmetics,[2] public opinion and marketing research, consulting, computers, and outdoor outfits. A brief description of the twin companies follows, and the size of each industry is highlighted in Table 3.1.

Cosmetics
In the cosmetics industry we studied ERIS SA and Dax Cosmetics. Both companies were started in the early 1980s, but operated as small craft shops at the peripheries of the market until 1989. In 1989, they started to develop. Eris became a clear leader in the late 1990s, controlling 17 percent of the cosmetics market. Eris is now one of the best brand names in Poland. Dax Cosmetics, by contrast, controls 4 percent of the market and has few well-established lines of products.

Public opinion
In the public opinion and marketing research industry we studied SMG/KRC SA and Demoskop. Both were started at the same time (in 1989) by students and employees of the sociology department at the University of Warsaw. And both focused primarily on public opinion polls. By 2000, SMG/KRC was a clear leader

Table 3.1   Pairs of analyzed firms in particular industries and their app. revenues in 1999

	Cosmetics	Public opinion and marketing research	Consulting	Computers	Outdoor
Leading firm	ERIS SA (20m USD)	SMG/KRC Ltd (12m USD)	WGK Ltd (3m USD)	Optimus SA (150m USD)	Alpinus SA (22m USD)
Peripheral firm	DAX Ltd (5m USD)	Demoskop (1.4m USD)	Management Focus Ltd (2.4m USD)	Baza Ltd (3m USD)	Janysport (250,000 USD)

with a wide range of products, a well-established brand, and revenues of more than twice its closest competitor, AC Nielsen. Demoskop established a good brand name but its revenues placed it at 11th place in the industry.

Consulting
We studied two such companies established in 1990: Warszawska Grupa Konsultingowa (WGK) and Management Focus. Neither of them was an industry leader when we began our research. However, they started with similar resources and opportunities, and attained different positions in the industry.

Computer
In the computer industry the clear leader is Optimus SA, best known for its dominant position in the personal computer market in Poland. When we began our research, it was a holding company with core offerings in the area of computers and IT integrated solutions. Its alter ego was Baza, which assembled and sold PCs and peripherals through a distribution network in Poland. Although Baza was almost a size of Optimus by the mid-1990s, it was on the verge of bankruptcy by the end of this decade.

Outdoor
In the outdoor (textiles and equipment) industry we studied a midsize company, Alpinus SA, which became a market leader and established a wide network of franchised stores in Poland. Its alter ego was a small, but interesting company – Janysport (after the name of the founder, Janyst) – which, despite some effort, never grew.

Data collection and analysis

We employed a case study approach for this study (Eisenhardt, 1989), and our main source of data was interviews and archival materials (e.g., company reports and industry analyses). The implementation of our study – involving both data collection and analysis – occurred over three phases.[3]

Phase 1
The first author began the study by conducting very general, open-ended interviews with the CEOs of each company. The purpose of these interviews was to gather general information about the industry and the organization. This general information was used in conjunction with our aforementioned literature review for the purpose of constructing a more refined interview protocol. These interviews were also used as an opportunity to identify initial managerial contacts within each company for subsequent interviews.

Phase 2
The first author and six MBA students conducted more focused and intensive interviews with managers during a six-month period. Managers who had extensive experience with the company – and extensive experience with the strategic development of

each firm – were chosen for interviews. Tenure of informants ranged from 5 to 11 years. Interviews lasted between 1 and 2.5 hours, and most (about 80 percent) were transcribed verbatim. When recording was not possible, researchers took extensive notes. Within each firm, between four and nine managers were interviewed, with the total number of interviews exceeding 50. In a few cases, follow-up interviews were conducted with managers or CEOs to explore and verify themes that emerged from analysis of the interviews.

The interview protocol was structured into four parts. The first part focused on changes in managerial perceptions of the environment and organization over time. Managers were asked questions about how they perceived the environment (as series of opportunities, threats, problems to be solved, etc.); how these perceptions have changed over time; how the firm cooperated with vendors and customers; and the nature of the firm's relations with competitors. Managers were also asked to describe the evolution of and reasons for the formalization, centralization, and standardization of their organizational operations.

The second part of our protocol concentrated on the actions and choices managers made. Specifically, we asked managers to identify which choices in the firm's history they considered to be strategic; how these choices were made (incrementally or in one big stroke); what the content of the choices was; and how the choices were evaluated retrospectively. Our knowledge of the organization's history through archival and other sources also prompted us to ask managers why they did not consider some choices that seemed strategically important from a theoretical point of view (e.g., decisions to build a factory or make acquisitions).

The third part consisted of a series of questions related to routines in the organization. We asked how routines were developed; what dimensions of organizational actions were formalized first and which were formalized later; what procedures could be bypassed and under what circumstances; and what areas in the organization were not regulated by procedures and why.

In the fourth and final part of questionnaire, we asked managers to indicate the most difficult events and "teaching points" in the history of the company; how they happened; and how they dealt with them. Managers were also asked to explain how these events changed the way the company was managed and if the influence of these events lasted over time (and for how long).

Phase 3
Once the interviews were transcribed, we coded them for recurring patterns or themes, and compared themes from each company with those that appeared in the others. We then compared the themes that emerged in our interview data with an analysis of public, archival data such as company reports and independent company and industry analyses. Data analysis occurred in an iterative fashion (Miles and Huberman, 1984): as themes emerged, we attempted to create theoretical explanations for their existence. These nascent theoretical explanations were retained, altered, or dropped as new data was analyzed. Ultimately, two main patterns of dominant logic development – each associated with a different performance outcome (e.g., industry winner or loser) – emerged. Before naming these patterns, we first illustrate them using a representative set of "twins" from our research.[4]

Findings: Representative Twins in the Public Opinion Industry

The firms in the public opinion industry are representative of the patterns of findings across the five sets of twins. These organizations conduct marketing and social research. The public opinion industry, as a whole, is growing worldwide owing to increasing demands for data among corporate and governmental organizations. In 1998, the total revenues in this industry were close to 4.9 billion USD in the United States, and close to 5.4 billion USD in the EU. The market for marketing data in Poland practically did not exist before 1989; research – primarily for political or academic concerns – was conducted by two state-funded centers. With the collapse of communism and transformation to a market economy, a market for social data and later for marketing data started to grow quickly. Between 1989 and 1990 nine research institutions were started, in 1992 there were 27 of them, and in 2000 this number reached 61. The market also grew – doubling or tripling every year, from 50,000 USD in 1990 to 700 million USD in 2000. During the time that we conducted our research, the market matured, with 15 firms generating 81 percent of revenues. Of these, a few firms – which were usually partially or completely owned by multinational research agencies – dominated. During the 1990s, the market also evolved. Whereas quantitative research dominated at the beginning of the decade, a balanced mix of "made to order" qualitative and quantitative research gained precedence during the middle of the decade; and by its end, there was fast growth in the area of syndicated research.

Demoskop SA (Figure 3.1) was founded in 1989 by two experienced sociologists as the first private, social research institute in Poland. They became well known during the time of the first free parliamentary elections in 1989 because they produced a surprisingly accurate forecast of the outcome of the elections. After the elections, they decided to establish a firm in the form of a civil partnership; and using their personal connections to government officials, they secured the first large order for research into social attitudes regarding changes occurring in Poland. The firm was not particularly well equipped in terms of assets, however. Their office was in the private apartment of one of the founders, and it did not have its own computer or software. Perhaps more importantly, they did not have a network of researchers collecting data. However, one of the founders' friends – an ethnographer – came up with an ingenious solution for this latter problem. He suggested using researchers employed at ethnographic museums – which were in every large city in Poland – for data collection (e.g., face-to-face and telephone). These museum ethnographers were young, well educated, and so badly paid that they were desperate to make extra money. The firm leveraged this idea and organized a network of these researchers to help deliver their first order. Slowly the firm got new orders – first through personal connections, and later through word of mouth – and eventually became the industry leader in terms of revenues and number of employees. Initially, most of the orders were for social and political research – thus allowing them to draw upon their core competencies. They also attributed their success to their ability to "read" and respond to their environment rather than their ability to foist new products and services on the market. As one manager noted, "This industry

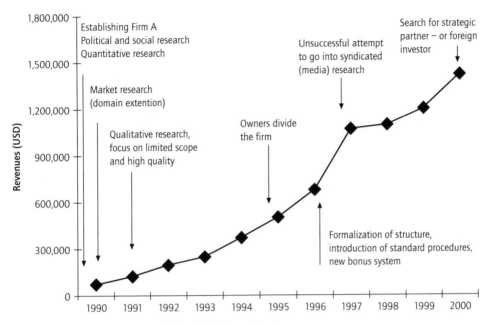

Figure 3.1 Key events in the evolution of the Demoskop SA

develops as environmental forces allow it rather than through its own ideas and products."

In 1992, however, they faced their first setback – one of the long-term contracts with a leading daily newspaper was not renewed. The newspaper cited the high costs that they were expected to pay for social indicator data as their main reason for terminating the contract. Only then did the firm notice that the public opinion market had become very competitive – several new firms had entered the market, and multinational FMCG companies fueled market growth with their research and promotional budgets.

There were several reasons why Demoskop had not participated in (and had not fully recognized) this growth. To begin with, because of their focus on social research and their reliance on quantitative methodology, they ignored the increasing demand for more qualitative research, such as focus groups. The founders – as former scientists – were well trained in quantitative research on large samples and did not recognize qualitative research as valid and meaningful. Moreover, they had hired people mainly from the sociology department at their former university where qualitative research was not recognized as very scientific. Finally, managers felt that their growth had suffered because the marketplace was engaging in "unfair practices." As one noted, "Unfair competition surfaced pretty early. Some firms performed low quality research at a low price, and others offered paybacks to promotional agencies for support. Finally, for some, we became a training school as they were stealing our best people."

As a result of their diminished market position, the founders made two important decisions for developing a market strategy of expansion in the latter part of 1992. First, they decided to extend the scope of their offerings. Second, they stressed the quality of their research methodology. In extending their scope, the founders decided to offer qualitative research services to their customers, to initiate an omnibus research survey (a large social- and market-oriented survey, administered on a monthly basis, on a representative sample), and to publish and sell monthly research reports. Following these initiatives, they purchased a license from the Thomas International Profiling System that enabled them to offer organizational (primarily HR) research services. These services were directed primarily at multinationals. In order to ensure high quality of research, the firm put more stress on research design, installed a multilayer control system of people performing surveys and research in the field, and became very active in industry associations lobbying for higher quality standards.

In terms of profitability, these decisions produced mixed results. For example, the omnibus survey and research reports proved to be costly, and the firm decided to drop the monthly research reports after only three months. However, the firm's revenues were growing. During this period, they employed more than twenty-five people, and moved to a new, modern office. For the first time, the founders hired a professional manager to manage finances.

The steady, but often slow, growth resulting from these moves was halted by a dispute between the founders: one founder wanted to retain the company's informal norms and procedures, while the other wanted to shift to a more professional and client-focused way of operating. As one founder noted, "I strived to analyze customer needs and adapt our projects to them, while he [the other founder] offered suggestions that were not answers to clients' demands but he considered them more interesting and important for buyers." Eventually their disputes about strategy, the level of formalization of the organizational structure, and proper behavior and dress code erupted into conflict. The founders decided to split the firm in 1995. The due diligence performed by the founder inclined to follow a more professional approach, who decided to take over the firm, showed that the firm was much more heavily indebted than he thought because the firm did not collect receivables well, and it had taken out large loans for the new office and for new equipment. Once he assumed control, however, the single founder started to formalize the firm's operations, and stressed the need to follow rules. The transition was not easy. As one of the respondents described it, "Some people did not follow rules and it created some problems and conflicts. Therefore we had to introduce control systems."

The founder hired a new CFO who introduced a formal system of accounting. He also brought on board a cash flow analyst. Along with these personnel changes, the founder implemented a new structure that centered on autonomous research teams and clear job descriptions. Regular, weekly meetings of managers and chief researchers were instituted in order to plan activities and divide jobs. A clear bonus system was established and human resource techniques like career planning and evaluations were introduced. As one manager noted, "We introduced formal job descriptions, a system of ranks and contracts with anti-competitive clauses."

While these changes stabilized the firm's performance, it did not increase the firm's market share or their profitability. The firm still focused narrowly on its

Figure 3.2   Key events in the evolution of SMG/KRC SA

strategy of delivering methodologically flawless public and marketing research. And the firm's lack of innovation and passive marketing – combined with these pressures for methodological purity in research – meant that their research was not always fast or useful. Moreover, they were still hampered by the belief that others in the market were not playing "fair." As a result, the company drifted until 2001 when a majority of shares were acquired by a foreign investor. Looking retrospectively, one of the executives commented, "High quality is not enough, a firm needs some aggressiveness and agility, even an element of yuppiness because customers have to be petted."

SMG/KRC SA (Figure 3.2) was created in 1990 by a group of students and young academics from the University of Warsaw. As one member recalled, "There were 12 of us in the first research team, and only two women. As a result of self-selection and gender bias, it was only natural that the atmosphere in SMG/KRC was pretty hilarious, even if with [some] elements of army barracks." The twelve started with social research performed for governmental agencies but very quickly – in 1991 – they exited this domain, citing a "lack of perspectives and limited profitability" as their main reasons. However, another reason was due to the fact that the founders were more at ease doing qualitative than quantitative research.

They then entered the field of marketing research – which they initially perceived as a more promising and less competitive market – with a public relations campaign targeted to inform the market (especially promotion agencies) about their existence and their resources. They offered to analyze the effectiveness of promotional agencies' promotional campaigns at a substantial discount in exchange for recommendations of their services to MNCs. They also decided to leverage their personal contacts with colleagues (and relatives) who were currently making careers in large firms. This did not prove difficult as many former students from the University were just beginning to work for MNCs, and many of them were launching spectacular careers. Using these personal contacts and the clients channeled by promotional agencies,

SMG/KRC started to grow rapidly. They built special facilities to conduct "focus sessions" for marketing research, and new people were employed. By late 1991, they were offering both qualitative and quantitative research to their clients.

These new hires, along with the expansion of offerings, changed the structure and culture of the organization. For example, a team of two researchers, one quantitative and other qualitative oriented, typically serviced clients. These teams formed the nucleus of a project structure. Regarding the culture, one manager described the change in this way: "We developed a culture of a consulting firm instead of just a data provider."

Rapid growth in the Polish consumer market and the entry of many foreign firms resulted in a growing demand for different types of public opinion research. As one informant noted, "There was enough work for everybody in the marketplace. The only problem was – who will work faster and make more money on it?" SMG/KRC adapted by hiring a group of specialists and extending their offerings even more (e.g., ad hoc surveys and quarterly omnibus research). Their fast growth forced an implementation of some formalization in SMG/KRC SA: they built a network of local researchers throughout the country, established local coordinators, installed procedures and timetables for survey execution, and implemented a basic wage system with high bonuses based upon revenues. After a short power struggle with one of the older founders, a founder with a strong personality and natural leadership ability – whose nickname was "the General" – became the president of the firm. He described his approach to management in the following way: "If I am asked what kind of martial style I represent – lion or fox, I would answer – donkey. I mix the stubbornness of a donkey with intuition and imagination." One of his first decisions was to limit brain drainage: Some of their best specialists were being hired away by the marketing departments of MNCs. An innovative answer to this problem was the establishment of a special executive search and selection department within the firm. The department had two functions. On one hand, they leveraged the strong ties – which resulted from the placement of its managers into MNCs – into research orders. On the other, this department worked as a deterrent to these MNCs and others who wanted to poach researchers from SMG/KRC. Specifically, the executive search department targeted each "poacher" as a potential source of labor from which to replenish SMG/KRC's own diminished talent pool. This department was run by the wife of the president, and eventually was spun off into a separate and highly successful firm using the same brand name.

In 1992, disaster struck. The whole department of quantitative research left the firm on a Friday and its employees established a new, competitive firm on Monday of the following week. The reasons for exit were many. Researchers from the qualitative department were getting higher bonuses because they controlled the client–firm interface; the quantitative department was located in a second-rate location outside of the main headquarters (which reinforced its "back-office" status); and despite bottom-up pressure for the establishment of an employee ownership plan, some owners strongly opposed it. As one manager put it, "The exit of an entire department worked as a 'cold shower' on everybody. We understood that in order to keep the best employees, we have to share."

A string of decisions and actions followed during the next two years. Key researchers and managers were offered either shares or large bonuses. The firm moved to a larger office that could house all the workers in the same premises. A new structure was implemented based upon autonomous teams that could better offer a full range of qualitative and quantitative services to external clients. One manager described the structural changes in this way:

> The leader of each team got wide responsibilities: they can hire and fire; they cannot offer bribes or dumping prices . . . Each of the leaders has their own vision of how to manage a team, how to manage projects. There are some general rules, but within these rules they have a lot of freedom. There is one general demand that over time was imposed upon leaders by the CEO of the company: Each has to be unique – either because of his own personality or because he was told to and planned to be . . . They have to be original and exceptional . . . they have to fight for prestige and identity because if you are unique you will always find clients that would like to work with you.

This new, empowering structure meant that every team was like a hologram image of the entire company. As such, managers felt that the company was more secure. As one of managers put it, "We are not a hostage to anybody nowadays. An important employee can leave us and it can even be a threat to sales – but the structure supporting the research teams does not rely on any particular person anymore."

In addition to these changes, normal procedures regulating internal competition were introduced, and decisions were decentralized at the team leader level. Experiments and new offerings were rewarded and researchers were proud that they created new markets. As the director of customer service explained, "Today every customer demands psychographic segmentation and research into the teenagers market, but it was us who created the need for it." The firm started to plan its activities, but framed its strategic objectives in the form of challenges. As one executive noted:

> Every year we put a different competitor into our strategic plan, one that we want "to hurt" the most in this year, just for sporty reasons, because it makes life interesting. We have to run with somebody, like in the games. Also, we put in the strategic plan new products that we want to introduce into the marketplace. We have to try new things because otherwise we get bored.

SKG/KRC also experimented with international expansion. This expansion was instrumental in the creation of CEMA (Central European Marketing Association): an alliance of top firms from the Czech Republic, Hungary, and the Slovak Republic. However, while the alliance proved to be moderately successful, the firm decided to exit it after three years because it provided partners with more orders than it got. In 1995, it joined an international research group, INRA, which was present in 22 countries. The firm obtained licenses from INRA for several products and, in exchange, participated in large international projects.

The firm remained incredibly active in the market and each team aggressively fought for new clients and contracts. It gained real market visibility after it won an important contract for a syndicated press panel and survey; it then obtained radio and television surveys, and organized a complex panel of consumers. These were important contracts as they involved cyclical, repetitive, large-sample research projects. In 1995, SMG/KRC became the market leader and has been growing faster than the market since then. In 1999, the founders decided that the opportunities resulting from its international cooperation with INRA were limited, and sold 47 percent of shares to a well-recognized international market research firm, Kantar Group. Looking retrospectively, one of the founders summed up the history of SMG/KRC in this way:

> Trust and optimism count. If somebody put us – these 12 key persons – on a ship tomorrow, we would sail where we had to. If we had been asked to build a house we would have done that too. Whatever we were doing together, we would manage well.

Conclusions

Both winners and peripheral firms in the Polish marketplace developed coherent dominant logics, but their elements and dynamic differed. These differences were illustrated in the public opinion firms, and the presence or absence of these differences across the other four "twins" is illustrated in Table 3.2.

Dominant logic of winning firms

In a nutshell, winners followed simple rules and used them to maximize their opportunities set. They were flexible and quick. The dominant logic of winners developed in a way similar to Siggelkow's (2000) *patch-to-patch* mode of behavior. That is, firms started with one strategic theme, and after it was well established they would experiment with the next theme, and so on. Under this process the dominant logic developed over time by adding new elements from those actions and choices that helped the particular firm to become successful in the marketplace. Winning firms also demonstrated consistent features across all of the main elements (perceiving/sensemaking, action/choice, learning, and codification of routines) that we identified as being central to dominant logics.

Perception/sensemaking
The most common characteristic of these dominant logics is an external orientation. Winners were focused on environmental trends and events, and perceived them as a set of opportunities to be leveraged. They focused on the marketing of their products and services, and this focus – combined with the creation of optional marketing strategies, experiments and innovations, and quick entries and exits – allowed them to create a standard and/or brand.

Actions/choices
Winners make many more choices that they see as being "strategic" than losers. For example, managers in SMG/KRC stressed eight strategic choices – new market

Table 3.2 Dominant logic of winners and peripheral firms

	Perception/sensemaking	Action/Choices	Learning	Codification of learning (routines)
ERIS SA (winner)	Opportunities dominate. Intuition and limited analysis as ways to analyze environment. Leaders note that a little "craziness" helps.	Few high-risk choices – investments in Swiss machinery, adapting firm to the needs of handicapped workforce, establishment of dedicated distribution, and moving up the market.	Few, very difficult situations used as teaching points.	Limited planning and budgeting – profit and app. budgets guide performance with the support of an IT control system. Strong role of leaders and core managers.
DAX Ltd (peripheral)	A mix of threats and opportunities. Company as trailblazer that competitors follow.	Two strategic choices: (1) export to Soviet Union, and later (2) imitation of foreign companies' offerings.	Limited learning – refocusing on Polish market after collapse of Russian market.	Precise marketing, sales, and production plans developed upon careful analysis. Strict budgeting.
SMG/KRC SA (winner)	Environment as a source of opportunities. Exploring and active shaping of an environment as a main focus.	Many strategic choices in terms of markets, products, organization, and internalization. Constant experimentation. Focus on development of market standards.	Specific and intense – both particular and general events trigger learning. Driven by leadership. Difficult situations converted into programs; institutionalizing changes to prevent repetitions of such problems.	Organizational structure of autonomous teams (hologram image) led by founders and most trusted managers, along with very limited formalization. Stressed innovation. Routine prices formalized.
Demoskop Ltd (peripheral)	Opportunities, threats, problems, and unfair competition.	Few choices that were mostly driven by internal pressures. Focus on high quality of research.	Major internal conflict converted into limited and focused teaching points.	Incremental, but systemic, formalization and standardization as ways of improving performance.
WGK Ltd (winner)	Opportunities driven, strong competition perceived as a problem to be solved through specialization.	Several major decisions – expansion from industry analysis to consulting, privatization, and investment banking.	Learning through analysis of good and bad experiences by founders – changes of strategy or structure followed.	Only routine processes (like bidding for jobs, accounting) formalized and standardized.
Focus Management Ltd (peripheral)	Balanced perspective of environment (both threats and opportunities).	Organic growth, focused on German companies.	Limited learning reported.	Selective formalization due to small size of the firm. Centralized decision making.

Table 3.2 (*cont'd*)

	Perception/sensemaking	Action/Choices	Learning	Codification of learning (routines)
Optimus SA (winner)	Focus on exploration and exploitation of new markets and challenges. Comparison with international companies in the industry emphasizes the good position of the firm.	Many strategic choices related to market, products, partners and organization – company perceived as a good incubator of new ideas and businesses. Being ahead of competition as a rationale for constant innovations.	Constant upgrading of products, diversification into IT integration, alliances with leading world companies, IPO and entry into e-business major source of learning.	Incremental but visible formalization leading to strategic planning and budgeting parallel to the process of handing over power to professional managers by the leader and founder of the firm in the late 1990s.
Baza Ltd (peripheral)	Volatile perspective on environment as a source of either many risks or threats.	Choice of markets and products made at the beginning of the firm. Replication of the business model in different regions of Poland, with limited adaptive changes.	No learning and analysis. Any difficulties blamed on suppliers, competitors, or personnel's lack of competence.	Limited but strict formalization. Centralized decision making.
Alpinus SA (winner)	Optimistic, proactive, "we can do it" attitude. Competitors as mobilizing force. Optimism "pumped into" organization.	String of explorative choices of partners, products, brands, distribution systems (franchise), outsourcing of production, etc. New initiatives get most of funding. Focus on intuitive thinking.	International and domestic competition as sources of new ideas. Experimentation through expansion of product lines and markets (200 variations) and organizational changes.	Strong leadership. General planning and programming approach transferred from experiences in leading climbing expeditions. Formalization on a "have to" basis; middle management expected to be creative but also to solve problems quickly through informal cooperation.
Janysport (peripheral)	Limited, balanced environmental perception and analysis.	Few choices; strong focus on implementation.	Selective learning focused on avoidance of threats.	Strong centralization of decisions; selective formalization – mainly of process of production.

entry, experimentation with new products and services, introduction of team-based structure, spin-off of a department specializing in executive search, introduction of an ESOP, a bid for risky and resource-intensive syndicated research, the establishment of an international network of agencies, and the choice of a foreign investor. Most of these actions were made quickly but with a reasonable amount of analysis and calculations.

Learning
Winners acknowledge easily that they went through traumatic situations and treated them as a learning experience. For SMG/KRC, these situations took the form of the exit of the entire quantitative analysis department and the loss of a major client. Traumatic or difficult experiences always triggered structural or procedural changes in winning firms, and were used as a *memento* or cultural saga to remind managers of possible problems and pitfalls.

Codification of learning/routines
In terms of routines, winners are absolutely pragmatic and opportunistic – they introduce routines only when it is necessary, and when they think that they will benefit from it. They did not impose rules for their own sake. By trial and error, these firms evolved into flexible but centralized organizations: Formalization and standardization are limited, and leaders can play a central role in bending the rules and procedures when needed.

The three simple rules of winners
It was the synthesis of links between and among these various elements of the dominant logic of winners that led to the successful implementation of their strategies (see Table 3.2). The process of dominant logic development can be summarized as *a process of developing simple rules that enabled them to make sense of a confusing, turbulent environment, and allowed them to make effective decisions.* We did not encounter in this research a comprehensive set of rules (e.g., how-to, boundary rules, priority, timing, exit, etc.) such as those proposed by Eishenhardt and Sull (2001). This may be due to the fact that in turbulent environments, no firm can know a priori what makes sense and what does not. Therefore winning firms developed their rules one by one ("patch to patch") by the design or wishes of their founders. These simple rules were modified, dropped, or added to in the wake of traumatic experiences and difficult choices.

To illustrate, SMG/KRC developed its rules at different moments in their history: the moment of inception when twelve young people decided to become entrepreneurs, when they decided to move from political to market research, after the painful exit of all their quantitative researchers. Similarly, Eris, the leader in the Polish cosmetics market, established some of its rules at the very beginning by making a decision that it would create a friendly environment for those with handicaps. It later decided to develop its own premium product lines and distribution network. Optimus, the leader in the computer market, developed some of its rules at its inception. These rules were an extension of its founder's attitude toward business – that it should be innovative and grow fast. Managers were encouraged to promote local initiatives and ideas. Initial profits were reinvested in the development of production facilities and distribution. Equity generated by the subsequent IPO was utilized to build new models, to invest in a brand that eventually became one of the top brands in Poland, and to diversify into IT services and e-business.

Generally these rules were developed and used by founders as key decision criteria, and they were used to craft their communication strategies both internally and externally.

In addition to commonality in the timing of rules (e.g., creating them one at a time), winning firms all tended to have the three interrelated rules that were central to their dominant logics: (1) be optimistic, (2) move fast, and (3) balance centralization and flexibility. With regard to the first rule, managerial perceptions of their market environment were very positive – it contained a set of opportunities to be followed, exploited, and leveraged. A natural extension of this optimism was an action orientation: Focusing on growth, taking challenges head on, and making room for initiative and errors were central features in the interviews of all the winning firms. Leaders in all the winning firms, since their inception, behaved like happy kids – they were constantly trying new things, they were making more (and more costly) mistakes than those in peripheral firms, and they were having fun. Of course, winners eventually mature. Each of the firms introduced routines that governed key processes, and most of them eventually hired professional managers to support the team of owners. However, they did not formalize or standardize their operations to the same degree as losing firms.

A second strategic rule of winners is, "Enter fast and then either grow fast or exit fast." In all the winning firms we observed, there was a passion to fight, to win, and to be a leader in many areas. Over time, SMG/KRC entered all possible markets for public opinion research – political, social, marketing, and all types of research – customized, syndicated, omnibus etc. Alpinus, the leading outdoor textiles producer, experimented with all possible product lines and types, even though most of them were ultimately unprofitable. And Optimus (the leading hardware and software provider) strived from the very beginning in 1990 to attain the position of a dominant firm in an industry, and used fast product upgrades, alliances with Microsoft, Intel, and Lockheed Martin, and constant extensions of its offerings (from hardware to software, consulting services, integration services, business-to-business services, etc.) as a means of keeping first-mover advantage.

To succeed in these diverse areas, founders established the boundaries of the strategic domain by their choice of industry, and then allowed managers to experiment with the option set. A key process in each of these firms was marketing. Given shifts in consumer preferences, it is perhaps logical that firms were structured in such a way as to allow their members to customize their activities toward a particular market niche. Winners also sometimes institutionalized innovations and experiments with special symbolic or financial rewards. Sometimes founders gave clear signals about the importance of innovation and creativity. For example, SMG/KRC introduced special awards for innovative thinking: Virtuti Researcheri, dinner with the president, one-month wage bonus, and so on.

Perhaps as importantly, these firms also had to trust their managers and know when to pull out of a market. At its inception, SMG/KRC quickly determined that its core business of providing social research for governmental agencies was not profitable, and left this market. It later broke off deals with CEMA and INRA when they were not profitable. Alpinus extended product lines and started to produce special out-door equipment for bikers and sailors, developed new brands and special "weekend lines" for ordinary customers, and even began a special offering for kids. None of them proved successful and Alpinus phased them out and decided to use only one brand – "Alpinus" – for all its products. It also focused on its traditional outdoor niche.

A third rule of dominant firms is "balance centralization and flexibility." As illus-trated above, this rule was partially accomplished by the winners' autonomous team-based structure. However, leadership was an important element in maintaining this balance as well. All winning firms had strong, and often charismatic, leaders. In all cases, leaders came from the team of founders, and each was clearly a symbol of the firms' successes. These leaders were (and are) used to being in charge – and they tended to control the most important activities of the firms since they were started. In all successful firms, leaders maintained a presence in all important aspects of the firm's operations. In SMG/KRC, the founders of the company ran most of the departments. In the cosmetics producer Eris, one of the founders, Irena Eris, ran the crucial R&D department, while her husband – in his role as CEO – focused on distribution and marketing. In Optimus, the computer producer, the CEO was a charismatic founder of the company who was behind all important choices and alliances that the firm entered. In Alpinus, the outdoor producers, three founders constituted an executive board and were responsible for production, distribution, and marketing. Middle managers are expected, however, to show initiative and to cooperate. As one manager noted, "We meet often in working groups but this is not regular. Life decides when we have to have a meeting." At the same time, in each of these firms the leaders spent considerable effort hiring and placing managers in other key positions, and gave these managers considerable freedom and responsibilities. They supported experiments and innovations, but killed them off quickly if they were not successful. They also periodically changed or modified the organization's structure. They supported the development of standard procedures, but did not allow bureaucracy to develop. They even broke their own rules, occasionally, and allowed others to do the same.

A dominant logic of peripheral firms

A surprising result of this research was the discovery that the dominant logic of companies that lost their chance in the marketplace – from almost the very beginning – was very coherent and nearly mature. Their evolution resembled a "*thin-to-thick*" pattern (Siggelkow, 2000). In a thin-to-thick pattern, the firm starts out with a consistent set of weakly developed strategic themes and elaborates them over time. In this case, dominant logics existed almost as a complete system from the firms' inception, and were refined and reinforced over time.

Perception/sensemaking
In terms of how they made sense of their environment, managers in peripheral firms acknowledged that they began with a "balanced" perception of the environment as

consisting of opportunities, threats, and problems to be solved. As they began to become industry leaders, however, they modified their vision and focused on threats and problems. Managers and founders of these firms became very good at spotting events and trends that might have a negative impact on the whole industry and on their firms. They also perceived competitors as ruthless or unfair, customers as fickle and not loyal, and the general business climate as bad.

Actions/choices

Managers in these firms did not label many decisions as strategic, and those that were labeled as such tended to be limited to choices made in the beginning of their life cycle. In spite of the influx of new competitors and other changes in their environment, owners of the peripheral firms stuck to their primary choices as predicted by organizational ecology theory (Hannan and Freeman, 1989). However, in turbulent markets, agility and experimentation pay off – whereas persistence and commitment do not. Janysport, for example, started as a producer of durable, high quality outdoor equipment and did not compromise its choice of strategy with new models or colors. Baza Ltd was established at the same time as Optimus but its owner did not change his strategic choice of establishing a limited product range (computers and peripherals) in spite of the market trend toward integrating computers, software, and services. Also, although his initial choice of structural configuration of the firm as a federation of regional offices enabled Baza to grow rapidly at the beginning, this structure eventually precluded synergies, cooperation, and the integration of effort. The structural configuration did not change even toward the end – only regional offices were closed in succession as the firm started to lose money in 1997. Dax Ltd made a strategic choice of focusing on the Russian market with a limited series of cheap but good-looking cosmetics and stayed with this choice until the Russian economic crisis of 1998. In the meantime, it lost any position and brand recognition in the core Polish market. Finally, Demoskop's long commitment to public and social research, and quantitative, "scientific" methodology, resulted in lost growth opportunities. It also led to a public image of being focused primarily on sociological and political research projects. In all of these cases, peripheral firms did have an articulated strategy. And the choices of markets and products – as well as sources of competitive advantage – were clearly made. At the same time, these strategies were not successful in the long run because they were rigid, driven by choices made at the inception of the firm, and were not adapted to changing environmental situations.

Learning

Perhaps the most striking feature of the peripheral firms researched is the fact that they rarely reported any difficult situations from their past. One would think that after a decade of limited growth or decline, the founders and managers could easily spot at least a couple of difficult situations and describe them as learning experiences. However, in most of the firms, respondents could not identify more than one of these events (and some firms could not identify any). When difficult situations were identified, they were rarely seen as learning opportunities. In the case of Baza, for example, any problems were blamed on the environment. In the cases of

Demoskop and Janysport, both firms reported conflicts between founders as their main "difficult experience." In the case of Dax, it was a crisis in the Russian market. However, peripheral firms did not learn easily, and they did not classify situations as learning experiences. If learning did occur, it was often used to confirm the existing dominant logic (e.g., reinforced the perception of the environment as hostile, and/or the correctness of earlier strategic choices). The owner of Baza powerfully summarized this lack of learning. When asked what insights he had gained – as the company was on the verge of bankruptcy – he stated bluntly, "I could not say that I learned anything important."

Codification of learning/routines
In terms of routines, losers formalized their activities and introduced standard operation procedures much faster than winners. Moreover, there was a clear difference between the formalization performed by winners and that of losers. Winners formalized only in areas that they felt they had to, for pragmatic or legal reasons – they standardized those activities that were critical in terms of overall productivity, or activities that had to be standardized by law (e.g., accounting). Peripheral firms formalized because they did not trust others: they centralized decisions when they did not trust management, they standardized activities when they did not trust employees, and set up certain procedures because they did not trust their competition to "play fair." Formalization was also used to prevent mistakes and errors. Managers in these organizations often used the metaphor of a machine to describe their operations. In the two cases of peripheral firms where we did not find significant formalization (see Table 3.2), we found that the reason for this lack of formalization was due to the fact that the founders were making nearly all of the important decisions, and thus found routines unnecessary.

Summary – happy kids and mature losers

This paper addresses the issue of dominant logic as a cognitive, strategic, operational, and evolutionary concept useful in studying how winners and losers built their strategies in turbulent markets. Studying the elements and dynamic processes of the dominant logics of outliers (winners and losers) that had almost equal chances for success in newly established markets in Poland, we came up with two major patterns of dominant logic. Winners do not have coherent strategies and rigid designs. They follow simple rules that enable them to create and leverage opportunities, influence the stream of events, establish standards, brand names, and publicity. Their dominant logic develops in a "patch-to-patch" fashion as they make strategic choices and learn from difficult experiences.

Peripheral companies develop differently. Their dominant logic develops in a "thin-to-thick" fashion, and therefore it is from the very beginning more coherent and consistent than in winning companies. But it is also more limited, rigid, and calculative. Peripheral companies do not follow the logic of opportunities but the logic of problem solving; they develop strategies and follow them tightly; they build centralized and formalized designs; and they eventually become slaves of their dominant logic by rationalizing failures and forgetting experiences.

Both winners and peripheral firms have a consistent dominant logic with cognitive and action/operational components, and most of the time they stick to their cognitive paradigm and managerial choices. However, the dominant logic of winners enables them to leverage opportunities and grow rapidly, while the dominant logic of losers limits their option set and make them inertial and unable to change. Firms that became peripheral develop an early solution in the form of a strategic choice (Child, 1997), and then tenaciously stick to that choice. Winners and peripheral firms were often able to indicate at least one difficult or traumatic experience in the history of the company (winners have much more of them); however, they treat them differently. In each case, winners made these events into a learning lesson, creating a new rule out of it (like starting ESOP after a group of specialists left the company, or making diversification an idea that was strictly forbidden after a disastrous diversification that looked like a sure winner). In peripheral companies these experiences were not leveraged in any way – at best, they were "bad memories" for everybody.

In a way, winners operated from the beginning like opportunistic and happy kids, using simple rules to experiment in the marketplace and search for such leveraging points as possible standards, brand names, publicity, fastest-growing segments, and internal flexible organization. Peripheral firms quickly developed full-fledged strategies and started to operate like mature and experienced companies. This dominant logic strengthens over time, irrespective of difficult experiences and visible limitation to growth – either because of owners' values, limited environmental scanning and therefore an erroneous evaluation of the company's position, formalization of operations, lack of learning, or plain inertia. It seems that managers at peripheral companies slowly learned how to live with limited options, and they tended to blame their predicaments on the environment or on the unfair and unjust strategies of winners.

Notes

1   To illustrate, Poland's gross domestic product (GDP) grew by almost 5.5 percent a year on average between 1993 and 1999. During that time, inflation declined by 7.5 percent and unemployment dropped to 10 percent. However, between 1999 and 2002 the economic growth rate slowed to less than 2 percent, inflation dropped to 2 percent, but unemployment increased to 18 percent.

2   The cosmetics industry did exist under the communist regime but many companies were so dependent upon exports to the Soviet Union that they either went bankrupt after 1989 or had such a limited reputation and product range that it did not create any barriers to entry.

3   Some of the data and an earlier version of the arguments reported here were used in K. Obłój, Simple rules and simple solutions: the dominant logic of high and low performers in turbulent environments, published in H. J. Stuting, W. Dorow, F. Claassen, and S. Blazejewski (eds.) (2003) *Change Management in Transition Economies: Integrating Corporate Strategy, Structure and Culture.* Berlin: Polgrave.

4   Owing to space constraints, we do not present findings from all five sets of twins. Rather, we chose one pair that illustrates the general trends across all of these sets.

References

Bettis, R. 2000: The iron cage is emptying , the dominant logic no longer dominates. In J. A. C. Baum and F. Dobbin (eds.), *Economics Meet Sociology in Strategic Management*. Stamford, CT: Jai Press, 167–74.

Bettis, R. A. and Prahalad, C. K. 1995: The dominant logic: retrospective and extension. *Strategic Management Journal*, 16, 5–14.

Cavaleri, S. and Obloj, K. 1993: *Management Systems*. Belmont, CA: Wadsworth.

Child, J. 1997: Strategic choice in the analysis of action, structure, organizations and environment: retrospect and prospect. *Organization Studies*, 18, 43–75.

Cote, L., Langley, A., and Pasquero, J. 1999: Acquisition strategy and dominant logic in an engineering firm. *Journal of Management Studies*, 36, 919–52.

Cyert, R. and March, J. G. 1963: *The Behavioural Theory of the Firm*. Englewood Cliffs, NJ: Prentice-Hall.

Eisenhardt, K. M. 1989: Building theories from case study research. *Academy of Management Review*, 14, 532–52.

Eisenhardt K. M. and Martin, J. M. 2000: Dynamic capabilities: what are they? *Strategic Management Journal*, 21, 1105–22.

Eisenhardt, K. M. and Sull, D. N. 2001: Strategy as simple rules. *Harvard Business Review*, January, 106–16.

Giddens, A. 1984. *The Constitution of Society*. San Francisco: University of California Press.

Gioia, D. and Poole, P. 1984: Scripts in organizational behavior. *Academy of Management Review*, 9 (3), 449–59.

Grant, R. M. 1988: On dominant logic, relatedness and the link between diversity and performance. *Strategic Management Journal*, 9, 639–42.

Hamel, G. and Prahalad, C. K. 1994: *Competing for the Future*. Boston, MA: Harvard Business School Press.

Hannan, M. and Freeman, J. 1989: Organizational ecology. Boston, MA: Harvard University Press.

Huff, A. S. 1982: Industry influences on strategy reformulation. *Strategic Management Journal*, 3, 119–31.

Jennings, D. and Seaman, S. 1994: High and low levels of organizational adaptation: an empirical analysis of strategy, structure and performance. *Strategic Management Journal*, 15, 459–75.

Kelly, D. and Amburgey, 1991: Organizational inertia and momentum: a dynamic model of strategic change. *Academy of Management Journal*, 34, 591–612.

Lampel, J. and Shamsie, J. 2000: Probing the unobtrusive link: dominant logic and the design of joint ventures at General Electric. *Strategic Management Journal*, 5, 593–602.

March, J. G. 1994: *A Primer on Decision Making: How Decisions Happen*. New York: Free Press.

March, J. G. 1996: Continuity and change in theories of organizational action. *Administrative Science Quarterly*, 41, 278–87.

Markus, H. and Nurius, P. 1986. Possible selves. *American Psychologist*, 41, 954–69.

Miler, D. 1999: Notes on the study of configurations. *Management International Review*, 2, 27–39.

Miles, M. B. and Huberman, A. M. 1984: *Qualitative Data Analysis*. Beverly Hills, CA: Sage Publications.

Miles, R. and Snow, C. 1978: *Organizational Strategy, Structure, and Process*. New York: McGraw-Hill.

Nelson, R. R. and Winter, S. G. 1982: *An Evolutionary Theory of Economic Change*. Cambridge, MA: Belnap Press of Harvard University.

Pettigrew, A. 1990. Longitudinal field research on change: theory and practice. *Organization Science*, 1, 267–92.

Porter, M. E. 1980: *Competitive strategy*. New York: The Free Press, 61–78.

Porter, M. E. 1996: What is strategy? *Harvard Business Review*, November–December, 61–78.

Powell, T. C. 1992: Organizational alignment as competitive advantage. *Strategic Management Journal*, 13, 119–34.

Prahalad, C. K. and Bettis, R. A. 1986: The dominant logic: a new linkage between diversity and performance. *Strategic Management Journal*, 7, 485–501.

Pratt, M. G. 2003: Disentangling collective identity. In J. Polzer, E. Mannix, and M. Neale (eds.), *Identity Issues in Groups: Research in Managing Groups and Teams, Vol. V*. Stamford, CT: Elsevier Science Ltd, 161–8.

Reynolds, P. D., Camp, S. M., Bygrave, W. D., Autio, E., and Hay, M. 2001: Global entrepreneurship monitor, 2001 executive report. London Business School.

Schein, E. 1985. Defining organizational culture. In *Organizational Culture and Leadership*. San Francisco, CA: Jossey-Bass, 3–15.

Siggelkow, N. 2000: *Evolution of fit*, Paper presented at the Strategic Management Society Conference, Vancouver, September.

Starbuck, W. H. 1983: Organizations as action generators. *American Sociological Review*, 48, 91–102.

Starbuck, W. H. and Milliken, F. 1988: Executives' perceptual filters: what they notice and how they make sense. In D. C. Hambrick (ed.), *The Executive Effect: Concepts and methods for studying top managers*. Greenwich, CT: JAI Press, 35–65.

Tushman, M. L., Newnam, W. H., and Romanelli, E. 1986: Convergence and upheaval: managing the unsteady pace of organizational evolution. *California Management Review*, XXIX (1), 29–43.

Walsh, J. P. 1995: Managerial and organizational cognition: notes from a trip down memory lane. *Organizational Science*, 6, 280–320.

Weick, K. 1995: *Sensemaking in Organizations*. Newbury Park, CA: Sage.

Zajac, E. J., Kraatz, M. S., and Bresser, R. F. 2000: Modelling the dynamics of strategic fit: a normative approach to strategic change. *Strategic Management Journal*, 21 (4), 429–55.

Zyglidopoulos, S. 1999: Initial environmental conditions and technological change. *Journal of Management Studies*, 36, 241–62.

CHAPTER 4

A Strategic Perspective on Capital Structure: The Implications of Trying to be an Innovator

Jonathan P. O'Brien

Keywords: Capital structure, strategy, innovation, performance.

Abstract

This paper argues that firm strategy should bear a significant influence on the capital structure choices made by managers. Maintaining lower leverage and greater cash balances gives the firm more financial slack (i.e., ready access to the funds necessary for financing projects). Within each industry, the basis of competition that each firm chooses will determine the strategic value to the firm of maintaining financial slack. The empirical analysis yields strong support for the proposition that financial slack should be a particularly critical strategic imperative for firms pursuing a competitive strategy premised on innovation. Furthermore, it is demonstrated that firms pursuing such a strategy that fail to recognize the value of financial slack are likely to suffer deleterious performance consequences.

Introduction

It would be a considerable understatement to say that the study of capital structure has inspired a prodigious amount of research within the field of finance. Yet, despite all this research, much is still unknown about how managers choose between debt and equity financing (Harris and Raviv, 1991). Early work on this subject, guided by Modigliani and Miller's (1958) assumption that the financing and investment decisions are separate processes, dismissed the potential for strategy to shed light on the "capital structure puzzle." However, ever since Jensen and Meckling (1976) acknowledged the potential for the investment and financing decisions to interact, the door has been opened for researchers to explore how competitive strategy influences capital structure.

Consistent with Balakrishnan and Fox (1993), we argue that the application of strategy may be most helpful in helping to understand intra-industry variation in capital structure. A critical assumption underlying many of the theories in strategic management is that in any given industry, the potential consumers are not homogenous, but heterogeneous with respect to their tastes and preferences. These different segments of consumers are best served by firms that operate under different strategies, as a firm that tries to serve all segments will serve all of them poorly (Porter, 1980). Therefore, within any given industry we are likely to observe variation in the strategies pursued by the industry incumbents. If competitive strategy guides the firm's investment decisions (Chandler, 1962), and the choice of investments can influence the choice of financing (Williamson, 1988), then we should expect that different capital structures best serve the needs of different strategies. Furthermore, this would suggest that we should expect to observe substantial variation in capital structures across firms within the same industry. Herein lies the potential for strategy to help clarify the capital structure puzzle. The most promising application of strategy to the topic of capital structure is not explaining inter-industry variations (as suggested by Harris and Raviv, 1991), but in explaining why we should expect to see as much variation within an industry as between industries.

In this paper, we focus on how a competitive strategy based on being an industry innovator will impact the capital structure decision. Previous research has adeptly pointed out that intense investment in R&D is associated with lower leverage because these investments create primarily intangible assets, which cannot serve as good collateral (Long and Malitz, 1985; Simerly and Li, 2000; Vicente-Lorente, 2001). This perspective suggests that the firm's absolute intensity of investment in R&D will influence capital structure. We take this argument a step further by proposing that financial slack (i.e., low leverage) should be a strategic priority for firms that are competing on the basis of innovation. Financial slack can help to sustain the competitive position of firms competing on the basis of innovation by helping to ensure: (a) continuous, uninterrupted investments in R&D; (b) that the funds necessary to launch new products are available when needed; and (c) that firms can expand their knowledge base through acquisitions when it is potentially beneficial to do so.

The arguments presented above suggest that maintaining a high degree of financial slack should be a strategic imperative for firms that are trying to compete on the basis of innovativeness. Accordingly, we propose that the strategic importance of financial slack to the firm should be proxied for, not by the firm's absolute intensity of investment in R&D (i.e., R&D expenditures scaled by total sales), but by the firm's relative intensity of investment in R&D (i.e., relative to other firms competing in the same industry). To illustrate this point, consider two firms competing in different industries that have the same absolute R&D intensity. If one of those firms has the highest R&D intensity relative to its industry peers, while the other has the lowest, it seems likely that the firm with the greater relative R&D intensity is attempting to compete within its industry on the basis of innovation, while the other firm is not.

Overall, we find strong support for our proposition that firms competing on the basis of innovation make financial slack a strategic priority. An implication of this result is that if capital structure *should* follow strategy, then we expect to observe significant performance penalties accruing to those firms that display misalignment between these two factors. Our empirical tests confirm this expectation, implying that failing to maintain sufficient financial slack can seriously inhibit a firm's ability to successfully implement a strategy premised on innovation.

Previous Research

Traditional explanations of capital structure

Modigliani and Miller (1958) first ignited interest in the study of capital structure when they proposed that given certain simplifying assumptions, the value of a firm is independent of its capital structure. This famous Proposition I violated the prevailing popular wisdom of the time, and thus generated a tremendous amount of controversy, debate, and research. However, the authors had never intended to suggest that capital structure is irrelevant in real-world applications (Miller, 1988). Rather, the authors simply wanted to investigate whether there could be any set of conditions, even in a frictionless world, where capital structure did not impact the value of the firm. Nevertheless, several years later Modigliani and Miller (1963) once again added fuel to the debate when they "corrected" their stance by relaxing the assumption of a tax-free world. Once the tax deductibility of interest payments is factored in, the value of the firm increases with leverage, even up to the point where it appeared that the optimal capital structure was all debt. Clearly, there had to be more to the story to explain why all firms weren't leveraged to capacity. That next piece to the puzzle came in the form of the costs of debt financing, which can act to at least partially offset the benefits of the tax deductibility of interest payments.

Among the costs of debt financing are the potential costs of financial distress. If the firm fails to meet its debt obligations, it does not simply pass seamlessly into the hands of creditors. Instead, going through bankruptcy causes the firm to lose some of its value. The magnitude of the costs of financial distress has been hotly debated, however, with estimates ranging from around one percent (Warner, 1977) to up to 20 percent (Andrade and Kaplan, 1998). Furthermore, these costs of financial distress might vary with the type of assets held by the firm (Long and Malitz, 1985), an important consideration that we will return to later.

Jensen and Meckling (1976) point out that there are also costs of debt that arise from the incentives that equity holders have to expropriate wealth from the bondholders. This induces the bondholders to demand protection via monitoring and bonding mechanisms. These monitoring and bonding mechanisms are not without cost, and thus they serve as another of the offsetting costs of debt financing. Although debt has its own agency costs, Jensen (1986) later pointed out that by virtue of reducing the amount of free cash flow over which managers have discretion, increased leverage may actually reduce the agency costs that arise from the divergent interests that exist between the managers and shareholders.

Motivated by the apparent benefits and costs of debt financing, a great deal of research has tried to determine if there is some optimal capital structure that balances the two opposing considerations (Jalilvand and Harris, 1984). However, the empirical evidence has been conflicting (Shyam-Sunder and Myers, 1999), and others have suggested that there might be no "optimum target" because there may actually be no tax benefits to debt (Miller, 1977). Furthermore, others have pointed out that there are other tax shields that can serve as substitutes for debt tax shields (DeAngelo and Masulis, 1980).

A slightly different perspective on capital structure came from Myers and Majluf (1984), who presented a "pecking order" model to explain corporate financing decisions. According to this model, there is no real optimum debt level for a firm. Instead, information asymmetries drive firms to prefer internal sources of financing (i.e., retained earnings) over external sources. According to this perspective, the firm generally uses internally generated cash flows to fund new projects. However, if retained earnings are not sufficient to cover the firm's financing needs, the firm will prefer to issue the least risky securities (i.e., debt) first, and the most risky securities (i.e., equity) only as a last resort. Thus, when profitability is low, debt levels may climb as the firm has to borrow money in order to fund projects. Conversely, when profitability is high, managers use excess cash flows to pay down debt. Some empirical tests have found strong support for the pecking-order model, at the expense of target adjustment models (Shyam-Sunder and Myers, 1999).

Early linkages of strategy to capital structure

The notion that there might be some relationship between a firm's "strategic" decisions and its capital structure first started to come to light with Jensen and Meckling's (1976) seminal paper on agency costs. The authors presented a simple model in which the owner-manager of a firm first issues debt, and then decides on which investments to make. Owing to the fact that the owner-manager has limited liability, the downside risk of the investment decisions will be borne more heavily by the bondholders than by the equity holders (i.e., the owner-manager). In the extreme cases where the firm will be unlikely to meet its debt obligations, all additional downside risk that is assumed by selecting riskier projects is borne by the bondholders. All of the corresponding upside risk (i.e., the outcome in the good state of the world), however, accrues to the equity holders. Thus, if investment decisions can be made after debt is issued, the equity holders are motivated to pursue riskier output strategies that will raise returns in the good states and lower returns in the bad states. As firms increase their debt load, the incentive to engage in riskier strategies also grows, since equity holders ignore any reduction in returns in the bankrupt states (Brander and Lewis, 1986).

If a firm has a coherent "strategy," then that strategy should guide its investment decisions. Thus, the realization that the debt level of the firm might affect the investment decision opened the door for researchers to examine other possible interactions between capital structure and corporate strategy. Although it wasn't the author's primary intention, the next major work to link strategy with capital structure reversed the causal inference, suggesting that that the firm's leverage might be influenced by its strategy. Titman's (1984) analysis of the effect of capital structure on a firm's liquidation decision revealed that a firm's capital structure might be a source of strategic value. Consistent with Jensen and Meckling's model of the agency costs of debt, Titman proposed that high levels of debt serve as a signal to the firm's customers that management is unconcerned with outcomes in the bad states and is engaging in riskier strategies. Accordingly, customers will judge such high debt companies to have a greater chance of going into bankruptcy and being liquidated. If the firm's products or services are of a nature such that customers care about whether the firm exists tomorrow (e.g., due to warranties or the ongoing servicing of durable goods), then customers will evaluate high leverage negatively. Therefore, opting for a low debt level can serve as a strategic advantage in that customers will prefer to buy from a producer with low leverage, all other things being equal.

Brander and Lewis' (1986) examination of capital structure from a game theoretic perspective was one of the first papers to explicitly give strong credence to the role of strategy in determining capital structure. The basic model the authors developed is a two-stage sequential duopoly game. In the first stage, both firms choose their respective financial structures. In accordance with Jensen and Meckling's (1976) agency costs model, one of the crucial assumptions here is that as firms take on more debt, they will have greater incentives to pursue riskier output strategies. In the second stage, the firms select their output, taking as given the financial structure chosen in stage 1. In this model, there is a random component to the aggregate market demand for the product that the two firms are producing. When the firms are selecting their output level in the second stage, they still do not know what that total market demand (and their corresponding marginal profit) will be. Under these uncertain conditions, choosing a high level of output (before demand is known) will constitute a riskier strategy than choosing a lower level of output. In Cournot oligopoly models, however, firms have an incentive to commit to producing large outputs, since this will induce their rivals to produce less. The result is a sequentially rational Nash equilibrium in both debt levels and output levels. In the first stage, both firms take on debt. Taking on debt in the first stage serves as a signal to the rival that the firm is committing to a riskier strategy, and thus a higher output level. The positive debt level chosen by both firms induces, as expected, both of them to produce at a higher (i.e., riskier) output level in the second stage than they would have selected in an all-equity world.

The model by Brander and Lewis (1986) demonstrated a relatively simple case in which financial structure would influence the output market equilibrium. Firms with foresight will take advantage of this fact to try to influence the output market in their favor. A firm that ignores the strategic effect of financial decisions will lose out and have a lower market value than the firm that took advantage of this knowledge. These strategic uses of financial structure are, however, purely predatory. The net effect when both firms use them is that both firms are worse off (Brander and Lewis, 1986). This end-game deleterious equilibrium was of little help to financial scholars attempting to explain the variations in capital structure observed empirically, as the rather simple conceptualization of strategy (primarily pertaining to the level of output) barely began to address the multifaceted ways in which strategy might interact with capital structure. However, this paper does serve as an important milestone in the literature because it was one of the first to explicitly recognize the role of strategy in influencing capital structure. The years following this paper saw a growing acceptance of the concept of "strategy" in financial theory. For example, the model of Sandberg et al. (1987) for selecting the degree of financial leverage claimed that leverage should be increased as long as it continues to have positive consequences and does not "impede the firm's ability to develop effective business strategy."

Modern strategic perspectives on capital structure

Although the subject of the relationship between strategy and capital structure had its origins in the finance and economics literature, the late 1980s saw a surge of interest in this topic within the field of strategic management. For several years, there had been a growing recognition within the strategy literature that "the valuation concepts which are the major preoccupation of the finance literature can also play a central role in strategy formulation" (Sandberg et al., 1987). However, strategy researchers had largely neglected the capital structure puzzle until Barton and Gordon's (1987) initial exploratory paper on the topic helped to stimulate interest within the field.

Barton and Gordon (1987) were interested in exploring whether the strategy literature might be able to fill in some of the gaps that existed in the financial literature regarding capital structure. As the authors pointed out, extensive theoretical and empirical work in the field of finance had failed to yield a consensus on not only which factors influenced capital structure, but also whether capital structure really has any effect on the value of the firm. They believed that the inclusion of strategic considerations would produce a "more eclectic and realistic assessment of the mechanism of the leverage decision at the firm level than the finance paradigm taken alone" (1987: 67). After an adept characterization of what remains unknown in the finance literature, the authors present several interesting propositions linking capital structure to the characteristics and desires of the firm's top managers. Perhaps the most important contribution of this paper is that it focused attention on the fact that both leverage and strategy affect the whole organization. If the firm exhibits strategic coherence (Porter, 1996), then we would expect these two factors to be related. Furthermore, the paper called for studying strategy at a richer level than mere decisions about output quantity or price, a call later reiterated in Harris and Raviv's (1991) extensive literature review of capital structure research. Although Barton and Gordon's (1987) suggestion of using diversification as a proxy for the overall corporate strategy was still a rather crude measure of strategy, it was still a big step forward.

Pursuing their theory that a managerial/behavioral perspective is necessary for understanding an individual firm's debt/equity mix, Barton and Gordon followed up their exploratory theoretical paper with an empirical investigation (Barton and Gordon, 1988). The authors divided a sample of firms drawn from Compustat into four discrete categories based on the firm's diversification strategy. For each strategy category, a measure of firm leverage (the ratio of owner's equity to invested capital)[1] was regressed on a number of different financial contextual variables, such as measures of firm size, profitability, capital intensity, risk and growth. The main effects for the financial contextual variables were generally either insignificant or very weak. However, one exception was profitability, which was strongly negatively related to leverage (as predicted by the authors). Furthermore, the authors also found some main effects for their measure of strategy, such that different "strategic orientations" tended to have different levels of debt. Despite some potential shortcomings,[2] this paper provided two valuable insights. First, we have perhaps our first empirical indication that strategy can help predict leverage. Second, although not predicted a priori, the authors detected multiple significant interactions between strategy and the financial contextual variables. This suggests that the relationship between certain financial variables and capital structure might be contingent upon the firm's strategy (Barton and Gordon, 1988). Thus, consideration of a firm's strategy holds the promise of being able to clear up some of the noise that might be obscuring the relationships tested in traditional empirical financial research.

The potential for strategy to serve as an effective predictor of capital structure really began to take form after the zeitgeist of the strategy literature shifted away from industry structure and towards firm heterogeneity. Theories that emphasized firm heterogeneity, such as Penrose's (1959) antecedents to the resource-based view of the firm, did not gain widespread popularity until the 1980s (Schendel, 1994). The key assumption that differentiated these theories from the previous work, which was largely based in IO Economics, was that factor markets are imperfect. Owing to these imperfections, firms will be heterogeneous with respect to the resources they control (Barney, 1991). The heterogeneity in resources available to the firm is what gives rise to variation in firm performance, and thus allows some firms to achieve a competitive advantage over other firms (Hunt, 1997).

The heterogeneity perspective on capital structure was exemplified in Williamson's (1988) application of the transaction costs economics framework to the capital structure puzzle. This perspective emphasized that debt and equity will be associated with different forms of governance structures. Bondholders can only seize control of the firm if it defaults on its debt or violates the covenants of the debt contract. Equity holders, however, have much broader rights than bondholders. Through their control over the board of directors, they can effectively intervene in strategic decisions if they deem it to be necessary. All other things being equal, managers will prefer the less interventionist control that is given to bondholders, and thus issue debt rather than equity. However, the type of governance opted for by the firm (i.e., debt or equity) will also be influenced by the characteristics of the assets that are deployed within the firm. In particular, the redeployability of the assets will bear a great influence on the governance decision. The more specialized the assets are, the less redeployable they are. Consequently, increasing the level of specialization of the assets makes debt more costly, as these assets are not able to serve as collateral as well as assets that are highly redeployable. The central idea here is similar to that of the model Shleifer and Vishny (1992) developed to analyze the liquidation value of the firm. Shleifer and Vishny (1992) focused on the degree to which assets are industry specific, and how this consequently affects the costs of financial distress. Williamson's (1988) model, however, leaves the door open for the assets deployed by the firm to be so specific that they might not even be of use to other companies in the same industry. This provides for the crucial link between the type of financing and the firm's business strategy (Balakrishan and Fox, 1993).

The theories in strategic management that stressed firm heterogeneity asserted that in order for a firm to gain a sustained competitive advantage in the marketplace, it must control resources which are valuable, rare, imperfectly imitable, and non-substitutable (Barney, 1991). These "resources" can broadly be classified into three forms of capital resources: physical, human, and organizational. Any resource that is valuable and rare may help a firm to improve performance for a short period of time, but in order to achieve a sustained competitive advantage it is crucial that competing firms not be able to effectively duplicate that resource (hence the imperfectly imitable and nonsubstitutable requirements). Thus, the emphasis in strategy was on develop-ing unique resources that might serve as a source of competitive advantage (Barney, 1991). Tailoring these resources to the firm's strategy and technology can help to reduce costs, improve quality, and enable the firm to differentiate its products and services from its competitors (Balakrishan and Fox, 1993). By their very nature, however, these firm-specific assets (which can take the form of R&D, brand name, etc.) are largely nontradeable and nonredeployable (Long and Malitz, 1985). This "asset specificity" problem results in high bankruptcy costs for firms with highly unique assets.

Building on the transaction costs literature, Balakrishan and Fox (1993) investi-gate whether measures of intangible capital contribute significantly to variation in leverage. If such a relationship were found, it would indicate that strategy, which can have an effect on the level of intangible assets in a firm, might help determine the capital structure. The interesting implication for the strategy literature to come out of this study regards firm heterogeneity. Using panel data and a variance com-ponents regression model, the authors demonstrated that industry and time effects were of little importance in determining capital structure in comparison to firm heterogeneity.

Subsequent research in the field of strategic management that addressed the issue of capital structure tended to emphasize the transaction costs economics framework. Kochhar (1996) compares agency theory and transaction costs economics in explain-ing the financing decision in leveraged buyouts (LBOs) and product diversification. While information asymmetries play an important role in both theories, agency theory assumes that these concerns can be adequately reduced through sufficient expenditures on monitoring devices. The result is that debt is used to reduce agency conflict and enhance firm value. Transaction costs economics, however, adds an extra consideration. The specificity of the assets will determine which form of governance is preferred. As discussed earlier, high specificity assets are not very redeployable, so the market demands a high rate of interest to fund such endeavors with debt, which leads to a "lemon problem," which leads to market failure (i.e., for debt). Thus companies will turn to equity financing for high specificity assets because it actually has the lower costs. Debt financing, which has lower governance costs, will be utilized for low specificity assets (for which the information asymmetries are not as great) because they can be used as collateral, leading to a lower interest rate (Kochhar, 1996). As evidence for the transaction costs economics framework, Kochhar (1996) contends that LBOs tend to occur in firms that have nonunique

assets and low investment opportunities. For these firms, the transaction costs economics viewpoint suggests that the low governance costs of debt will be preferred to
equity financing. Kochhar and Hitt (1998) also found support for a transaction costs
economics perspective in the financing decision made by firms that are diversifying.
Perhaps more interestingly, by using a three-stage least-squares estimation technique, the authors were able to show that the relationship between the firm's
financial strategy and its corporate diversification strategy is a reciprocal, dynamic
one in which each influences the other.

An interesting extension of the transaction costs perspective was developed by
Simerly and Li (2000), who incorporated elements of agency theory and environmental dynamism to propose that a firm's capital structure should be matched to its
environment. The authors propose that although there are benefits to debt, obligations to creditors place certain constraints on firms that equity financing does not.
Thus, too much debt may hinder either the firm's ability or its willingness to make
the strategic investments in areas such as research and development that are critical
to maintaining competitive advantage. Furthermore, since lenders are generally risk
averse, the strategic costs of debt financing will increase with the level of uncertainty
in the firm's environment. Thus, equity financing will be more appropriate in contexts where there is a great deal of uncertainty about the firm's environment. The
authors found strong support for their proposition that a mismatch between the
firm's leverage and its environment will encumber performance.

Linking innovation to capital structure

Although innovation has long been a prominent topic in the field of strategy, there
has been relatively little effort to explicitly examine how a strategy based on innovation might impact capital structure. One of the few exceptions is Jordan et al.
(1998), who investigated the relationship between capital structure and strategy
using a variant of Porter's (1980) generic strategy typology. The authors found that
a strategy based on innovation was associated with the lowest level of debt, while
firms pursuing a cost-leadership strategy had the highest levels of debt. Similarly,
Vicente-Lorente (2001) found that R&D investments that are characterized by a
high degree of specificity or opaqueness are associated with lower leverage. The
negative association between R&D expenditures and leverage is not new, for R&D
has been a common control variable in finance studies examining the determinants
of capital structure since Long and Malitz (1985) suggested that these investments
create intangible assets that will likely suffer from market failure. Since the intangible
assets created by R&D investments cannot be efficiently traded on the open market,
they cannot serve as effective collateral and thus cannot support a high level of debt.
The interesting finding to emerge from the Vicente-Lorente (2001) study was that
some R&D investments are less specific than others, and thus more capable of
supporting debt.

The linkage between R&D intensity and leverage raises an interesting, yet apparently unexplored, question. If R&D is negatively related to leverage simply because
those investments create intangible assets that are incapable of supporting much
debt, then why does R&D intensity remain a significant predictor of leverage even

after the firm's tangible assets ratio has been controlled for (e.g., Hovakimian et al., 2000)? Furthermore, as an ongoing expense, R&D tends to be a minor line item for all but a few outliers. Across all business segments in the sample tested in this paper, annual R&D expenditures averaged less than 2 percent of sales. Thus, it seems quite possible that a firm's intensity of investment in R&D serves as more than just a proxy for the "stock of strategic resources such as innovative capabilities" (Vicente-Lorente, 2001: 162).

Theory and Hypotheses

We propose that R&D intensity, when appropriately modeled, taps into a dimension of strategy that goes beyond the mere creation of intangible assets. The R&D intensity of a firm, relative to its industry rivals, indicates the strategic importance of innovation to a firm. Certainly, large expenditures on R&D do not guarantee that a firm will be an effective innovator. However, firms that invest in R&D at a much higher rate than their competitors are most likely *trying* to compete on the basis of innovativeness. Strategy researchers have previously proposed that in order for a firm to be an effective innovator, it must maintain sufficient "slack resources" (Bourgeois, 1981; Damanpour, 1991; Nohria and Gulati, 1996; Singh, 1986; Zajac, Golden and Shortell, 1991). Furthermore, this need for slack resources may manifest itself in the form of financial slack (Bromiley, 1991; Cyert and March, 1963; Davis and Stout, 1992). One of the most prominent ways in which the need for financial slack will manifest itself is a relatively low leverage ratio. A conservative financial structure (i.e., low leverage) gives the firm more financial slack because potential lenders will see the firm as a safe debt instrument and provide it with ready access to lines of credit (Brealey and Myers, 1996: 501).

Obviously, a firm that is trying to compete on the basis of innovation must continue to innovate in order to stay alive in the marketplace. This need to continue innovating makes financial slack an essential complement to the effective implementation of a strategy based on innovativeness for three primary reasons. First, as Froot et al. (1993) point out, cash flow volatility can potentially jeopardize the investments in R&D that the firm must continue to make in order to maintain its competitive position. Maintaining a smooth continuous rate of R&D investment is critical to innovators because "maintaining a given rate of R&D spending over a given time interval produces a larger increment to the stock of R&D know-how than maintaining twice this rate of R&D spending over half the time interval" (Dierickx and Cool, 1989: 1507). Thus, R&D expenditures cannot be allowed to fluctuate with the firm's potentially volatile cash flows, and financial slack helps to provide insulation against cash flow volatility and ensures that investments in R&D are maintained even during sour times.

A second reason why financial slack is an essential component to competing on the basis of innovation concerns product launches. Being an *effective* innovator requires more than just developing new products, it requires getting those products to market. Financial slack can help ensure that the firm has the financial resources required to launch new products as soon as they are ready. Finally, some strategy researchers have suggested that firms may use acquisitions to expand their stock of knowledge (Huber, 1991; Kogut and Zander, 1992; Karim and Mitchell, 2000). Thus, firms competing on the basis of innovation may sometimes enhance their competitive position through acquisitions. Obviously, sufficient financial slack can assist the firm in making the acquisitions that it deems necessary in a timely fashion.

Fundamentally, we agree with previous research that has argued that firms with a large stock of intangible assets will not be *able* to borrow as much.[3] Moreover, we accept simple firm-level R&D intensity as a valid proxy for the stock of intangible assets (although we believe there may be better proxies for this construct). However, we add to this argument by positing that firms attempting to compete on the basis of innovation will make financial slack a strategic imperative, and hence the firm will not *want* to borrow as much. The strategic importance of innovativeness to the firm will manifest itself not in absolute R&D intensity of the firm, but rather in the R&D intensity of the firm relative to others in its industry. Thus, while absolute R&D intensity may proxy for the stock of some intangible assets, relative R&D intensity will proxy for the strategic importance of innovativeness to the firm. The strategic importance of innovativeness, and hence R&D, to the firm will vary within an industry depending on whether each firm is attempting to be an innovator, a fast follower, or a low-cost mass producer. In certain industries, firms may have to invest fairly heavily in R&D just to be a low-cost mass producer. However, for these firms, innovation is not their competitive strength and thus financial slack will not be as important as it will be for the firm that is attempting to be the innovator. On the basis of these arguments, we propose the following two hypotheses.

Hypothesis 4.1a: Consideration of a firm's intensity of investment in R&D, relative to industry peers, will improve our ability to predict leverage versus a model that considers only absolute R&D intensity.

Hypothesis 4.1b: A firm's intensity of investment in R&D, relative to industry peers, will be negatively associated with leverage.

It is worth pointing out that there are strong empirical precedents to lead us to believe that hypothesis 4.1b will be supported. Although R&D intensity is usually computed as a simple firm level ratio, some studies have computed it as deviation from industry average and found a strong negative relationship between R&D and leverage (e.g., Hovakimian et al., 2000). However, to the best of our knowledge, the two different measures of R&D intensity have never been directly pitted against one another while introducing more precise controls for the present stock of intangible assets. Under such circumstances, we believe that the best interpretation for a strong relationship between the relative measure of R&D and leverage relates to the strategic importance of financial slack to the firm.

Our proposition that some firms make financial slack a strategic imperative has some interesting implications for the previously discussed pecking-order model. This model predicts a negative relationship between profitability and leverage, since firms pay down debt when profits are high and borrow when profits are low. While firms for whom financial slack isn't a strategic imperative might mechanistically follow the pecking order, we predict that the relationship will not be as strong for firms pursuing a strategy of innovation. An increase in profitability will not necessarily induce the innovators to reduce leverage. If the company has been true to its strategic orientation and has maintained comfortable financial slack, they may be more likely to divert the cash received from increased profitability to other uses.[4] Shyam-Sunder and Myers (1999) recognized the potential for such interactions with strategy when they acknowledged that "we doubt the pecking order would do as well for a sample of growth companies investing heavily in intangible assets." Thus, we expect the pecking-order model to more accurately characterize firms for whom innovation is not a strategic priority, and hence we predict an interaction between relative R&D intensity and profitability in regards to their effect on leverage.

Hypothesis 4.2: The negative relationship between profitability and leverage will be weaker for firms that attempt to compete on the basis of innovation.

As Opler et al. (1999) point out, there is more to financial slack than excess debt capacity. Credit lines could get cancelled precisely when the firm needs them the most. Thus, the need for financial slack is likely to manifest itself in another form: large cash balances. A firm that requires a lot of financial slack will feel comfortable holding a little more debt if they can also hold a little more cash. Debt and cash can be thought of as "two faces of the same coin," simultaneously serving as both substitutes for each other and also as different measures of the same theoretical construct (financial slack). Using time-series and cross-sectional tests, Opler et al. (1999) found support for a static tradeoff model of cash holdings. Moreover, strong growth opportunities (which induce larger information asymmetries), riskier cash flows, large R&D/sales ratios, and low net working capital/assets ratios were all economically significant predictors of large cash holdings. In regard to R&D intensity, we expect to find similar results as Opler et al. However, we anticipate that holding large cash balances will be a greater strategic priority for firms attempting to compete on the basis of innovation. Thus, adding R&D intensity relative to one's industry peers will improve the fit of the model.

Hypothesis 4.3a: Consideration of a firm's intensity of investment in R&D, relative to industry peers, will improve our ability to predict cash holdings versus a model that considers only absolute R&D intensity.

Hypothesis 4.3b: A firm's intensity of investment in R&D, relative to industry peers, will be positively associated with cash holdings.

Simerly and Li (2000) found that a mismatch between capital structure and the level of environmental dynamism can negatively impact the value of the firm. Similarly, we predict that there will be performance implications for firms that are trying to be innovative and yet fail to properly appreciate the strategic value of financial slack. The preceding arguments have posited that firms that are trying to compete on the basis of innovation maintain financial slack because failing to do so may hinder the effectiveness of their strategy. Therefore, we predict a negative interaction between strategy and leverage in regard to their impact on firm performance, indicating that firms that pursue innovation while maintaining high leverage will pay a price for this misalignment.

Hypothesis 4.4: There will be a significant negative interaction between relative R&D intensity and leverage with regards to their impact on firm performance.

Research Methods

Data sources

Most of the variables used in this study were derived from the Compustat industrial and business segment databases, commonly referred to as Compustat I & II, respectively. Compustat contains detailed financial information at the level of the firm for all companies that file reports with the Securities and Exchange Commission (SEC). This population primarily encompasses all public companies in the US, although it also includes some non-public firms as well as some foreign firms. Since the computation of some of the variables used in this study required a value for the market value of the firm, we restricted our sample to just those firms that had public equity (approximately 80 percent of all observations). Furthermore, since all of the independent variables were lagged one year, we also excluded any firms that were listed for only one year. Thus, our initial sample included all 16,358 firms that were listed in both Compustat I and II for at least two years between 1980 and 1999 (the years for which segment data was available), and encompassed 110,117 firm/year observations. Occasional missing variables and our requirement that the independent variables be lagged one year resulted in approximately 92,000 observations being used in the final analyses.[5] Thus, our data represent a large sample of firms operating in many different industries over a significant period of time. Our primary concern with using Compustat for this study was the potential variability in SIC coding at the four-digit level from year to year (see Davis and Duhaime, 1992). Therefore, for all variables and analyses described in this paper, we aggregate industries up to the three-digit SIC code level in order to attenuate any potential variability in year-to-year coding.

The only variable used in this study that was not constructed from Compustat data was a proxy for the dynamism of the firm's primary industry. This variable was constructed from stock return data derived from the University of Chicago's Center for Research in Security Prices (CRSP) database.

Dependent variables

Our analyses test three different groups of models, each of which employs a different dependent variable. The first group of models tests hypotheses 4.1a, 4.1b and 4.2, which relate to strategic influences on firm-level leverage. The dependent variable for these models was Leverage, which is computed by dividing the book value of debt by the total market value of the firm. The total market value of the firm was calculated as the book value of debt plus the market value of equity plus the carrying value of preferred stock. The second group of models tested the relationship between strategy and cash holdings posited by hypotheses 4.3a and 4.3b. The dependent variable for these models, Cash, represents the dollar value of all cash and short-term investments held by the firm, divided by the book value of assets. The final set of models examines the performance implications of a mismatch between strategy and capital structure, as posited in hypothesis 4.4. Since all the firms in the sample were public companies, we chose to use the firm's market-to-book ratio as the dependent variable.[6] The primary strength of this proxy for performance, which corresponds very closely to Tobin's q (Chung and Pruitt, 1994), is that it incorporates all future expectations regarding firm performance. The variable Market:Book was calculated by dividing the market value of the firm by the book value of total assets.

Because we were concerned about potential nonlinearities in the relationships between the dependent variables and the independent variables, a Box-Cox procedure was employed in order to determine if a nonlinear transformation of the dependent variables might be appropriate. For all three variables, a transformation approximately equivalent to the natural log was found to maximize model fit. Thus, Leverage, Cash, and Market:Book were all transformed by taking the natural log.[7] All three dependent variables were derived from Compustat I data.

Independent variables

The primary independent variable of theoretical interest is our proxy for the importance of innovation to the firm's strategy. As discussed above, we believe that the importance of innovativeness to a firm's strategy will manifest itself not in the absolute magnitude of R&D intensity, but rather in the firm's R&D intensity relative to industry rivals. The variable Innovation serves as our proxy for the relative R&D intensity of the firm, and is constructed as follows. First, we compute R&D intensity for every business segment listed in Compustat II from 1980 to 1999 by dividing total segment level R&D spending by total segment sales.[8] Next, we compare the R&D intensity of each segment to all business segments competing in the same industry (as defined by the segment's primary SIC code). All business segments competing in an industry are then assigned a percentile score based on their R&D

intensity. For single segment firms, this percentile score represents the firm-level value for the variable Innovation. For firms operating in multiple industries, the firm-level value for Innovation was computed by taking the weighted average (by segment assets) of the percentile scores assigned to each segment.[9] Higher scores on this variable indicate that a firm invests more heavily in R&D than its industry rivals, and thus is more likely to be attempting to compete in the basis of innovativeness.

All three sets of models tested also included a number of firm-level control variables that previous research had linked to the dependent variables. The variable R&D Intensity, as measured by simple firm-level expenditures on R&D divided by sales, provides a more traditional proxy for the intensity of investment in R&D. Similarly, Advertising Intensity represents the firm-level investment in advertising scaled by total firm sales. We control for the Size of the firm by including the natural log of total book value of firm assets. The variable Profitability controls for firm-level accounting profitability, as measured by return on assets (i.e., operating income before depreciation and amortization divided by the book value of assets). The firm's Capital Intensity is calculated by dividing the book value of total firm assets by total firm sales. Finally, the variable Tangible Assets controls for the firm's ratio of tangible assets to total assets by dividing total property, plant and equipment by the total book value of assets.

In addition to the firm-level control variables described above, we also controlled for some industry-level factors that might impact both financial slack and firm performance. We control for the performance of the firm's primary industry with the variable Industry Performance. For each firm, this variable is calculated by summing the total market values of all firms in the industry, excluding the focal firm, and dividing by the total book value of assets for all those firms included in the numerator. Firms competing in higher market-to-book industries may have more growth opportunities available to them, even if they have thus far failed to capitalize on them.

In addition to industry performance, we also controlled for industry-level volatility with the variable Dynamism. To construct this proxy, we first gathered monthly stock return data for all firms listed in CRSP between 1950 and 2000. We then formed value-weighted portfolios for each industry (according to each firm's primary industry) in order to derive monthly returns for each industry. Our measure of total industry uncertainty was then produced by modeling the returns data for each industry as a generalized autoregressive conditional heteroskedasticity (GARCH) process (Bollerslev, 1986). The GARCH model produces a time-varying estimate of the conditional variance about the trend in the dependent variable. This approach is advantageous since it allows environmental dynamism to vary each period, and it only considers the variance about any trends in the data.[10] Specifically, we employed the GARCH-in-mean, or GARCH-M model, which is parameterized by two values that specify the number of lags for the squared error terms and the number of past variances to be included in the computation of the current variance. We used a one-period lag on both parameters (i.e., a GARCH-M[1,1] model), which generally provides excellent fit for modeling returns volatility (Solnik, 1996: 81). Diagnostic checks of the GARCH models based on the residuals and parameter significance indicated that the parsimonious GARCH-M(1,1) model provided better fit than alternative lag specifications.

All variables described above were calculated on an annual basis, and all independent variables were lagged one year in order to avoid potential endogeneity with the dependent variables. As described below, all models also included the lag of the dependent variable. The variable Leverage was also included as an independent variable in the performance models. Furthermore, tolerance statistics were computed on the data in order to determine whether any of the relationships between the independent variables were strong enough to necessitate excluding any variables from the model.[11] The analysis revealed that all tolerances were well above the most conservative commonly used cutoff of 0.01 (Neter et al., 1996: 388). Descriptive statistics for the dataset are given in Table 4.1.

Analysis

Since our dataset contains multiple observations per firm, the potential confounding influence of unobserved heterogeneity due to firm level effects is a concern. Three of the most common approaches for modeling this type of panel data are: (a) random-effects models; (b) fixed-effects models; and (c) including lagged dependent variables as predictor variables. Random-effects models are most appropriate when the sample contains observations that are a random draw from a specified population (Johnson, 1995). Such is not the case here, as the sample includes all members of a given population (i.e., all public firms). Moreover, random-effects models are highly susceptible to the biases that can result from failing to include in the model all explanatory variables that influence the dependent variable, which is very likely to occur in any non-experimental study (Allison, 1994; Johnson, 1995). Although a fixed-effects model would be preferable to a random-effects design, it also has a critical shortcoming with regard to the present study. Because the firm fixed effects capture all factors that are constant within a firm over time, these models cannot produce stable estimates for variables that are either invariant or display little change within a firm over time (Allison, 1994; Johnson, 1995). If firms show a fair degree of stability over time with respect to their strategy, then variables attempting to measure strategy should not be included in the model. Therefore, it was determined that the best approach for modeling the panel data, in this context, would be the inclusion of lagged dependent variables as predictor variables in OLS regression models. This technique is particularly appropriate for the current study because all dependent variables are essentially stock variables, rather than flow variables. Thus, the past levels of these variables are likely to have a causal influence on the present levels, which makes the use of lagged dependent variables the preferred approach for the modeling of panel data (Finkel, 1995).

Results

Leverage models

The results of the least squares regressions that were used to test the hypotheses 4.1a, 4.1b and 4.2 are reported in the first three models of Table 4.2. Model 1 presents the base model that excludes the effect of Innovation, our measure of

Table 4.1 Descriptive statistics

	Mean	Std Dev	Min	Max	(1)	(2)	(3)	(4)	(5)	(6)	(7)	(8)	(9)	(10)	(11)
(1) Leverage	-1.42	0.72	-2.30	0.10											
(2) Cash	-1.61	0.61	-2.50	0.10	-0.480										
(3) Market:Book	0.08	0.85	-9.44	6.50	-0.441	0.354									
(4) Innovation	0.53	0.18	0.07	0.99	-0.158	0.206	0.176								
(5) R&D Intensity	0.47	16.07	-0.03	3,309	-0.031	0.058	0.051	0.069							
(6) Advertising Intensity	0.02	1.19	0.00	332.5	-0.012	0.017	0.022	0.007	0.008						
(7) Size	4.40	2.24	-5.30	13.09	0.318	-0.267	-0.220	-0.127	-0.028	-0.015					
(8) Profitability	0.06	0.38	-28.59	24.27	0.076	-0.108	-0.103	-0.169	-0.099	-0.026	0.251				
(9) Capital Intensity	5.65	104.89	0.01	12,270	-0.029	0.040	0.024	0.026	0.375	0.041	-0.029	-0.036			
(10) Tangible Assets	0.32	0.25	0.00	1.00	0.335	-0.345	-0.061	-0.069	-0.016	-0.009	0.202	0.065	0.000		
(11) Indus. Performance	1.05	0.64	0.00	16.93	-0.251	0.253	0.328	0.039	0.054	0.002	-0.143	-0.111	0.025	-0.143	
(12) Dynamism	0.01	0.02	0.00	1.35	-0.010	0.008	-0.002	-0.012	-0.003	0.008	-0.038	-0.010	-0.001	-0.023	0.025

All correlations with absolute value of $r > 0.007$ are significant at $p < 0.05$

Table 4.2   Regressions for leverage and cash models

Dependent Variable:	Leverage Models			Cash Models	
	Model 1 Leverage	Model 2 Leverage	Model 3 Leverage	Model 4 Cash	Model 5 Cash
Intercept	−0.290***	−0.245***	−0.245***	−0.402***	−0.455***
	(0.006)	(0.007)	(0.007)	(0.005)	(0.006)
Lagged Dep. Var.	0.821***	0.818***	0.818***	0.753***	0.748***
	(0.002)	(0.002)	(0.002)	(0.002)	(0.002)
Innovation	–	−0.089***	−0.088***	–	0.083***
		(0.007)	(0.007)		(0.007)
Innovation X Profit.	–	–	0.035**	–	–
			(0.012)		
R&D Intensity	0.000	0.000	0.000	3.E-04***	3.E-04***
	(0.000)	(0.000)	(0.000)	(0.000)	(0.000)
Advertising Intensity	0.000	0.000	0.000	−0.004***	−0.004***
	(0.001)	(0.001)	(0.001)	(0.001)	(0.001)
Size	0.011***	0.011***	0.011***	−0.008***	−0.007***
	(0.001)	(0.001)	(0.001)	(0.001)	(0.001)
Profitability	−0.014***	−0.020***	−0.042***	0.010***	0.015***
	(0.004)	(0.004)	(0.008)	(0.003)	(0.003)
Capital Intensity	0.000	0.000	0.000	0.000+	0.000+
	(0.000)	(0.000)	(0.000)	(0.000)	(0.000)
Tangible Assets	0.103***	0.102***	0.102***	−0.072***	−0.073***
	(0.006)	(0.006)	(0.006)	(0.005)	(0.005)
Industry Performance	−0.021***	−0.022***	−0.022***	0.029***	0.030***
	(0.002)	(0.002)	(0.002)	(0.002)	(0.002)
Dynamism	0.077	0.066	0.065	−0.047	−0.037
	(0.056)	(0.056)	(0.056)	(0.050)	(0.050)
n	91,234	91,234	91,234	92,618	92,618
R-Square	0.7087***	0.7092***	0.7093***	0.6433***	0.6439***
F	–	142.14***	8.98**	–	156.33***

F statistic is for the improvement in R-square versus the previous model
Standard Errors for coefficients are in parentheses.
$+ p < 0.10$; $* p < 0.05$; $** p < 0.01$; $*** p < 0.001$ (all tests are two-tailed)

relative R&D intensity. While the results for most of the control variables were pretty much as might be expected, there were a couple of surprises. Curiously, after controlling for all the other factors included in the model, simple R&D intensity does significantly influence firm leverage. Furthermore, environmental dynamism and advertising intensity also failed to significantly influence leverage.

In model 2 we add our proxy for the importance of innovation to the firm's competitive strategy. As predicted by hypothesis 4.1a, the addition of this variable significantly improves the fit of the model ($p < 0.001$), and the coefficient on Innovation is negative, as predicted by hypothesis 4.1b. Although in the presence of the other control variables, absolute R&D intensity does not appear to be related to firm leverage, firms that compete on the basis of innovation do tend to maintain lower leverage. Presumably the additional financial slack afforded by maintaining low leverage may be beneficial in implementing a strategy premised on innovation.

Model 3 adds in the interaction between Innovation and profitability. As predicted by hypothesis 4.2, this interaction is positive and it significantly improves the fit of the model ($p < 0.01$). Firms that mechanistically follow the pecking-order model of corporate finance will display a negative relationship between profitability and leverage. However, this relationship is weakened amongst firms that compete on the basis of innovation. Apparently, for these firms the potential benefits of financial slack make it a strategic priority, and hence they have a greater tendency to depart the less deliberate pecking-order model.

Cash models

Model 4 in Table 4.2 presents the base model for cash holdings, while model 5 adds in the variable Innovation. The effects of most of the control variables were in accord with expectations, although surprisingly advertising intensity had a negative influence on cash holdings. As predicted by hypothesis 4.3a, the addition of Innovation significantly improves the fit of the model ($p < 0.001$) and the coefficient is positive, as predicted by hypothesis 4.3b. Thus, large cash holdings appear to function as a form of financial slack, similar to low leverage. Pursuing an innovative strategy apparently makes it more advantageous to hold the cash that might be necessary to fund R&D and launch new products.[12]

Performance models

The potential impact of a misalignment between firm strategy and capital structure is tested in the models of Table 4.3. Model 2 shows that adding Innovation to the base model significantly improves the fit of the model ($p < 0.001$), and that pursuit of an innovative strategy is generally associated with improved performance. The interaction between Innovation and leverage, which is added to model 3, significantly improves the model fit ($p < 0.001$). As predicted by hypothesis 4.4, the interaction is negative. Apparently, firms that maintain high leverage while attempting to pursue a strategy based on innovation incur a significant performance penalty.[13]

In addition to being statistically significant, the interaction between Innovation and leverage in Model 3 is also very economically consequential. By taking the derivative of the regression equation with respect to Innovation, we can determine that the slope of the relationship between Innovation and Market:Book is only 0.0001 at high (i.e., 95th percentile) leverage, while it is 0.23 at low (i.e., 5th percentile) leverage. This relationship is depicted graphically in Figure 4.1, which plots the contribution that following a strategy of innovation makes towards firm performance for three different levels of leverage. For firms that maintain high leverage, pursuing a strategy based on innovation offers no performance benefits. However, firms that pursue such a strategy while maintaining a high degree of financial slack experience appreciably enhanced performance.

Table 4.3   Regressions for performance models

Dependent Variable:	Performance Models		
	Model 1 Market:Book	Model 2 Market:Book	Model 3 Market:Book
Intercept	−0.197***	−0.283***	−0.191***
	(0.007)	(0.009)	(0.015)
Lagged Dep. Var.	0.766***	0.762***	0.760***
	(0.002)	(0.002)	(0.002)
Innovation	–	0.160***	−0.027
		(0.010)	(0.026)
Innovation X Leverage	–	–	−0.113***
			(0.015)
R&D Intensity	0.001***	0.001***	0.001***
	(0.000)	(0.000)	(0.000)
Advertising Intensity	−0.003+	−0.003+	−0.003+
	(0.001)	(0.001)	(0.001)
Size	0.001	0.002*	0.002*
	(0.001)	(0.001)	(0.001)
Profitability	−0.015**	−0.004	−0.001
	(0.005)	(0.005)	(0.005)
Capital Intensity	−5.E-05**	−5.E-05**	−5.E-05**
	(0.000)	(0.000)	(0.000)
Tangible Assets	0.052***	0.055***	0.058***
	(0.007)	(0.007)	(0.007)
Leverage	−0.056***	−0.052***	0.005
	(0.003)	(0.003)	(0.008)
Industry Performance	0.061***	0.063***	0.063***
	(0.003)	(0.003)	(0.003)
Dynamism	−0.112	−0.094	−0.090
	(0.072)	(0.072)	(0.072)
n	91,234	91,234	91,234
R-Square	0.6403***	0.6414***	0.6416***
F	–	270.36***	58.4***

F statistic is for the improvement in R-square versus the previous model
Standard Errors for coefficients are in parentheses.
+ $p < 0.10$; * $p < 0.05$; ** $p < 0.01$; *** $p < 0.001$ (all tests are two-tailed)

Discussion and Conclusions

The results of this study have strong implications, both theoretical and practical, for the field of strategic management. First, our findings help to substantiate the view that capital structure is not just a function of exogenous industry, regulatory, and product market factors, but also of a firm's strategy and its basis for competition

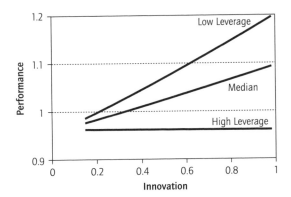

Figure 4.1 Influence of innovation and leverage on performance
The x-axis of this figure plots Innovation from low (0.1) to high (1.0), while the y-axis displays the predicted market-to-book ratio, which is the exponent of the dependent variable *Market:Book* from Model 3 of Table 4.3. Three series are plotted, representing firms with low (5th percentile), median, and high (95th percentile) leverage. All other variables are held constant at their mean.

within an industry. This suggests that capital structure cannot be treated as either exogenous or irrelevant to strategy. Managers who fail to realize that an inappropriate capital structure can hamper the effectiveness of a firm's strategy, and thus hinder a firm's ability to compete, may suffer significant performance consequences. Furthermore, the results of this paper also suggest that when researchers include R&D expenditures in their models, they should give careful consideration to exactly what construct they are attempting to model. Depending upon whether the intention is to assess strategy or the stock of some intangible assets, different specifications may be required. Finally, although the model used here was a rather crude test of the pecking-order model, it does have strong implications for research in that area. The relevance of the pecking order may be underestimated if researchers don't account for the fact that whether or not a firm follows the pecking order may depend on its business strategy.

The results of this study may also have implications for research in agency theory, once again suggesting that firm strategy may be a critical moderator. For example, Jensen and Murphy (1990) suggested that managers may prefer to increase firm leverage in order to increase the risk of the firm's common stock and hence to increase the value of their stock options. While this may be true of non-innovator firms, for whom increased leverage may not compromise the effective implementation of firm strategy, we doubt if it would hold for innovator firms. For firms attempting to compete on the basis of innovation, diminished financial slack may undermine their strategy, and hence have a net negative effect on the value of the stock options.

In their extensive review of the literature on capital structure, Harris and Raviv (1991) probably underestimated the potential contributions of strategy when they suggested that the consideration of strategic variables might "help in explaining inter-industry variations in capital structure." However, using concepts such as firm and resource heterogeneity, some researchers have made significant progress in helping to explain variation in capital structure within an industry (e.g., Balakrishnan and

Fox, 1993; Barton and Gordon, 1988; Ginn et al., 1995; Jordan et al., 1998). This study builds upon the prior work linking strategy to capital structure by suggesting that we *should expect to see* almost as much variation in capital structure within an industry as between industries. Competitive priorities must be determined in relation to those firms that are perceived as competitors (i.e., other firms in the same industry). If strategy really does impact capital structure, then its usefulness lies in explaining the variation we observe within an industry. The results of this study lend credible support to this argument by demonstrating that a firm's gross expenditures on R&D may not be as important in determining capital structure as the strategic importance of innovation to the firm.

The central premise of this paper is that pursuing a strategy of innovation makes financial slack more desirable. It is worth noting, however, that a negative relationship between leverage and pursuit of a strategy of innovation might also be predicted on the basis of the financial risk borne by shareholders. As a firm relies more heavily on debt financing, the firm's common stock becomes riskier. If we assume that either shareholders or managers have risk preferences, then following a riskier strategy (such as innovation) could potentially prompt the firm to maintain lower leverage. However, this perspective would struggle to explain the strategy by profitability interaction, the tendency for innovator firms to hold more cash, and the results of the performance models. Thus, we believe that the strategic value of financial slack to innovator firms is the most compelling explanation of the body of results presented here.

It is also worth speculating whether the results of this study might vary with the overall health of the stock market. Because managers tend to avoid issuing equity in a depressed market, all managers may perceive equity financing to be a less attractive alternative in a bear market. In firms for whom financial slack is not a strategic priority, managers may turn more to debt financing, enticed by low interest rates. Although innovator firms may rely on debt financing a little more than they previously did, they should still maintain much lower leverage than non-innovators because slack is still crucial to their strategy, and equity financing is even more unattractive than it used to be. Thus, in down markets, average leverage will likely be higher, but the relationships should be largely unaltered. Likewise for the performance models, the average market-to-book ratio may be smaller in down markets, but the relationships should still hold.

Although the arguments presented in this paper have suggested that a firm's choice of competitive strategy will affect its subsequent capital structure, it may also be the case that a firm's current financial structure limits the range of viable strategies open to the firm. A highly levered firm with factories geared towards large-scale, low-cost output would be faced with quite a challenge if it decided that it needed to become an innovator. Thus, it is highly likely that there is a dynamic relationship between strategy and capital structure, such that each affects the other. As previously discussed, Kochhar and Hitt (1998) did find evidence of such a reciprocal relationship. Future research examining the causal structure of this relationship in more detail may provide empirical evidence by demonstrating that a managerial decision to shift competitive strategy is followed by the predicted change in capital structure. Similar to the work of Rajagopalan and Finkelstein (1992), recently deregulated industries might serve as a fertile context for such investigations.

Future research on this topic may also try extending the scope of strategies considered beyond innovativeness. For example, most of the arguments presented could logically be extended to encompass firms that attempt to differentiate through brand image or advertising. Another promising avenue of research concerns the potential interactions between strategy, capital structure and environmental dynamism. Although we did not find dynamism to be a significant predictor of leverage, other studies have (Simerly and Li, 2000). A logical extension of this paper would be to propose that financial slack will be particularly important to firms that are attempting to be innovators in highly dynamic environments. Further research in this area is clearly indicated.

Acknowledgments

I am grateful to Timothy Folta and Yoon-Suk Baik for their helpful comments on earlier drafts of this manuscript.

Notes

1   Care must be used in interpreting these results since the ratio of owners' equity to invested capital will be inversely related to more traditional measures of leverage.
2   Since this study utilized a cross-sectional design, causality could not be confidently established.
3   In addition to making traditional debt financing less available, following a strategy of innovation may also make project financing less viable. Project financing generally requires that the project be "physically isolated from the parent and that it offer the lender tangible security" (Brealey and Myers, 1996: 695). A key strategic investment comprised heavily of intangible assets is unlikely to meet these criteria, thus enhancing the value of financial slack to such a firm.
4   In fact, if there are tangible benefits to debt, we may actually see some innovator firms increase their debt load in response to increased profitability if they are confident the increase is sustainable.
5   The precise number fluctuated from model to model in accordance with sporadic missing data items.
6   Firm-level market:book ratio has also sometimes been used as a proxy for the firm's growth opportunities. However, only those growth opportunities that the firm is expected to be able to take advantage of will be impounded into the firm's current market:book ratio. Thus, we feel firm level market:book is more aptly considered a holistic proxy for performance than a proxy for all opportunities open to the firm.
7   Before the transformation, a small constant (i.e., 0.1) was added to Leverage and Cash in order to make all values positive (as recommended by Neter et al., 1996: 132).
8   If R&D was missing, it was assumed to be equal to zero. This is consistent with common practice in financial research (e.g., Opler et al., 1999; Minton and Schrand, 1999) and theoretically justifiable. Since 1975, firms have been required to expense and disclose virtually all R&D expenditures (White et al., 1994: 397). Thus, missing values for R&D are likely the result of negligible expenditures. Furthermore, as Himmelberg et al. (1999)

report, excluding firms from the analysis that do not report R&D expenditures biases the sample towards firms which make intensive investments in R&D.

9   Ideally, we would prefer to conduct all analysis at the level of the segment. However, since debt resides at the level of the firm, and not the business segment, it was necessary to aggregate up to the firm level.

10   Common regression models provide estimates of the variability about a trend, but don't allow that variance to change over time. Alternatively, simple variances computed for individual time intervals fail to account for trends in the data, which will increase the measured variance but may not constitute an element of uncertainty if they are predictable.

11   The tolerance statistic is the reciprocal of the variance inflation factor, another commonly used diagnostic test for potential multicollinearity problems (see Neter et al., 1996: 386–8).

12   In an unreported regression model, we found a significant negative interaction between Innovation and profitability. This would suggest that although most managers may follow a pecking-order type model of cash holdings, the managers of firms competing on the basis of innovation are less inclined to do so. In another model, the interaction between simple R&D intensity and profitability failed to reach significance.

13   In an unreported regression, the interaction between leverage and the simple measure of R&D intensity failed to reach statistical significance.

References

Allison, P. D. 1994: Using panel data to estimate the effects of events. *Sociological Methods and Research*, 23, 174–99.

Andrade, G. and Kaplan, S. N. 1998: How costly is financial (not economic) distress? Evidence from highly leveraged transactions that became distressed. *Journal of Finance*, 53, 1443–93.

Balakrishnan, S. and Fox, I. 1993: Asset specificity, firm heterogeneity and capital structure. *Strategic Management Journal*, 14, 3–16.

Barney, J. B. 1991: Firm resources and sustained competitive advantage. *Journal of Management*, 17, 99–120.

Barton, S. L. and Gordon, P. J. 1987: Corporate strategy: useful perspective for the study of capital structure? *Academy of Management Review*, 12, 67–75.

Barton, S. L. and Gordon, P. J. 1988: Corporate diversification and capital structure. *Strategic Management Journal*, 9, 623–32.

Bollerslev, T. 1986: Generalized autoregressive conditional heteroskedasticity. *Journal of Econometrics*, 31, 307–27.

Bourgeois, L. J. 1981: On the measurement of organizational slack. *Academy of Management Review*, 6, 29–39.

Brander, J. A. and Lewis, T. R. 1986: Oligopoly and financial structure: the limited liability effect. *The American Economic Review*, 76, 956–70.

Brealey, R. A. and Myers, S. C. 1996: *Principles of Corporate Finance*. New York: McGraw-Hill.

Bromiley, P. 1991: Testing a causal model of corporate risk taking and performance. *Academy of Management Journal*, 34, 37–59.

Chandler, A. 1962: *Strategy and Structure: Chapters in the History of the American Industrial Enterprise*. Cambridge MA: MIT Press.

Chung, K. H. and Pruitt, S. W. 1994: A simple approximation of Tobin's q. *Financial Management*, 23, 70–4.

Cyert, R. M. and March, J. G. 1963: *A Behavioral Theory of the Firm*. Englewood Cliffs, NJ: Prentice-Hall.

Damanpour. F. 1991: Organizational innovation: a meta-analysis of effects of determinants. *Academy of Management Journal*, 34, 555–90.

Davis, G. F. and Stout, S. K. 1992: Organization theory and the market for corporate control: a dynamic analysis of the characteristics of large takeover targets, 1980–1990. *Administrative Science Quarterly*, 37, 605–33.

Davis, R. and Duhaime, I. M. 1992: Diversification, vertical integration, and industry analysis: New perspectives and measurement. *Strategic Management Journal*, 13, 511–24.

DeAngelo, H. and Masulis, R. 1980: Optimal capital structure under corporate and personal taxation. *Journal of Financial Economics*, 8, 3–29.

Dierickx, I. and Cool, K. 1989: Asset stock accumulation and sustainability of competitive advantage. *Management Science*, 35, 1504–11.

Finkel, S. E. 1995: *Causal Analysis with Panel Data*. Newbury Park, CA: Sage Publications.

Froot, K. A., Scharfstein, D. S., and Stein, J. C. 1993: Risk management: Coordinating corporate investment and financing policies. *Journal of Finance*, 48, 1629–58.

Ginn, G. O., Young, G. J., and Beekun, R. I. 1995: Business strategy and financial structure: an empirical analysis of acute care hospitals. *Hospital and Health Services Administration*, 40, 191–200.

Harris, M. and Raviv, A. 1991: The theory of capital structure. *Journal of Finance*, 46, 297–355.

Himmelberg, C. P., Hubbard, G., and Palia, D. 1999: Understanding the determinants of managerial ownership and the link between ownership and performance. *Journal of Financial Economics*, 53, 353–84.

Hovakimian, A., Opler, T., and Titman, S. 2000: The debt–equity choice. *Journal of Financial and Quantitative Analysis*, 36, 1–24.

Huber, G. P. 1991: Organizational learning: the contributing processes and the literatures. *Organization Science*, 2, 88–115.

Hunt, S. D. 1997: Resource-advantage theory: An evolutionary theory of competitive firm behavior. *Journal of Economic Issues*, 31, 59–77.

Jalilvand, A. and Harris, R. 1984: Corporate behavior in adjusting to capital structure and dividend targets: an econometric study. *Journal of Finance*, 39, 127–45.

Jensen, M. C. 1986: Agency costs of free cash flow, corporate finance, and takeovers. *American Economic Review*, 76, 323–9.

Jensen, M. C. and Meckling, W. H. 1976: Theory of the firm: managerial behavior, agency costs and ownership structure. *Journal of Financial Economics*, 3, 305–60.

Jensen, M. and Murphy, K. 1990: Performance and top management incentives. *Journal of Political Economy*, 98, 225–64.

Johnson, D. R. 1995: Alternative methods for the quantitative analysis of panel data in family research: pooled time-series models. *Journal of Marriage and The Family*, 57, 1065–85.

Jordan, J., Lowe, J., and Taylor, P. 1998: Strategy and financial policy in UK small firms. *Journal of Business Finance and Accounting*, 25, 1–27.

Karim, S. and Mitchell, W. 2000: Path-dependent and path-breaking change: reconfiguring business resources following acquisitions in the US medical sector, 1978–1995. *Strategic Management Journal*, 21, 1061–81.

Kochhar, R. 1996: Explaining firm capital structure: the role of agency theory vs. transaction cost economics. *Strategic Management Journal*, 17, 713–28.

Kochhar, R. and Hitt, M. A. 1998: Research notes and communications linking corporate strategy to capital structure: diversification strategy, type and source of financing. *Strategic Management Journal*, 19, 601–10.

Kogut, B. and Zander, U. 1992: Knowledge of the firm, combinative capabilities, and the replication of technology. *Organization Science*, 3, 383–97.

Long, M. and Malitz, I. 1985: The investment-financing nexus: some empirical evidence. *Midland Corporate Finance Journal*, 3, 53–9.

Miller, M. H. 1977: Debt and taxes. *Journal of Finance*, 32, 261–75.

Miller, M. H. 1988: The Modigliani–Miller propositions after thirty years. *Journal of Economic Perspectives*, 2, 99–120.

Minton, B. A. and Schrand, C. 1999: The impact of cash flow volatility on discretionary investment and the costs of debt and equity financing. *Journal of Financial Economics*, 54, 423–60.

Modigliani, F. and Miller, M. 1958: The cost of capital, corporation finance and the theory of investment. *American Economic Review*. 48, 261–97.

Modigliani, F. and Miller, M. 1963: Corporate income taxes and the cost of capital: a correction. *The American Economic Review*, 53, 433–43.

Myers, S. C. and Majluf, N. S. 1984: Corporate financing and investment decisions when firms have information that investors do not have. *Journal of Financial Economics*, 13, 187–221.

Neter, J., Kutner, M. H., Nachtshem, C. J., and Wasserman, W. 1996: *Applied Linear Statistical Models*. Chicago, IL: Irwin.

Nohria, N. and Gulati, R. 1996: Is slack good or bad for innovations? *Academy of Management Journal*, 39, 1245–64.

Opler, T. C., Pinkowitz, L., Stulz, R., and Williamson, R. 1999: The determinants and implications of corporate cash holdings. *Journal of Financial Economics*, 52, 3–46.

Penrose, E. T. 1959: *The Theory of Growth of the Firm*. Oxford: Blackwell.

Porter, M. E. 1980: *Competitive Strategy*. New York: The Free Press.

Porter, M. E. 1996: What is strategy? *Harvard Business Review*, November–December, 3–20.

Rajagopalan, N. and Finkelstein, S. 1992: Effects of strategic orientation and environmental change on senior management reward systems. *Strategic Management Journal*, 13, 127–42.

Sandberg, C. M., Lewellen, W. G., and Stanley, K. L. 1987: Financial strategy: planning and managing the corporate leverage position. *Strategic Management Journal*, 8, 14–24.

Schendel, D. 1994: Introduction to "Competitive organizational behavior: Toward an organizationally-based theory of competitive advantage." *Strategic Management Journal*, 15, 1–4.

Shleifer, A. and Vishny, R. 1992: Liquidation values and debt capacity: A market equilibrium approach. *The Journal of Finance*, 47, 1343–66.

Shyam-Sunder, L. and Myers, S. C. 1999: Testing static tradeoff against pecking order models of capital structure. *Journal of Financial Economics*, 51, 219–44.

Simerly, R. L. and Li, M. 2000: Environmental dynamism, capital structure and performance: a theoretical integration and an empirical test. *Strategic Management Journal*, 21, 21–49.

Singh, J. V. 1986: Performance, slack and risk taking in organizational decision making. *Academy of Management Journal*, 29, 562–85.

Solnik, B. 1996: *International Investments*. Reading, MA: Addison-Wesley Publishing.

Titman, S. 1984: The effect of capital structure on a firm's liquidation decision. *Journal of Financial Economics*, 13, 137–51.

Vincente-Lorente, J. D. 2001: Specificity and opacity as resource-based determinants of capital structure: evidence for Spanish manufacturing firms. *Strategic Management Journal*, 22, 157–77.

Warner, J. B. 1977: Bankruptcy costs: some evidence. *Journal of Finance*, 32, 337–48.

White, G. I., Sondhi, A. C., and Fried, D. 1994: *The Analysis and Use of Financial Statements*. New York: John Wiley & Sons.

Williamson, O. 1988: Corporate finance and corporate governance. *Journal of Finance*, 43, 567–91.

Zajac, E. J., Golden, B. R., and Shortell, S. M. 1991: New organizational forms for enhancing innovation: the case of internal corporate joint ventures. *Management Science*, 37, 170–84.

CHAPTER 5

The Value of Managerial Learning in R&D

Paolo Boccardelli, Alessandro Grandi, Mats G. Magnusson, and Raffaele Oriani

Keywords: Competence, knowledge, technology, managerial, learning effect.

Abstract

This paper explores the impact of the R&D learning curve on performance by capturing its effect on the development and utilization of organizational capabilities. In R&D project work, managerial competencies are used to influence the exploration and exploitation of technological competencies. Competencies of selection and creation play a crucial role in leading the evolutionary traits of technological competencies, while those of combination and transformation enable organizations to efficiently embed technological competencies in products and processes. The combined use of these competencies over time influences R&D outcome in terms of innovation and efficiency. By studying R&D projects in two business units at Ericsson, the relationship between the R&D learning curve and performance is investigated.

Introduction

This paper investigates the role played by the R&D learning curve and organizational capabilities with respect to the relationship between R&D activities and firm performance.

In this study the basic assumptions come from resource- and knowledge-based theoretical backgrounds, which allow us to argue that knowledge and intellectual capital held by a firm positively affect its economic performance.[1] Accordingly, the larger the accumulation of knowledge and the larger the easy-to-use intellectual capital, the larger the potential to gain economic rents. Authors who have empirically tested this relationship earlier, referring to the market value of the corporation as performance indicator, have generally found a positive impact of the R&D capital on the value of the firm (Hall, 1999).

However, R&D efforts do not automatically convert into new technological knowledge and innovative products. Indeed, a significant area of uncertainty concerns the outcome of R&D projects. Even though this discussion has reached a relevant standing in the recent discourse of scholars as well as practitioners, we argue that most of the contributions have left out an important aspect of the relationship between this type of intangible assets and companies' performance, namely the role of organizational "assets" in mobilizing knowledge and intellectual capital.

The present paper aims to demonstrate that organizational capabilities can be considered an intangible asset which enables a fruitful flow of intellectual capital and knowledge within organizations, thereby significantly affecting their rate of knowledge development as well as their economic results. The achievement of outstanding performance is thus a result of a superior capability of fusing different types of tangible and intangible assets.[2]

We also shed new light on the empirical analysis. A limitation to several previous studies is that they use the R&D investments as a measure of knowledge capital and do not consider that different companies can manage R&D activities in different ways. Other studies examine the management of the R&D process at the firm level, but do not relate it to the firm's economic performance (Boccardelli and Magnusson, 2000). Therefore, the consideration of the role of organizational capabilities and the effort to measure them provides a possibility to capture firm-specific effects affecting the potential value of R&D investments.

Based on the above, we can note that R&D performance depends on a number of firm-specific factors. First of all, the results of R&D projects are substantially affected by the existence of complementary assets allowing the commercialization of the new products and services (Teece, 1986). Secondly, managerial capabilities allow a better selection and orientation of R&D activities on one side, and a more effective recombination and transformation of the knowledge generated on the other side, leading to a tighter fit between new products and market requirements (Iansiti and Clark, 1994). Thirdly, managerial choices determine a different balance between basic and applied research that, in the perspective of the trade-off between exploration and exploitation (March, 1991), can critically impact on the overall performance of a firm's R&D investments (Iansiti, 2000). Taken together, this points to the need for a more detailed understanding of the role played by managerial capabilities in converting R&D investments into products and services generating economic results.

Theoretical Background

A fundamental point of the resource-based theory of the firm is that superior rents come from firms' possession of specific resources that are unique, scarce and difficult to imitate (Rumelt, 1984, 1987, 1991; Wernerfelt, 1984; Dierickx and Cool, 1989; Barney, 1986, 1991; Amit and Schoemaker, 1993; Peteraf, 1993). Accordingly, the coordination and integration of resources induce the development of core competencies (Nelson and Winter, 1982), and competitive margins stem from the part of internal capabilities that are overlapped with the "strategic industry factors" (Amit and Schoemaker, 1993).

The Dynamic Capabilities approach develops this theoretical framework, focusing on the process of competence transformation. Indeed, as strategic industry factors are not immutable and stable over time, the success of a company critically depends on the ability to change and reorient its core competencies in order to face new environmental challenges (Teece et al., 1997).

This distinction suggests that two sub-roots can be individuated in the resource-based perspective (Boccardelli and Magnusson, 2000). The first one is that of organizational rents, originally studied by Penrose (1959) and developed by Nelson and Winter (1982). It focuses on the role of integration and coordination of specific resources in the creation of isolation mechanisms and barriers to imitation processes. The second one, which we refer to as entrepreneurial rents, claims that superior rents arise from innovation and change over time (Teece et al., 1997).

The twofold perspective of the RBV of the company is well depicted by the bathtub metaphor provided by Dierickx and Cool's contribution (1989). In their work Dierickx and Cool pointed out that economic rents stem from stocks and flows of valuable resources. Stocks are those assets from which, to a large extent, companies gain profits at each moment in time, while flows are those resources that convey the existing asset configuration and modify it in order to match evolutionary traits of competitive arenas.

Focusing on the technological dimension of the corporate assets, we can provide some fundamental definitions. Technological competencies are sets of knowledge and skills related to the fields of science and technology. Managerial capabilities, on the other hand, are a set of structures, resources, mechanisms, and procedural as well as managerial knowledge, which enable the implementation of strategic decisions on evolutionary paths of technological competencies. The last of these allows a dynamic configuration of company competencies.

Managerial competencies in this sense can be considered strategic assets to exploit in order to lead operational and organizational activities, which, aiming to dynamically configure companies' assets, pick up existing competencies, use and modify them, and search for new ones. Put in a straightforward manner, they are directed to combine, select, transform, and create technological competencies.

Competencies of combination of different technological competencies into products and applications, by means of the mechanisms of coordination and integration (Clark and Fujimoto 1991; Leonard-Barton et al., 1994; Henderson and Cockburn, 1994; Iansiti and Clark, 1994), aim to combine different technological resources, such as human capital and knowledge in terms of technologies, into coordinated activities for the development of new competencies or applications. Therefore, they perform in a twofold perspective: the organizational perspective, which allows the coordination of heterogeneous activities and resources (Clark and Fujimoto, 1991), such as product development teams or project management methods; and the knowledge perspective, which enables the integration of different components of technological knowledge into more complex and systematic ones (Iansiti and Clark, 1994), such as product platforms or families.

Capabilities of selection of competencies to be kept and those to be abandoned (Garud and Nayyar, 1994) aim to shed light on technological assets that cannot be used in the existing project portfolio, because they are not yet ready to use for the marketplace or their residual potential does not justify any further investment. In that sense, competencies of selection drive those managerial decisions which define and constitute shelves of innovation and which are directed to technological out-sourcing. These competencies allow companies to concentrate development resources on specific innovation tasks for the marketplace, and they can boost performance because they release resources from the development of old and well-defined competencies.

Capabilities of transformation of original competencies into new ones that fit better with the new business environment (Smith et al., 1996; Lei, 1997; Teece et al., 1997) are used to perform the necessary incremental development to obtain and launch on the market numerous applications from the same technological basis. They drive those processes that, starting from the existing technological knowl-edge basis, perform the necessary changes and developments to adapt products and applications to the new challenges emerging from the competitive landscape. As technological competencies primarily consist of human capital and technological knowledge, competencies of transformation mainly reside in the capability of the R&D resources to incrementally learn and evolve.[3] Hence they can be analyzed by considering all the structures and mechanisms that enable researchers and designers to learn from a specific knowledge basis, such as research collaboration and problem-solving methods.

Finally, competencies of creation of radically new technological knowledge (Nonaka, 1994; Nonaka and Takeuchi, 1995) represent capabilities to develop path-breaking innovation, which sometimes destroy the existing knowledge base and replace it with a new one. The purpose of creative competencies is to assist in bringing about radically new products and processes. Often, this implies finding ways of breaking away from established ideas to create room for the application of new perspectives – which can be facilitated by the use of strong metaphors that do not fit with existing frames of interpretation (Nonaka, 1994; Nonaka and Takeuchi, 1995), or by exposure to individuals or communities holding different perspectives (Boland and Tenkasi, 1995). In this sense, the openness of research and development teams to external communities, like those of marketers, suppliers, and customers, can be interpreted as a means to foster and amplify their creativity, which entails a major capability to reach disruptive outcomes in innovation activities.[4]

Following these definitions and by taking into account the distinction between stock- and flow-resources (Dierickx and Cool, 1989), technological competencies will be considered differently depending on whether an analysis of firms' resource- and competence-configurations is made from a static or a dynamic point of view. In a static view, technological competencies are those assets from which com-panies to a large extent gain sources of economic rents. This statement derives directly from the traditional argumentation of RBV (Rumelt, 1984; Wernerfelt, 1984; Dierickx and Cool, 1989; Barney, 1991). Switching to a more dynamic perspective of resource-configuration, technological competencies must also be con-sidered as flow-components which improve, modify and, sometimes, destroy the previous assets in search of a new configuration that better matches the environmental dynamics (Teece et al., 1997; Boccardelli and Magnusson, 2000; Eisenhardt and Martin, 2000). Managerial competencies play a crucial role in supporting and leading the evolutionary traits of technological competencies, but they also work in the static realm as they allow the organization to embed competencies in products and processes.

Therefore, at each moment in time, the potential economic performance of a firm's technological knowledge depends on two related aspects of managerial capabilities: the creation, orientation, and reconfiguration of knowledge-flows in a way that is consistent with the environmental evolution, and the coordination and integration of existing knowledge assets. The first aspect mainly aims at the exploration of new technological competencies, while the second might be considered to aim primarily at the exploitation of existing technological competencies. Given this, the entire body of technological knowledge, as well as the managerial capabilities, should be considered when investigating the relation between R&D efforts and economic performance.

Consequently, the management of the whole set of knowledge is a firm- and even unit-specific capability, which should be fostered in order to improve R&D performance. This requires that managerial competencies evolve over time by means of an accumulation of knowledge in the memory of the organization (Walsh and Ungson, 1991). This organizational memory consists of previously experienced stimulus–response relationships that are potentially useful in present decision-making processes. The organizational memory exists not only at the individual level but also at the organizational level, e.g., in terms of organizational routines stored as procedural knowledge (Cohen and Bacdayan, 1996). The speed of the needed knowledge accumulation depends on the complexity, and consequently on the nature, of the task to be performed (Boccardelli and Grandi, 2001).

Project complexity poses significant challenges to the organization and, thereby, often implies the emergence of new or modified managerial practices. Indeed, in their contribution Brown and Duguid (1991) define a relationship between working, learning, and innovation within communities of practice. A similar phenomenon can be grasped when observing R&D operations, the nature of which is truly knowledge based. Accordingly, the interaction of communities of practice in knowledge-based operations generates new patterns of managerial and procedural knowledge.

We therefore argue that R&D projects with a more explorative focus, measured in terms of higher project-innovation rate, higher complexity, and higher number of new technologies, require a different type and intensity of managerial tools and mechanisms than do projects strictly focusing on bringing forward new products based on existing technological competencies.

The R&D activities ultimately affect the firm's performance. Previous literature has tried to define specific causal relations between R&D investments, knowledge, and economic results of the corporation (Clark, 1987; McGrath et al., 1996). Empirical studies show a positive relationship between the R&D capital and the market value of the firm (Hall, 1999). We are, however, still lacking a deeper understanding of the factors shaping this relationship over time and across different firm-specific contexts.

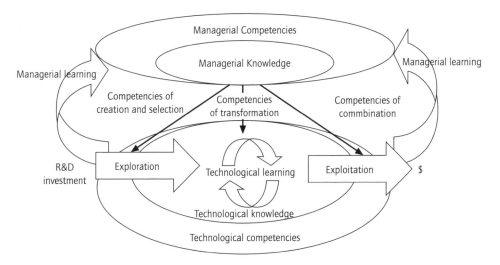

Figure 5.1 A resource-based model for R&D work, incorporating the use and development of managerial competencies

An Analytical Model Relating the R&D Learning Curve and Firm Performance

In this contribution we maintain that the relations between R&D efforts and economic performance are mediated by a set of managerial competencies affecting R&D projects. In Figure 5.1, a schematic illustration of this relationship is presented.

R&D operations allow the organization to improve its potential of gaining outstanding economic performance throughout the development of technological competencies and knowledge. However, R&D performance depends on a set of managerial capabilities divided into two categories: those of selection and creation, and those of combination and transformation. The selection/creation group is applied to the development of different technological options, and the combination/transformation group to the management of development projects providing the marketplace with new products and services.

In order to successfully develop new products and applications, the firm has to continuously link evolving customer needs to product specifications (Clark and Fujimoto, 1991; Iansiti and Clark, 1994). In this context, both selection/creation and combination/transformation competencies affect the performance of R&D investments, even though in different ways. The first group links the flows of new knowledge to the evolution of product markets; the second induces the exploitation of existing technological knowledge for new product commercialization. In other words, the goal of the selection and creation competencies is to develop valuable technology-based options for the future, while the tasks of combination and transformation competencies are more closely aligned to the extraction of the potential value embedded in the current stock of technology.

The technological knowledge base available to a company mainly derives from the technological path explored earlier by the firm. This pathway may contain both attempts at organic generation of new technology and the use of external technology sources, i.e., the selection of exogenously developed knowledge assets. The aggregate technological knowledge available constitutes a pallet of knowledge assets that can potentially be exploited through its conversion to marketable products and services. The knowledge stock thereby generated can be further refined, however, in order to better suit the needs regarding input to, e.g., new product development projects. This transformation, implying incremental changes of previously developed knowledge, is here referred to as technological learning.

Although the difference between learning and the creation of new knowledge is difficult to demarcate, as learning always consists of the development of some new piece of knowledge for at least one individual, it appears logical to differentiate between radical and incremental processes of knowledge creation, because they display different characteristics and call for different management approaches (Nonaka, 1994). Transformation also relates to the diffusion of new technology within a firm. In order to make new technology available to new product development projects, it has to be packaged in a more readily available format, and this may also be considered a learning process.

The use of competencies of combination, e.g., through the use of management tools that integrate different sets of stored information or coordinate people from different functions, is a key to the fruitful utilization of the knowledge assets available to the firm. To convert existing technological knowledge into goods and services in a timely and efficient manner is the main objective of the deployment of these competencies of exploitation.

The relative importance of the different competencies depends upon the particular knowledge dynamics of the industry, eventual biases toward long-term or short-term performance, and the heterogeneity of the technologies used. The more rapid the knowledge development, the more necessary it becomes to explore new technological paths in order to avoid obsolescence of the knowledge that is kept. Likewise, a rapidly changing technological field also calls for a stronger focus on updating already existing technological knowledge. The stronger the demands for short-term performance, the more focus will need to be put on combination efforts.

Creating new technology is a time-consuming endeavor and it is hardly viable to give much attention to this issue in an industry where firms are evaluated strictly on their weekly or monthly economic performance. Furthermore, such a bias in focus ought to favor acquisitions over organic development as a source of new technology. Finally, it can be argued that heterogeneity of technology may incur difficulties in the combination of R&D activities. The more diverse the knowledge base that needs to be used, the more difficult it becomes to manage its transformation into products and services.

When the firm utilizes the different categories of competencies mentioned, and receives feedback from this use, managerial learning can take place. As Boccardelli and Grandi (2001) have pointed out, the nature of the projects undertaken influences this kind of learning. More complex projects pose more substantial challenges to management and thereby potentially lead to an increased learning effect. The larger is the accumulated managerial knowledge, the greater is the potential to manage either effectively or efficiently the entire body of technical knowledge to be used in R&D activities.

Taken together, the reasoning above leads us to ask how different managerial capabilities should be developed and used by firms in order to allow them to reconfigure technological competencies over time in a way that provides them with an increased potential to harvest economic rents.

Research Settings, Data and Methods

The analytical model explained above has been applied within an explorative study in the telecommunications industry, which, in terms of technological issues, is characterized by heterogeneous business areas. In this industry, strategic industry factors seem to be based on two key issues: the accumulation of wide and deep technical competencies, and the capabilities to develop and launch new products and systems in the marketplace at an increasing pace.

This heterogeneity characterizing the telecom industry gave us the opportunity within the same company to analyze two R&D settings operating in different business areas. The first is that of Ericsson Microwave Systems (EMW), which initially was operating in the military communications equipment business and, recently, when the defense business started to decline, began to develop solutions for civilian markets and, particularly, radio links for telecom operators. This last market is characterized by highly volatile trends owing to the competitive pressures transferred by operators to the suppliers of technological infrastructure. The emergence of new operators has, in fact, resulted in a large opportunity for microwave technologies, as microwave radio-links allow newcomers to reduce time to service and the total amount of investment in infrastructure. In this business area, technological knowledge has been accumulated over the past 15 years and there has been a substantial change in the technical contents of products and processes (see Table 5.1).

In Table 5.1, TECHCOMP measures the amount of technological competencies used and developed in the series of the projects investigated. TECH-CHANGE indicates the differential value of technological competencies of each project compared with their average value in the sequence of projects running within the same setting. TIME is a variable count measuring the passing of time after each project.[5]

Table 5.1   The accumulation and change of technological competencies over time in EMW and RDS

	EMW TIME	RDS TIME
TECHCOMP	0.491	0.096
TECH-CHANGE	0.303	0.001

Source: adapted from Boccardelli and Magnusson (2000)

The results in Table 5.1 show a positive correlation between the time variable and the development and use of technological competencies, and also between the time variable and the change in the technical contents of the project. The interpretation is that, in EMW, technological competencies have both been accumulated and changed over time.

The second business unit is that of Ericsson Telecommunications Italy, which provides the global sub-product Unit Residential Services (RDS) with technical solutions and products for a worldwide market. This site has for the past 20 years contributed to the development of the AXE system, the Ericsson switching system for PSTN and ISDN services. The RDS unit was, in fact, born just three years ago. The analysis has, therefore, focused on a sequence of projects running within those parts of the Italian subsidiary of Ericsson that later came to constitute the RDS unit. In the switching-system business, market trends are essentially stable and it appears very complex and costly to boost market share. The main customers of equipment suppliers are traditional telecom operators, which after monopolistic market conditions have had to cope with competitive pressures emerging from newcomers. In order to compete with new actors using alternative and potentially more effective technologies, the former monopolists demand continuous innovations and changes to the basic technological platform, which for this part of Ericsson lies in the development environment of AXE switching systems. In this area, technological change today mainly concerns the software component needed to develop the platforms and architectures that enable the launch of new services to end-markets. This business area, in contrast with radio-links, has for the past 15 years displayed stable and mature technological knowledge, strictly related to the AXE architecture in terms of the product, and to the development environment of AXE systems in terms of programming tools and environment.[6]

Table 5.1 shows that over the past 15 years the wealth of technological knowledge embodied in the AXE systems has remained fundamentally the same and only minor changes have resulted from a sequence of projects that has incrementally developed technological competencies. Neither TECHCOMP nor TECH-CHANGE shows any correlation with TIME, which means that in the period investigated the accumulation of technical competencies was limited and that only insignificant changes took place in terms of the technological competencies of RDS. In RDS the main technological asset to be exploited was generated at the beginning of AXE history. Nowadays, the RDS unit mainly designs new software blocks and functionality for the market.

Data Collection and Methods

For this study we have used data from a larger research project (Boccardelli and Magnusson, 2000) that we are conducting in order to investigate the factors allowing an operationalization of the dynamic resource-based perspective within technology-based organizations. The whole data set is divided into three main classes: data on technological competencies, data on managerial competencies, and data on performance (see Table 5.2).

Table 5.2   Grouping the data set used in the empirical analysis

Variable	Measurement	Data Source
	Technological Competencies	
Technological Competencies developed and used in each project TECHCOMP	• Measured as the number of granted/filed patents and the number of technologies used/developed within the project • 4 items resulted in 1 index by considering a vectorial sum	Internal Docs
Innovation Rate generated by each project INNOVA	• Measured as the percentage of projects modified or totally new subsystems • The ratio between the number of modified or new subsystems and the total number of subsystems	Internal Docs
	Managerial Competencies	
The use of Competencies of Combination COORDMECH	• Presence and usage intensity of coordination mechanisms • Measured by 5 items indicating the intensity of usage of a specific tool; the resulting index has been measured as the mean • Presence and usage intensity of integration mechanisms	Questionnaire
INTEMECH	• Measured by 5 items indicating the intensity of usage of a specific tool; the resulting index has been measured as the mean	
The use of Competencies of Selection	• The relevance of strategic committees for selecting those competencies to be kept and those to be abandoned	Questionnaire
SELECT	• Selection Ratio between outsourced and internally developed technologies • Measured by 2 items indicating the intensity of usage of strategic committees and the selection ratio; the resulting index has been measured as the mean	Internal Docs
The use of Competencies of Transformation TRANSFORM	• Presence and usage intensity of learning mechanisms and patterns • Measured by 15 items indicating the usage intensity of a specific tool; the resulting index has been measured as the mean	Questionnaire
The use of Competencies of Creation CREATION	• The presence and usage intensity of knowledge creation mechanisms • Measured by 5 items indicating the usage intensity of a specific tool; the resulting index has been measured as the mean	Questionnaire
	Economic Performance	
SALES	• The amount of sales related to the development projects investigated	Internal Docs

Seventeen new product development projects (eight in EMW and nine in RDS), covering a period of 15 years, have been studied. Project performance data, e.g., lead times and man-hours, as well as technical and managerial data, in terms of technologies used in the project and utilization of various management methods and tools, have been collected for all the investigated projects.

In order to capture quantitative data, both internal documents and a questionnaire have been used. Internal documents have been used to collect technical and project performance data, as well as data on economic performance of the two business units. First, we used in-depth interviews with R&D managers and project managers to examine the evolution of the technical competencies over time and the overall streams of R&D projects that had been performed. Thereafter, questionnaire forms were circulated to project managers, in order to map the evolution of the managerial competencies adopted in the projects. Put briefly, the questionnaire had been designed to estimate the presence and relative usage intensity of organizational tools, mechanisms and factors supporting different managerial competencies applied in the R&D projects. Altogether, we collected data related to 40 items, which were then grouped in eight different variables, as shown in Table 5.2.

Data from the projects investigated have been simply correlated in order to get indications for relationships between the measured variables. The significance of the resulting analyses has not been tested at a general model level. However, since the data come from the population of innovation projects in each investigated setting for a period of about 15 years, the results can arguably be interpreted as a quantitative format of explorative and descriptive analyses of two different cases. In combination with the understanding of the studied phenomena that has been generated through the qualitative part of the study, a number of research propositions have been developed.

Research Findings

In Table 5.3, correlations between managerial competencies and economic performance are shown. A number of observations can be made from the displayed results. First of all, we can observe that a growth in project complexity, measured by the innovation rate of the development tasks, has a positive correlation with economic performance in a dynamic setting, like that of EMW, while it appears negative in a stable environment like that of RDS. This result can be viewed as a further confirmation of the logical rules of the game for the two different kinds of competitive arenas described earlier.

Proceeding to the analysis of managerial competencies, we note that in EMW competencies of combination, characterized by mechanisms of coordination and integration, and competencies of creation positively affect economic performance. In RDS, on the contrary, we find that competencies of transformation and selection are positively correlated with economic performance.[7]

The results in Table 5.3 fit very well with the analytical model presented above, since they show that in a dynamic environment competencies of creation mobilized in the early and more research-focused stages of the development process positively

Table 5.3   Pearson correlations between managerial competencies and economic performance

	EMW SALES	RDS SALES
INNOVA	0.655	−0.489
COORDMECH		
INTEMECH	0.513	−0.129
SELECT		
TRANSFORM	0.858**	0.245
CREATION		
	−0.065	0.494
	0.247	0.712*
	0.628*	0.037

** Significant at the 0.01 level (1-tailed).
* Significant at the 0.05 level (1-tailed).

affect economic performance; i.e., by generating new technologies, more effective ways of developing products and services can be achieved. In stable settings, on the contrary, primarily competencies of selection and transformation can improve economic performance. As there are normally few new technologies emerging in stable settings, the possibility to improve performance in this situation is a question of utilizing available technological knowledge in the most efficient way possible. This highlights the importance of making the right decisions regarding the use of outsourcing in order to leverage internal as well as externally available technology in the best possible way. However, even more important in a stable setting appears to be the effort to continuously update the existing technological base through incremental learning, as revealed by the strong positive correlation between competencies of transformation and economic performance.

Moving to the market-driven development stages, which are the closest to the launches of new products in the marketplace, we see that in the stable setting integrative capabilities, which provide a part of competencies of combination, positively affect economic performance, even though this correlation does not appear to be very strong and significant. In a dynamic environment, all types of competencies of combination (both coordination and integration mechanisms) show a positive correlation with economic performance, but integrative capabilities play a major role in the task of exploiting the existing knowledge base. A reasonable explanation for this difference is that there is a larger need for rapid development projects in more dynamic settings, as the continuous launching of new products is the key to gaining market shares. As technology performance normally increases quickly in these fields, the possibility to develop products based on new technology, outperforming already existing solutions, is a matter of integrating new technology in a rapid manner. In stable settings, the possibilities to gain advantages from this way of working are much more limited, and competition is not necessarily as related to new product development as it is in dynamic settings, but may be more intense in, for instance, the areas of marketing and distribution.

These results appear consistent with the settings analyzed and might be worth a deeper analysis, where the effect of R&D activity on the performance is not only a result of specific stocks of resources, but also depends on how managerial capabilities are used to reconfigure the stock of resources. In a synthesized form, this can be summarized in the following research proposition.

> Proposition 5.1. The effect of R&D projects beginning at time t on the firm's economic performance at time $t+k$ depends on both selection/creation and combination/transformation competencies applied by the management of the projects.

Besides this, in previous research it has been shown that coping with complexity in R&D work induces an accumulation of procedural and managerial knowledge throughout a sequence of projects (Boccardelli and Grandi, 2001). Accordingly, we argue that the more complex are the R&D tasks, the greater will be the accumulation of managerial knowledge to be applied in future R&D cycles. Consequently, it can be claimed that a "learning effect" improves the capability of the R&D organization to produce knowledge flows.

Whereas in the manufacturing area many contributions seek to investigate experience and learning effects, few investigations focusing on R&D work have been undertaken.[8] We therefore aim to show that as managerial competencies drive evolutionary paths and exploitation tasks of technological assets, a learning effect working on managerial factors enables the organization to speed up the conversion of the stored technological knowledge into applications ready to use in the marketplace. This learning effect consequently should provide the capacity to manage future R&D cycles in a way that improves the development and accumulation of technological flows, allowing the company to cope with more complex and challenging projects in a better way. As a consequence, it generates a larger potential of economic performance by means of a greater stock of technological assets to exploit.

Wishing to explore the impact of this learning effect, we now investigate the relation between task uncertainty and differences in the use of managerial competencies. In order to do this, five variables (COORD-CHANGE, INTE-CHANGE, SELECT-CHANGE, TRANSF-CHANGE, CREA-CHANGE) have been used to measure change in the use of managerial competencies over time from one project to the next. The values of the five variables are calculated as follows:

$$\text{COORD-CHANGE} = \text{COORDMECH}_t - \text{COORDMECH}_{(t-1)}$$
$$\text{INTE-CHANGE} = \text{INTEMECH}_t - \text{INTEMECH}_{(t-1)}$$
$$\text{SELECT-CHANGE} = \text{SELECT}_t - \text{SELECT}_{(t-1)}$$
$$\text{TRANSF-CHANGE} = \text{TRANSFORM}_t - \text{TRANSFORM}_{(t-1)}$$
$$\text{CREA-CHANGE} = \text{CREATION}_t - \text{CREATION}_{(t-1)}$$

Following what has been stated above, we should expect that the higher the values of innovation rate, the higher the values of the variables measuring the change in the use of managerial competencies.

Table 5.4 Correlations between the accumulation of managerial knowledge and innovation rate

	EMW INNOVA	RDS INNOVA
COORD-CHANGE	0.025	0.359
INTE-CHANGE	−0.064	0.018
SELECT-CHANGE	0.001	0.437
TRANSF-CHANGE	0.277	−0.017
CREA-CHANGE	0.598*	−0.159

* Significant at the 0.05 level (1-tailed).
Source: adapted from Boccardelli and Grandi (2001)

Table 5.4 shows different results for mature and dynamic business environments, respectively. A higher rate of innovation is positively correlated with a growth in the use of competencies of combination (COORD-CHANGE) and selection (SELECT-CHANGE) in RDS, and with competencies of transformation (TRANSF-CHANGE) and creation (CREA-CHANGE) in EMW. These results show that technological contingencies, like innovation rate or task uncertainty, call for different solutions if they emerge from different business environments, and hence they affect the accumulation over time of managerial competencies in a heterogeneous manner. This implies that task uncertainty is selectively correlated with the accumulation of managerial knowledge.

In order to explore the role of managerial learning for economic performance in more detail, though, we must take into account the relations between variables measuring the change in competence utilization and those estimating economic performance (Table 5.5).

In Table 5.5, the change in competence utilization that occurred between t and $t+1$ has been correlated with performance variables at time $t+2$. The specific time lag chosen depends on project lead times that characterize the two investigated settings. The results in Table 5.5 show that in a dynamic setting like that of EMW, the

Table 5.5 Correlations between the accumulation of managerial knowledge and economic performance

	EMW SALES	RDS SALES
COORD-CHANGE	0.086	0.252
INTE-CHANGE	0.611*	−0.474
SELECT-CHANGE	0.672	−0.093
TRANSF-CHANGE	0.394	0.418
CREA-CHANGE	0.968**	0.175

* Significant at the 0.05 level (1-tailed).
** Significant at the 0.01 level (2-tailed).

change in the use of all different types of managerial competencies is positively correlated with economic performance.[9] In a stable environment, like that of RDS, a positive correlation can only be found with competencies of transformation.[10] Based on this, we suggest that in dynamic settings managerial learning, generated by the growth in the utilization of competencies to manage a specific R&D project, potentially improves the capability of the project itself to provide economic value for the company, while in stable environments the unique managerial learning positively affecting economic performance might be that coming from the use of competencies of transformation.

Combining this with the results in Table 5.4, we see that it is possible that only in dynamic settings can a growth in project complexity and scope boost economic performance through the managerial learning occurring in the use of competencies of creation.

Furthermore, the radical difference between the roles played by managerial learning in dynamic and stable business settings, respectively, calls for a differentiated view of the effect that this kind of learning has on economic performance. While the impact of managerial learning appears to be substantial in dynamic settings, its effect in stable environments could be questioned. This appears fully logical, as the potential benefits of managerial learning come from the need to reconfigure the company's technological competencies, and consequently the need to hold the appropriate managerial competencies to do this in a fruitful manner. When there is limited need to perform these changes, managerial learning also becomes less important. This leads us to formulate our second and third research propositions as follows.

Proposition 5.2. The effect of R&D projects at time $t+k$ on the firm's economic performance at $t+j$ (with $j>k$) depends on the accumulation of managerial knowledge resulting from managerial learning at time t.

Proposition 5.3. The effect of R&D projects running in dynamic settings at time $t+k$ on the firm's economic performance at $t+j$ (with $j>k$) depends on project complexity and scope of the project performed at time t.

Discussion and Managerial Implications

This paper aims to shed new light on the relationship between R&D investments and firms' performance. While previous research has emphasized the role played by intangible assets for the economic performance of companies on a general level, we argue that the firm-specific capability to carry out R&D projects is a hitherto overlooked factor enabling companies to extract the value embedded in technological assets.

Reliance on the resource- and competence-based perspective as the theoretical background allows the creation of a powerful link between R&D operations and the competencies that can be exploited in the marketplace. In particular, the distinction between stock- and flow-resources provides the resource-based framework with a dynamic view, which enables us to better understand the evolutionary paths of companies' competencies. Finally, the consideration of the role played by managerial factors (Boccardelli and Magnusson, 2000) allows investigation of the other sorts of competencies conditioning the extent to which the technological base may be converted into value for the firm. This idea is far from new, being at the core of the seminal work of Penrose (1959), though with a few exceptions (e.g., Teece et al., 1997) it has not been given much room in the recent resource-based discourse. In previous research, as a matter of fact, we argue that the exclusion of managerial factors is a severe shortcoming of the existing theories on firms' competencies, and we claim that these factors' inclusion is a highly urgent issue in order to turn the resource-based approach into managerially useful practice (Boccardelli and Magnusson, 2000).

By studying the development and use of both technological and managerial competencies in a series of R&D projects in two different business settings, it has been possible to perform an empirical exploration of the relationship between resources and economic performance with a more complete analysis of firm-specific factors than has been the case in earlier studies. Based on the theoretical development and the empirical findings of the study, a number of observations can be made. First of all, we note that the possession of firm-specific assets in terms of technological competencies is not sufficient to convert resources into value for the marketplace, but that a company also needs a set of managerial competencies that reconfigure its technological knowledge base. While other authors have tried to address the same objective (e.g., McGrath et al., 1996), we believe that our contribution provides a deeper analysis and permits a more straightforward generation of actionable guidelines for practitioners.

Furthermore, this paper continues the development of a dynamic perspective for the management of companies' competencies. Indeed, the analysis of a "learning effect" in managerial competencies enables us to observe the accumulation of managerial and procedural knowledge rather than the development of other sorts of resources. Taking the accumulation of stocks of resources as the main phenomenon explaining the dynamics of firms' resources (Dierickx and Cool, 1989), our investigation stresses the fact that companies must also accumulate capabilities on how to manage tasks. This approach seems even more important considering that the accumulation of assets to exploit in the marketplace, such as technologies, increases the complexity to be managed in future operations. The "learning effect" in managerial competencies therefore enables companies to uphold operational performance in more complex and dynamic domains. From a theoretical standpoint, the paper also contributes to a better understanding of the whole cycle of research and development, since the developed analytical model does not simply relate economic performance to the inflows of resources in R&D operations, but describes the mechanisms that influence the exploration and exploitation of knowledge taking place in R&D.

A few managerial implications can also be grasped from this study. First of all, it provides clear guidelines in high-technology settings for what concerns the management of evolutionary traits of competencies. As a matter of fact, the differences affecting the investigated settings indicate that business areas dominated by heterogeneous scenarios require extremely different managerial decisions and resources. This diversity comes from the different nature of R&D operations characterizing these settings.

By showing the role of the R&D learning effect we have empirically illustrated a frequent claim in the management of innovation, namely the relevance of complex and risky projects for R&D organizations. Indeed, we show that engaging in complex, explorative and innovative projects results in accumulation not only of technological but also of managerial knowledge. An implication of this is that companies, when selecting and managing innovation projects within a portfolio, should take into account the learning effect that comes from accumulation of managerial knowledge in the organizational memory. This gives rise to some questions regarding the substantial downsizing that is taking place at present in the depressed telecom industry. In order to adhere to the demands for significant and rapid cost reduction, numerous companies have chosen to reduce investments in R&D activities in general, and in more explorative initiatives in particular. This tendency towards a stronger focus on exploitation of already generated technological knowledge ought to imply also a shift in the usefulness of different managerial capabilities, possibly entailing a change in the competitiveness of certain firms' resource positions. Another related issue is the reluctance shown by venture capitalists to invest in new start-up companies in the telecom industry. By not providing means for exploring new technologies and businesses, the amount of complementary assets that can be used to exploit the technological competencies of telecom operators and infrastructure providers is limited, which at least in a shorter time span potentially reduces performance at the industry level.

The study has some limitations as well, for it is evidently very difficult to test and assess some variables like technological and managerial knowledge, which are at the core of this model, in a historical perspective. It appears clear that the empirical part of this study can be improved by increasing the number of R&D projects studied and the number of business sectors covered, in order to allow a more thorough test of the propositions – which could validate the conclusions at a more general level. However, since the investigation covers a substantial number of the involved units' new product development projects during a 15-year period, these considerations can certainly be applied to the business settings analyzed. Furthermore, it would be useful to test the analytical model through the use of multiple correlation so as to obtain more robust findings.

Notes

1 This consideration derives from the original contribution of Dierickx and Cool (1989) regarding the distinction between stock- and flow-resources. In this perspective, knowledge and intellectual capital can be considered stock-resources to exploit in the marketplace.

2   Different kinds of intangible assets can be traced inside the corporation (see, for example, Hall, 1992): R&D-related assets (technological knowledge, patents, licenses, trade secrets), market-related assets (trademarks, reputation, brands), relational assets (alliances, partnerships, networks), human capital (employees' know-how and skills), and managerial competencies. Recent literature has increasingly explained the firm's competitive advantage on the basis of the exclusive ownership and control of particular categories of intangible assets (Amit and Schoemaker, 1993; Teece, 2000; Lev, 2001).

3   Both experiential (Rosenberg, 1982; Brown and Duguid, 1991; Wheelwright and Clark, 1992) and cognitive (Argyris and Schön, 1978) learning are considered in the concept of competencies of transformation.

4   Open source movement in the software industry and the interpretation of product development as a synthesis of two opposed forces, centripetal and centrifugal (Sheremata, 2000), can be theoretically referred to as the concept of creative chaos (Nonaka, 1994; Nonaka and Takeuchi, 1995) and competencies of creation.

5   See Table 5.2 for an explanation of the variables used.

6   During these 15 years one single substantial change occurred in 1994 with the development of *application modularity* technology, which changed the software architecture of the AXE system.

7   The correlations between economic performance and TRANSFORM in EMW, and economic performance and INTEMECH in RDS, are positive but very low.

8   Among others, Pruett and Thomas (2000) argued that learning manifested in automobile design is a function of a design's cumulative production volume.

9   As regards competencies of combination, we found a positive relation only for integrative capabilities.

10 A positive but very low correlation can be also found with competencies of combination and creation.

References

Amit, R. and Schoemaker, P. J. 1993: Strategic assets and organizational rent. *Strategic Management Journal*, 14, 33–46.

Argyris, C. and Schön, D. 1978: *Organizational Learning*. Reading, MA: Addison Wesley.

Barney, J. B. 1986: Strategic factor markets: expectations, luck, and business strategy. *Management Science*, 32 (10), 1231–41.

Barney, J. B. 1991: Firm resources and sustained competitive advantage. *Journal of Management*, 17, 99–120.

Boccardelli, P. and Grandi, A. 2001: *Competence exploitation and competence development: the dynamic fit between organisational design and task uncertainty in R&D projects – results from a study in the telecom industry*. Paper presented at the R&D Management Conference 2001 "Leveraging Research & Technology," February 7–9, 2001, Wellington, New Zealand.

Boccardelli, P. and Magnusson, M. G. 2000: *Dynamic mastering of technological and managerial competencies – some preliminary results from a study in the telecom industry*. Paper presented at the Strategic Management Society 20th Annual International Conference, Strategy in the Entrepreneurial Millennium: New Winners, New Business Models, New Voices. British Columbia University, Vancouver, Canada, October 15–18, 2000.

Boland Jr., R. J. and Tenkasi, R. V. 1995: Perspective making and perspective taking in communities of knowing. *Organization Science*, 6 (4), 350–72.

Brown, J. S. and Duguid, P. 1991: Organizational learning and communities-of-practice: toward a unified view of working, learning and innovation. *Organization Science*, 2, 40–57.

Clark, K. B. 1987: Investment in new technology and competitive advantage. In D. J. Teece (ed.), *The Competitive Challenge. Strategies for Industrial Innovation and Renewal*. Cambridge, MA: Ballinger, 59–81.

Clark, K. B. and Fujimoto, T. 1991: *Product Development Performance: Strategy, Organization and Management in the World Auto Industry*. Boston, MA: Harvard Business School Press.

Cohen, M. D. and Bacdayan, P. 1996: Organizational routines are stored as procedural memory: evidence from a laboratory study. In M. D. Cohen and L. S. Sproull (eds.), *Organizational Learning*. Thousand Oaks, CA: Sage Publications, 403–29.

Dierickx, I. and Cool, K. 1989: Asset stock accumulation and sustainability of competitive advantage. *Management Science*, 35, 1504–11.

Eisenhardt, K. M. and Martin, J. A. 2000: Dynamic capabilities: what are they? *Strategic Management Journal*, 21, 1105–21.

Garud, R. and Nayyar, P. R. 1994: Transformative capacity: continual structuring by intertemporal technology transfer. *Strategic Management Journal*, 15, 365–85.

Hall, B. H. 1999: Innovation and market value. NBER Working Paper # 6984, Cambridge, MA.

Hall, R. 1992: The strategic analysis of strategic resources. *Strategic Management Journal*, 13, 135–44.

Henderson, R. and Cockburn, I. 1994: Measuring competence? Exploring firm effects in pharmaceutical research. *Strategic Management Journal*, 15, 63–84.

Iansiti, M. 2000: How the incumbent can win: managing technological transitions in the semiconductor industry. *Management Science*, 46, 169–85.

Iansiti, M. and Clark, K. B. 1994: Integration and dynamic capability: evidence from product development in automobiles and mainframe computers. *Industrial and Corporate Change*, 3, 557–605.

Lei, D. T. 1997: Competence-building, technology fusion and competitive advantage: the key roles of organizational learning and strategic alliances. *International Journal of Technology Management*, 14 (2/3/4), 208–37.

Lev, B. 2001: *Intangibles. Management, Measuring and Reporting*. New York: Brookings Institution Press.

Leonard-Barton, D., Bowen, H. K., Clark, K. B., Holloway, C. A., and Wheelwright, S. C. 1994: How to integrate work and deepen expertise. *Harvard Business Review*, Sept.–Oct., 121–30.

March, J. G. 1991: Exploration and exploitation in organizational learning. *Organization Science*, 2, 71–87.

McGrath, R. G., Tsai, M. H., Venkataraman, S., and MacMillan, I. C. 1996: Innovation, competitive advantage and rent: a model and a test. *Management Science*, 42, 389–403.

Nelson, R. R. and Winter, S. G. 1982: *An Evolutionary Theory of Economic Change*. Cambridge, MA: Belknap.

Nonaka, I. 1994: A dynamic theory of organizational knowledge creation. *Organization Science*, 5, 14–37.

Nonaka, I. and Takeuchi, H. 1995: *The Knowledge Creating Company*. New York: Oxford University Press.

Penrose, E. T. 1959: *The Theory of the Growth of the Firm*. New York: John Wiley.

Peteraf, M. A. 1993: The cornerstones of competitive advantage: a resource-based view. *Strategic Management Journal*, 14, 179–91.

Pruett, M. and Thomas, H. 2000: *Organizational learning and product designs: a model and empirical test*. Paper presented at the Strategic Management Society 20th Annual International Conference, Strategy in the Entrepreneurial Millennium: New Winners, New Business Models, New Voices. British Columbia University, Vancouver, Canada, October 15–18, 2000.

Rosenberg, N. 1982: *Inside the Black Box*. Cambridge: Cambridge University Press.

Rumelt, R. P. 1984: Toward a strategic theory of the firm. In R. Lamb (ed.), *Competitive Strategic Management*. Englewood Cliffs, NJ: Prentice Hall, 556–70.

Rumelt, R. P. 1987: Theory, strategy and entrepreneurship. In D. Teece (ed.), *The Competitive Challenge*. Cambridge, MA: Ballinger, 137–58.

Rumelt, R. P. 1991: How much does industry matter? *Strategic Management Journal*, 12, 167–85.

Sheremata, W. A. 2000: Centrifugal and centripetal forces in radical new product development under time pressure. *Academy of Management Review*, 25, 389–408.

Smith, K. A., Vasudevan, S. P., and Tanniru, M. R. 1996: Organizational learning and resource-based theory: an integrative model. *Journal of Organizational Change Management*, 9, 41–53.

Teece, D. J. 1986: Profiting from technological innovation. *Research Policy*, 15, 285–305.

Teece, D. J. 2000: *Managing Intellectual Capital. Organizational, Strategic and Policy Dimensions*. Oxford: Oxford University Press.

Teece, D. J., Pisano, G., and Shuen, A. 1997: Dynamic capabilities and strategic management. *Strategic Management Journal*, 18, 509–33.

Walsh, J. P. and Ungson, G. R. 1991: Organizational memory. *Academy of Management Review*, 16, 57–91.

Wernerfelt, B. 1984: A resource-based view of the firm. *Strategic Management Journal*, 5, 171–80.

Wheelwright, S. C. and Clark, K. B. 1992: *Revolutionizing Product Development: Quantum Leaps in Speed, Efficiency and Quality*. New York: Free Press.

Appendix B Questionnaire submitted to project managers responsible for each product development project

For each of the mechanisms/tools below, respondents were asked to tick the most appropriate alternative on a five-point scale ranging from "Not at all" to "To a very large extent," to the following questions:

In the development of the product did you use . . .
1. product development teams with members from different functional departments?
2. information systems (e.g., intranet, e-mail, video-conference, etc.)?
3. meetings?
4. committees?
5. project management methods?
6. other coordination/integration mechanisms and tools, like . . .

7. databases?
8. information systems?
9. scientific journals and publications?
10. meetings with suppliers and customers regarding R&D issues?

11. collaboration with research institutes, universities, and so forth?
12. other resources and patterns of learning, like . . .

13. problem definition?
14. storytelling (formal or informal sessions in order to share past experiences)?
15. brainstorming or idea generation?
16. learning forums and meetings?
17. prototypes?
18. definition of objectives?
19. failures analysis?
20. the definition of heuristic decision trees?
21. the definition of probabilistic decision trees?
22. feedback mechanisms?
23. other learning mechanisms, like . . .

24. product planning sessions?
25. R&D–top management committees?
26. Multi-project management (coordination of a number of relative projects)?
27. diffused and shared vision of a developed product?
28. product platforms?
29. modularization?
30. marketing research?
31. customer requirements specifications (formal tools and methods for defining customer requirements, e.g., Q.F.D.)?
32. customer/user involvement in the development work?
33. customer-oriented prototypes (prototypes used and evaluated by customers)?
34. product clinics?
35. other marketing tools and mechanisms, like . . .

Furthermore, the respondents were asked the following questions:

36. How many times during the project did top management meet with the project manager in order to discuss the value of a technological competence? (1, 2, 3, 4, more than 4.)
37. In the selection of project members the most considered criterion was that
     a. all the competencies available in the R&D unit were present.
     b. all the competencies, which according to forecasts should be necessary, were present.
     c. team members were used to working with the specific project leader.
     d. team members who had contributed the most to the development of a certain knowledge base were present.
     e. others:

CHAPTER 6

The Paradox of Social Capital: Structural, Cognitive and Relational Dimensions

Linda F. Edelman, Michael Bresnen, Sue Newell, Harry Scarbrough, and Jacky Swann

Keywords: Social capital, value creation, intellectual capital.

Abstract

Recently, there has been a shift in the way management scholars view the firm, from traditional models that are based on ideas of opportunism and the internalization of transaction costs (Williamson, 1975), to newer theories that argue for a socialized perspective (Kogut and Zander, 1992). One of the key components of these new theories is social capital. In this paper, we question the popular contention that the accumulation of social capital has a positive and proportionate effect on the performance of projects in organizations. To do this, we use data collected from over 38 interviews of senior and middle managers in five firms operating in different industrial sectors in the United Kingdom. Our findings indicate that while social capital has many beneficial effects with respect to information access and retrieval, there are also a host of less-beneficial aspects to utilizing social capital, which are under-explored in the current empirical literature. Implications and future research directions are discussed.

Introduction

Recently, there has been a shift in the way management scholars view the firm. While traditional models of the firm are based on opportunism and the internalization of transaction costs (Williamson, 1975), more recently, scholars have argued for a more socialized perspective in which "organizations are social communities where individual and social expertise is transformed into economically useful products and services" (Kogut and Zander, 1992: 384). These newer theories argue that the firm's principal purpose and source of competitive advantage is the creation and dissemination of firm-specific knowledge (Kogut and Zander, 1996). The implication here is that if firms can develop the capabilities for creating and sharing knowledge, then these capabilities can be used to generate organizational advantage. In other words, organizations can create value from their organizational knowledge, not merely appropriate it (Moran and Ghoshal, 1996).

Implicit in this new perspective of the firm is the notion of social capital. Social capital is defined as "the sum of the resources, actual or virtual, that accrue to an individual or group by virtue of possessing a durable network of more or less institutionalized relationships of mutual acquaintance and recognition" (Bourdieu and Wacquant, 1992: 119). Unlike financial or human capital, which can be possessed by a large number of people, social capital is unique. It resides in the structure of relationships between or among actors, making it a resource that does not lie with one individual, but instead is jointly owned (Coleman, 1988; Putnam, 1995). Therefore, it is a potential source of differential advantage.

Nahapiet and Ghoshal (1998), in their comprehensive review of the conceptual literature on social capital, divide the construct into three distinctive but clearly interrelated dimensions: structural, cognitive, and relational. Structural social capital refers to the ways in which motivated recipients gain access to actors with desired sets of knowledge or intellectual capital. Structural social capital is principally studied using a network approach. In network research, the frequency of contact and distance among actors in a particular firm or organizational field are plotted to form a web-like diagram illustrating actor interaction patterns. The objective in much of this research is to determine the central nodes of the network, in that developing an awareness of the location of critical communicators is helpful in understanding communication patterns as well as resultant organizational behaviors such as power positioning and knowledge flows (Brass and Burkhardt, 1992).

In contrast, cognitive social capital recognizes that exchange occurs within a social context, which is both created and sustained through relationships (Nahapiet and Ghoshal, 1998). Similar to the notion of community of practice (Brown and Duguid, 1991), cognitive social capital refers to the shared meanings that are created through stories and continual dialogues within a specific, often clearly defined group. These shared meanings are self-reinforcing in that participation in the community is contingent upon an a priori understanding of the context coupled with repeated contribution to these ongoing dialogues.

Finally, the relational aspects of social capital are concerned with the underlying normative dimensions that guide exchange relationship behaviors. Norms exist when the socially defined right to control an action is not held by the individual actor, but instead is held by others (Coleman, 1990). Therefore, norms represent a degree of consensus, and hence are a powerful albeit fragile form of social capital (Nahapiet and Ghoshal, 1998). Social capital theorists are particularly concerned with norms of cooperation and control such as trust and reciprocity. Specifically, norms of trust lead to enhanced cooperation, which in turn lead to increased trust. Similarly, reciprocity is an obligation, which when satisfactorily fulfilled, can lead to further reciprocal arrangements.

Nahapiet and Ghoshal (1998) argue that these three dimensions of social capital: structural, cognitive, and relational, operating in conjunction with the opportunity for combination and exchange in the organization, result in the creation of new intellectual capital. New intellectual capital is associated with organizational effectiveness and value creation.

In addition to Nahapiet and Ghoshal (1998), other conceptual work has examined social capital effects. Walker, Kogut and Shan (1997) discuss the role played by social capital in network formation while Burt (1997) posits that there are significant information and control effects in the use of social capital for senior managers. Leana and Van Buren (1999) discuss how employment practices can foster or discourage the development of potentially beneficial social capital in firms. Most recently, Adler and Kwon (2002) present a comprehensive framework for understanding social capital. In their framework, the use of social capital depends upon a wide range of factors, including the nature of the task, the existence of complementary resources, and the systems of norms and beliefs surrounding the interpretation and symbolic impact of social capital assets.

Many empirical studies that use a social capital perspective have recently appeared in the strategic management literature. These studies have tended to focus on the role of social capital in exchange relationships across organizations (Chung et al., 2000; Tsai, 2000), as well as within organizational activities such as resource allocation (Bouty, 2000), innovation (Tsai and Ghoshal, 1998), CEO compensation (Belliveau et al., 1996), and managing change (Gargiulo and Benassi, 2000). Another stream of social capital research has examined the interconnectedness of human and social capital in project-based firms (DeFilippi and Arthur, 1998) and in work/race relations (James, 2000). Finally, recent research has examined the differences in social capital usage across national boundaries (Burt et al., 2000).

In all of these studies, the accumulation of social capital is predominantly viewed as having a positive and proportionate effect on performance – in short, "more is better." Thus, the emphasis in current research is primarily on the benefits that can be achieved by participating in social networks and on the deliberate construction of social relationships for achieving these benefits. However, as Locke (1999) notes, this one-sided perspective has several limitations. He argues that loss of objectivity, the role of individual discovery, and the direction of the causal linkages between social and intellectual knowledge creation are all critical flaws in current social capital theory. According to Locke (1999), loss of objectivity is a function of actors becoming deeply embedded in an existing network. This can lead to the exclusion of new actors or ideas that are potentially beneficial. Locke (1999) cites the Japanese banking crisis as an example of a network in which the actors were more concerned with solidarity than they were with sound banking practices.

The role of individual discovery refers to the inherently individual process of knowledge creation. Locke (1999) argues that knowledge is created individually through the integration of sensory material through reason. In addition, he suggests that organizational advantage is not derived primarily from social interaction but instead that value creation is an intellectual process. Therefore instead of a feedback loop between social capital and intellectual capital as argued by Nahapiet and Ghoshal

(1988), Locke (1999) suggests that knowledge is discovered at the individual level and that this is followed by dissemination at the social level and eventual routinization at the organizational level.

In addition to Locke (1999), Adler and Kwon's (2002) framework suggests that there are risks to using social capital. They argue that these risks stem from three main dangers. First, there is the possibility of "over-investment" that transforms an asset into a liability (e.g., where actors become over-committed to specific relationships). Second, there is the possibility of generalized negative consequences of strong and beneficial localized ties (e.g., the exclusionary effects of cliques in organizations). Third, there is the possibility that the solidarity benefits of social capital may so tightly embed an actor into a particular relationship that it restricts the free-flow of new ideas and in the case of firms, it restricts the adoption of new innovations that are generated outside the network.

Finally, Burt (1992), Hansen (1999), and Adler and Kwon (2002) all argue that there is a cost dimension to utilizing social capital. Building and maintaining strong social capital ties requires an investment in time and energy that may not be justified given the potential for information redundancy. While non-redundant contacts can lead to information that is novel or new, redundant contacts take as much time to maintain and may not lead to new information (Hansen, 1999).

In this paper, we critically examine many of the underlying assumptions made in most conceptual and empirical studies relating to social capital. While we agree with the fundamental premise outlined in Naphiet and Ghoshal (1998), which is that social capital is a vital component in the creation of intellectual capital and hence may be a source of differential advantage, in our paper we strive to provide a more holistic perspective on social capital. We do this by exploring its beneficial as well as less-beneficial aspects. Instead of an unmitigated, "more is better" approach, as is so popular in much of the current scholarly research, our findings suggest that using social capital has both benefits and drawbacks. By this we mean that social capital is simultaneously *beneficial* in that it provides access to knowledge that may otherwise be difficult to obtain and concurrently *detrimental* in that it encourages the use of local search behaviors over other more extensive and potentially beneficial information-gathering activities. Thus, social capital extends relationships as well as channels further networking opportunities. The underlying premise of our work is that a balanced perspective is needed to fully understand the ramifications of using social capital, for, as our findings indicate, social capital is a necessary, but not sufficient, condition for value creation in organizations.

Research Design and Methods

This investigation was an exploratory qualitative study of five unrelated projects, for the purpose of understanding the processes by which project-based learning was created and disseminated. The unit of analysis was the project. What the five case studies give us that other research designs cannot is an intensive investigation of processes, which reveals the common patterns among projects. The limits of qualitative research involving a small set of cases are well documented: We do not know if the findings from this inquiry can be generalized to a larger population. The value of the research instead lies in its ability to provide insights through rich detail, and to provide directions for future inquiries.

The project did not begin as an exploration of social capital effects. Our initial mandate was to use interviewing techniques to better understand the ways in which projects take their learning and transfer it to other like groups within the firm. Interviewing was chosen as the method of investigation because there is a strong indication in the organizational learning and knowledge transfer literatures that the context in which the transfer occurs is extremely important in the transfer process (Argote, 1999; Szulanski, 1996). Only after the interviews were completed and the data analyzed did we realize the important and often contradictory effects that using social capital in the context of project-based learning had on the organizations in our sample.

Sample criteria and selection

Companies were chosen based on industrial sector. Five diverse industrial sectors are represented in the data: telecommunications, pharmaceuticals, health-care, construction, and social services. A sixth sector, automotives, was initially included in the research design, but the company declined to participate owing to a large number of internal changes that were ongoing during the time-frame of our inquiry. These particular industrial sectors represent a substantial portion of the UK economy, and hence were identified as critical for inclusion in a UK-based cross-sector inquiry. For the purposes of this study, the organizations are called Teleco, Drugco, Healthorg, Constructco, and Servicesorg. All organizations are located in the United Kingdom. In addition, all of the organizations have UK headquarters, with the exception of Drugco, which is a subsidiary of a US pharmaceutical company. All of the organizations are large, well-established companies, in that they each have been operating for at least 30 years and on average employ over 50,000 people.

While all of the organizations were in mature industries, they were operating under relatively different environmental conditions. For example, Teleco was operating in a rapidly changing environment and was undergoing a major cultural transformation as it moved into Internet-based competition. Conversely, Constructco's and Servicesorg's environments were relatively stable. Drugco and Healthco were also operating under conditions of some uncertainty, but these were due mainly to regulatory and other governmental changes and so were not of the radical proportion that Teleco was experiencing.

In each company, a specific project was chosen as the focus of our investigation. Projects were chosen by the organization, based on a set of guidelines set by the research team. Since we were interested in generic project-based learning issues, we asked each organization to provide us with a typical project. We also recognized the difficulties in comparing projects at different phases of their life cycle (Leonard-Barton, 1990), and so we requested a mature project that was well established in the organization. To our surprise, we found that projects and project teams are not synonymous, and that in two of the five cases the core project members consisted of one or two actors. These projects stand in contrast to two other projects that have

Table 6.1   Project characteristics

Project within organization	Number of individuals on the core project team	Project duration	Project complexity*
Teleco Technology-watch project	2	Ongoing	High complexity. Multiple sources and types of knowledge were used to assess the viability of a new technology. Project outcomes were not clearly defined.
Drugco Prostate cancer treatment	5	Ongoing	Moderate complexity. Team members have different experiences; however, within the project there were clearly identifiable goals.
Healthorg Cataract diagnosis and treatment process	7	Finite Project nearly completed	High complexity. Team was comprised of members from different disciplines each with separate concerns abut the project's outcome.
Constructco Role of the Regional Engineering Manager	12	Ongoing	Low complexity. Team members have similar backgrounds and experiences. Project has clearly identifiable goals.
Serviceorg Change in organizational structure	8	Finite Project nearly completed	Low complexity. Team members have similar backgrounds and experiences and were working on a clearly identifiable goal.

* To measure complexity, we follow Winter (1987), who defines complexity as comprising multiple types of firm competencies.

an identifiable team structure, identifiable objectives, and finite time duration. One project was a hybrid of the two project types. Table 6.1 summarizes the characteristics of the projects.

Data collection

In all cases, we began our investigation with an introductory meeting with a senior manager, administrator, or the director of the organization. While this person had a general familiarity with our interest in project-based learning, it was necessary for us to acquaint him with the particulars of this inquiry, and to help him in identifying an appropriate team for us to investigate. In each case save one, a suitable project was determined at this initial meeting. In one instance, a project was identified that failed to meet our "typical project" criteria and so another project was chosen. In conjunction with the introductory meetings, archival data was collected about each organization to help the researchers to understand the organizational context in which the project was situated. Web sites were accessed when available, and written documentation in the form of financial reports and/or press releases was requested.

After our initial meeting, we met next with the various project managers. It was at this meeting that we learned the details of each project as well as the names and contact information of the project members. To the extent that it was available, we collected archival project documentation, to include project process charts as well as sets of minutes from previous project meetings. Subsequent to this meeting, we met with numerous members of the project team. In all cases, the project team members interviewed had different roles and responsibilities on the team, thereby providing us with a holistic perspective on project-based learning for that group. In addition, regularly scheduled meetings were held within the research group to discuss our findings to date. In total, we interviewed thirty-eight individuals in five organizations over a seven-month period, logging thirty-five total interview hours. Table 6.2 provides a breakdown of the number of individuals interviewed per organization and the length of each contact.

While the interviews varied in length from one-half hour to over two hours, on average each interview lasted for approximately one hour and fifteen minutes. One of the interviews was conducted in a group format and included four different project participants. At another organization, researchers attended a project meeting. In all cases, every attempt was made to have two researchers present at each interview, although this was not possible for every interview owing to scheduling difficulties. Before each interview, interviewees were sent a letter describing the objectives of the research project and outlining the subject of the interview. At each interview, the researcher gave a brief example of knowledge transfer to help the respondent understand the general phenomenon of interest. All interviews were structured to maximize the respondent's ease of response. Respondents were asked initially to describe their role in the project, then to discuss the process of knowledge creation and transfer first within the project and then between other projects and other organizations, if applicable. At each interview, numerous open-ended questions were asked to encourage respondents to relate stories of how knowledge was created and transferred within and across similar projects within the organization.

Table 6.2   Interview summary

Organization	Interviews	
	Number of interviews	Hours spent interviewing
Teleco	8	8
Drugco	5	5
Healthorg	9	7.5
Constructco	8	7
Serviceorg	8	7.5
Total	38	35

All interviews followed a pre-designed interview protocol. The protocol included questions about the facilitators and barriers to knowledge transfer among project teams. Questions in the interview protocol were developed based on an extensive review of the knowledge management literature, a workshop in which senior managers from each of the five companies discussed project-based learning issues, as well as from the backgrounds in knowledge management research of the various research team members. Appendix C presents the interview protocol.

Data analysis

As is typical in inductive studies, writing the five case studies was an iterative process in which the data was constantly revisited (Eisenhardt, 1989; Yin, 1989). To aid in data consistency, the interview data was initially coded based on a coding scheme developed by the research team. Data coding was also an iterative process in which the research fellow and the team's four principal investigators searched the data for regularities and patterns and then recorded these key words and phrases to represent topics or themes which became the categories for further study (Bogdan and Biklen, 1992). It was our analysis of these themes that led us to focus on the contradictory role played by social capital. Within each category, if inconsistencies occurred among the data that was collected, third-party sources were consulted for clarification. Triangulation across the different sources of primary and archival data revealed a high level of data consistency.

After each case study was complete, the data was re-analyzed to develop the conceptual insights presented in this paper. While there were no preconceived hypotheses at the outset of the inquiry, patterns emerged from the data reflecting the contradictory nature of social capital in the creation and transfer of project-based learning.

Validity and transferability of the findings

Recent research suggests that the processes involved in analyzing qualitative data have motivated a change in the traditional frameworks that were used for ensuring the data validity (Erlandson et al., 1993). Specifically, to ensure validity, qualitative data must be checked against the criteria of credibility and transferability. In this study, data categories were identified across the five case studies to ensure data credibility. In addition, after each case study was written, it was sent to the individual project manager as a way to ensure the validity of the findings. Findings were also discussed with company representatives at a project workshop meeting held near the conclusion of the project. Data transferability was addressed by employing an interview protocol for collecting data that utilized questions that were primarily drawn from the existing literature. In relying on previous literature to ground our inquiry, future researchers examining project learning could potentially apply the interview protocol to their own studies.

Findings and Discussion

Characteristics of social capital

Unlike economic capital that resides in people's bank accounts and human capital that is inside their heads, social capital is found in the nature of personal relationships. To possess social capital, a person must be related to others, and it is those others, not he or she, who are the actual source of his or her advantage (Portes, 1998: 7). Social capital can be thought of as "know-who"; it is about everyone you now know, everyone you knew, and everyone who knows you even though you do not know them (Burt, 1992). Unlike financial and human capital, which can be possessed within an acceptable range by a large number of people, social capital is unique. As such, it is a source of differential advantage.

The emergent nature of social capital is one of the principal factors leading to the positive and unintentional negative effects that social capital can have on actors and organizations. Social capital is unplanned in that it arises out of day-to-day interactions. This implies that at the outset of any innovative activity, the social capital benefits accruing to a particular team or project are unable to be anticipated. Therefore, the social capital effects on both intellectual capital creation and subsequent innovation activities are likely to be unclear.

In addition, owing to its emergent nature, social capital cannot be readily appropriated as an organizational resource. This is due in part to the unique nature of social capital in that it is not held by individuals or organizations but instead is found in the nature of relationships. Therefore, the characteristics of social capital correspond closely to social interaction patterns, such as physical proximity, occupational affiliation, mutual interests, and informal relationships rather than to any planned strategic approach of a particular organization. However, social capital is an important part of organizational capital and hence it will reflect the historical evolution of organizational strategy.

Nahapiet and Ghoshal (1998) emphasize the beneficial effects of social capital on the creation of intellectual capital. In their work, they identify three dimensions of social capital: structural, cognitive, and relational. Our study of inter-project learning builds on their conceptual work. However, our findings indicate that a more equivocal perspective on social capital is needed to more fully understand the role that it plays in organizational value creation. While not invalidating Nahapiet and Ghoshal's (1998) analysis of social capital's important role in the development of intellectual capital, our findings indicate that social capital can operate in the organization in both beneficial and less-beneficial ways. Tables 6.3 and 6.4 present illustrative examples from the five cases in our inquiry.

Structural social capital: knowledge redundancy

The structural dimension of social capital in project-based learning refers to the ways in which motivated recipients gain access to actors with desired sets of knowledge or intellectual capital (Nahapiet and Ghoshal, 1998; Portes, 1998). The fundamental notion here is social networks, which refers to economic action that is seen as

Table 6.3 Benefits of social capital: examples from five case studies

Cases → Benefits	Teleco	Drugco	Healthorg	Constructco	Servicesorg
Structural Access to information which is difficult to otherwise obtain	Extensive use of internal personal social networks for information gathering.	Company developed networks across different medical professional groups as well as networks of internal salespeople.	Used networks to obtain information about cataract diagnosis and treatment processes in other NHS Trusts.	Network of regional engineering managers provides access to engineering knowledge and expertise.	Personal networks crucial to bringing in external knowledge needed for reorganization effort.
Cognitive Dialogues of shared meaning	Remaining employees have long-term employment and so had worked with numerous colleagues through several reorganization schemes.	Within sales group, success stories are important in creating and maintaining motivation levels. Efforts to share success stories was ongoing.	Created a community of disparate professionals that then created their own shared meanings.	Shared engineering language facilitates communication.	Similar training and backgrounds among employees led to ease of communication.
Relational Facilitates access for exchange, i.e., trust, reciprocity	Radical change in corporate culture has reinforced a strong sense of cohesion among remaining employees. High levels of trust and reciprocity.	Created context in which different medical professionals learned to trust one another. Lead to the establishment of a standing forum dedicated to cross-functional medical issues.	Created context in which disparate professionals learned to trust one another.	Strong norms of reciprocity and trust among regional engineering managers and between REM and construction site managers.	Consultative ethos of organization helped to dispel anxiety about changes in work processes.

Table 6.4 Drawbacks of social capital: examples from five case studies

Cases → Drawbacks	Teleco	Drugco	Healthorg	Constructco	Servicesorg
Structural Redundancy of contacts makes it difficult to obtain novel information	Significant employee exodus leaving gaps in internal organizational knowledge, individual networks incomplete leaving valuable knowledge inaccessible.	Numerous restructuring efforts have left gaps in knowledge networks.	Process reengineering team comprised of eye-care specialists who previously worked with the Trust.	Redundancy of information as all regional engineering managers have similar backgrounds and experience	Lack of staff movement to other career offices limited exchange of information among them.
Cognitive Exclude new sources of knowledge from outside the community.	Remnants of preexisting organizational culture created some lingering resistance to adopting externally generated innovations.	Highly politicized atmosphere in which all new knowledge was disseminated not only to the relevant sales employee but also to his/her country manager.	Overcome professional perceptions and barriers to create new knowledge.	Network of regional engineering managers used to the exclusion of other sources of knowledge within or outside company.	Staff had little experience in working in other sectors, thereby limiting their diversity of experience and ideas.
Relational Norms of control and compliance leading to resistance to innovation.	Remnants of former "make not buy" culture leading to "not invented here" attitudes with respect to information gathering efforts.	Internal resistance by country managers to innovative sales practices led to resistance to their adoption.	Overcome resistance by administrative staff to proposed changes in cataract process.	In regional offices, remnants of the former more adversarial corporate culture made carrying out the coordination part of the job of REM difficult.	Radical change in customer service orientation led to lack of trust towards managers by career counselors.

embedded in an ongoing set of relationships (Granovetter, 1992). It is these relationships, which are often established for other purposes, that form channels to information that is otherwise difficult to retrieve.

According to Burt (1992), networks provide three forms of benefits: access, timing, and referrals. Access refers to receiving information as well as knowing who can use that information, while timing refers to the timeliness of the information and referrals is about using the network as an information filter. The logic here is that information is asymmetrically distributed within and across organizations. Social networks provide access to unevenly distributed information. As with the more general concept of social capital, with networks more contacts are perceived to be better. To this end, employees who desire stronger networks join more committees, and attend more social functions.

Yet, paradoxically, network size is a mixed blessing. Burt (1992) reminds us that it is the diversity of the network, not its absolute size, that is important. Non-redundant contacts can lead to information that is novel or new. Conversely, while redundant contacts can lead to the same new information as non-redundant contacts, they take more time to maintain than do their non-redundant counterparts. For example, one of our cases, Constructco, developed and implemented a "regional engineering manager" (REM) role in its civil and build divisions in each of the regions that it services. While these regional engineers used each other as a first point of information when they had a problem to solve, each REM tended to be privy to the same sorts of information and therefore was less helpful with difficult or unusual problems. REMs were forced to expand their search for new knowledge outside the company, leading to problems with inaccurate information and subsequent sub-optimal solutions. These drawbacks of social capital were also apparent in the Servicesorg case in that low levels of turnover meant that employees were a homogeneously socialized group, whose social network was predominantly within the organization rather than outside it.

Similarly, Hansen (1999) highlights the effort required to maintain strong relationships, arguing that the efficacy of these in terms of knowledge transfer depends on the knowledge type. Strong ties, he suggests, may only be effective when the knowledge to be transferred is complex and largely tacit. However, explicit knowledge transfers easily through weak-tie relationships. This suggests that when organizations need to transfer complex non-articulable knowledge, the effort required to maintain strong, often redundant ties is well spent; however, such efforts are not justified when the knowledge to be transferred is explicit.

Just as organizational learning may lead to "competence traps" (Levitt and March, 1988), social capital may produce "relationship traps" in that utilizing social capital can lead to a disincentive to engage in other information search patterns. Consider, in our research, Teleco, whose mandate is to change from a traditional telephone provider to a dominant player in the revolutionary telecommunications marketplace. To do this, Teleco has changed its technology strategy from "make to buy." Instrumental to that change is the technology watch group, which identifies companies that are involved in technologies that Teleco perceives as necessary to its future development. The technology watch team uses its extensive internal network to draw upon the expertise of Teleco employees when deciding if a company is potentially interesting. However, using this internal network is potentially problematic in that it limits the other types of search activities that might be employed. For example, within Teleco a sophisticated information retrieval system was developed to help overcome local search behaviors, but according to one project manager, it is not widely utilized because people prefer *"personal e-mail, the coffee point, and meetings."*

> Proposition 6.1 Structural social capital provides access to asymmetrically distributed information; however, it can create disincentives to engage in other information search activities.

Cognitive social capital: social closure

While the structural dimensions of social capital refer to the ways in which knowledgeable actors are accessed, the cognitive dimensions recognize that exchange occurs within a social context that is both created and sustained through ongoing relationships (Nahapiet and Ghoshal, 1998). As such, meaningful communication is sustained through the ongoing dialogue of shared meanings among parties to the exchange (Boisot, 1995). This perspective is similar to Brown and Duguid's (1991) concept of community of practice. In a community of practice, knowledge is constructed as individuals share ideas through collaborative mechanisms such as narration and joint work. Within such communities shared means for interpreting complex activity are constructed, often out of conflicting and confusing data. It is this *process* of constructing meaning that provides organizational members with identity and cohesiveness.

While the concept of a community of practice is intuitively appealing, the boundaries of such communities are defined as much by those whom the community excludes as by who is included. Therefore, while social capital can tie projects within organizations together, it can also exclude new sources of knowledge that come from outside the boundaries of the social network (Portes, 1998). This is not to say that community boundaries operate as impermeable barriers to knowledge exchange; instead, by this we mean that the boundaries of a social network are continually re-created through repeated interactions among community members. Consider Drugco, for example, which is introducing a radical new procedure to treat prostate cancer that avoids the deleterious effects of both surgery and traditional beam radiation. The new procedure requires the expertise of both radiologists and urologists. In this case, the company has found that the difficulty in selling their product does not lie in getting patients to accept the new procedure, but instead lies in getting these two disparate groups of doctors, with different professional identities, to work together.

> Proposition 6.2. Cognitive social capital creates binding ties among community members; however, it can exclude new sources of information that originate outside the boundaries of the social network.

Relational social capital: resistance to change

The relational dimensions of social capital refer to the underlying normative structure of exchange. Coleman (1988) describes norms as "a powerful but fragile form of social capital." As such, norms have a significant influence on the exchange process in that they open up access to actors for exchange as well as ensuring the motivation to engage in such actions (Putnam, 1993; Nahapiet and Ghoshal, 1998).

Resources obtained through social capital have the character of a gift. As such, they are subject to norms of reciprocity. Under reciprocity norms, donors provide privileged access to resources in the expectation that that they will be fully repaid in the future, although the timing and form of repayment are unspecified at the time of the exchange. Reciprocity norms govern the relationship between the Regional Engineering Manager and the construction site managers in Constructco. In this organization, the REM must rely on the site managers to carry out his initiatives. The site managers are willing to do this, even though it means extra work, because they know that the REM is acting as an advocate for them in the home office.

While the norm of reciprocity is the most widely discussed norm with respect to social capital, other normative dimensions are also important. For example, social capital can form unwritten norms of control and compliance among actors in projects, thereby limiting beneficial innovative organizational activities in that they deter individuals from deviating from established norms. For example, Servicesorg introduced an innovative way of managing clients when they moved into their new flagship building. Instead of having individual offices and client appointments, counselors were now expected to hot-desk and to work with clients on a walk-in basis. Counselors were extremely resistant to this change, even though it meant better service to their various constituents. Traditional perceptions about client service had to be changed before this new innovative approach was fully accepted.

Alternatively, consider Healthorg, which was reengineering the process through which cataracts were diagnosed and treated. Traditional normative assumptions about the role of medical professionals needed to be altered before the new process was accepted. Through extensive contact and consensus building, perceptions about professional qualifications were slowly modified and preexisting normative barriers to innovation were broken down (Newell et al., 2003).

Proposition 6.3. Relational social capital creates strong normative forces that enable the exchange relationship; however, they also deter individuals from deviating from expected behaviors.

Interaction effects – structural, cognitive, and relational dimensions

In addition to the potentially detrimental individual effects of social capital on firm innovation activities, findings from our inquiry suggest that there are important interaction effects that further limit the effectiveness of utilizing a social capital approach. Our data indicates that in situations where an established group has developed a shared set of understandings and, along with that, strong norms of trust and reciprocity, then they are also likely to have developed both strong and multiple social linkages. While these strong links may facilitate knowledge movement within the group, they can also create strong and potentially damaging barriers around the group, shielding them from possibly beneficial knowledge and information that is outside the boundaries of the defined community. This strong barrier both makes it difficult for community members to access information outside the focal group and may also lead to skewed perceptions in that once individuals are fully indoctrinated into the community, they may not perceive the possible benefits of search activities outside of the group boundaries.

The "not invented here" syndrome (Katz and Allen, 1982), which is well documented in the new product development literature, illustrates the potentially damaging interaction effects between structural, cognitive, and relational social capital elements. Katz and Allen (1982) found that in research and development situations, strong social norms, reinforced by shared experiences, led to a propensity for internal information search as well as a lack of acceptance of new ideas that were generated from outside the group. Over time, this reluctance to accept externally generated ideas led to an overall decline in the level of firm innovation. Similarly, in our study, one of the principal barriers that Healthorg encountered in the development of their new cataract procedure was the need to overcome a general lack of trust of other healthcare professionals in different professional disciplines. Once the consultants broke their norms of professionalism and accepted that the optometrists were capable of handling their enhanced role in the new cataract diagnosis and treatment procedure, and in turn, the optometrists found the consultants willing to communicate with them on the diagnosis and treatment of non-routine situations, then the new cataract process could be implemented fairly smoothly.

Interaction effects can work to the detriment of an individual firm's efforts toward value creation. Remembering that social capital is a resource that is jointly held, then as organizational members leave the focal firm, they take their preexisting social capital ties with them. Individual network ties with current and former colleagues, which are based on norms of trust and reciprocity as well as on a common affiliation with a broader community of professionals, are often stronger than individual ties to the focal organization (Liebeskind et al., 1996). This can lead to potentially valuable information leaving the focal firm through these preexisting social networks. For example, in Teleco, there is ongoing radical organizational change, and many employees are leaving the focal firm to seek out more fulfilling, and potentially more lucrative positions in other aspects of telecommunications. This employee exodus, while potentially beneficial to Teleco in that it contains the possibility of using social networks to bring in new ideas and information that are external to the organization, can also be potentially damaging in that preexisting social capital ties may cause vital organizational knowledge to be disseminated outside the firm's boundaries.

Proposition 6.4. The structural, cognitive, and relational dimensions of social capital interact. These interaction effects magnify the benefits and the drawbacks of utilizing social capital.

Future Research and Conclusions

There is no doubt that social capital is an important dimension in value creation. Social capital is vital in the search and discovery of valuable and potentially difficult to obtain information. While we recognize the value of social capital, in this paper we argue that as a value creation mechanism, social capital has not been fully explored. Indeed there are instances where social capital can impede value creation by leading to aborted exploration processes, the exclusion of new knowledge, and organizational disincentives for adopting new innovations. We further argue that these less-beneficial aspects of social capital interact with each other, and that these interaction effects are mutually reinforcing, thereby creating powerful forces that have the ability to destroy any beneficial value creation efforts brought by using social capital.

These potentially deleterious effects of social capital may have been more transparent in our research because it was conducted across multiple industrial sectors. Our findings suggest that the social context in which social capital is created has a significant impact on its form and usage. For example, in a number of our cases we examined the attempted creation of intellectual capital across professional boundaries. As suggested by a social capital approach, these boundaries were powerful barriers to developing shared meanings as well as developing norms of cooperation and trust. While the actors in our cases did make substantial headway in overcoming many of these preexisting professional barriers, the new and significantly more efficient processes that they created were unable to be replicated in, for example, other health services. This was due in part to existing professional barriers in other national health trusts, which prevented the adoption of a cross-functional approach to health-care.

We suggest that future empirical research should continue this cross-sectoral approach to examining social capital. It seems quite likely that social capital is utilized differently in dynamic versus benign environments, large versus smaller organizations, or in short- versus longer-term projects (i.e., across different project life-cycle dimensions). In addition, research indicates that knowledge creation and dissemination is facilitated when homologous actors participate in the process (Lincoln and Miller, 1979; McPherson and Smith-Lovin, 1987; Marsden, 1988), thereby suggesting important and currently unexplored interaction effects between human and social capital.

Finally, we argue that current research should move beyond the qualitative and exploratory methods of this inquiry to include both grounded case study as well as deductive theory-testing techniques. Value creation activities are critical to the development of an organizational advantage, and hence the development of a more generalizable and predictive model of the paradoxical effects of social capital would therefore be of great scholarly value.

In addition to adding to scholarship, developing an enhanced perspective on social capital has several managerial implications. Given its paradoxical effects, we counsel managers to beware of adopting a narrow perspective on social capital. Our study suggests that while social capital is a potential source of differential advantage, it can also grossly hinder the value creation process by limiting trust, excluding new ideas, and providing sub-optimal solutions to problems. Moreover, as the contrast between our study and previous work indicates, the drawbacks to using social capital are not readily identifiable from any preexisting pattern of social relations. Instead, the beneficial as well as less-beneficial effects of social capital are principally contingent on the nature of the task, rather than inherent in the social network (Hansen, 1999). Therefore, we suggest that the drawbacks to utilizing social capital discussed here would be less problematic in relation to well-specified tasks, where appropriate forms of social capital can be predetermined and thereby cultivated appropriately. However, with innovation processes, the ex ante uncertainty of the scope and scale of the innovation task makes this a priori determination much more difficult. Therefore, it is this type of task where we would expect to find the unintended less-beneficial consequences of utilizing social capital to be most likely to appear. Since innovation processes often rely extensively on using social capital, we suggest that, paradoxically, social capital is the most difficult to manage when it is most critical to the competitive advantage of the firm.

Acknowledgment

The Engineering and Physical Sciences Research Council (EPSRC) provided funding and support for this project.

References

Adler, P. S. and Kwon, S.-W. 2002: Social capital: prospects for a new concept. *Academy of Management Review*, 27 (1), 17–40.

Argote, L. 1999: *Organizational Learning: Creating, Retaining and Transferring Knowledge*. Boston: Kluwer Academic Publishers.

Belliveau, M., O'Reilly C., and Wade J. 1996: Social capital at the top: effects of social similarity and status on CEO compensation. *Academy of Management Journal*, 39 (6), 1568–84.

Bogdan, R. C. and Biklen, S. K. 1992: *Qualitative Research for Education*. Boston: Allyn and Bacon.

Boisot, M. 1995: *Information Space: A Framework for Learning in Organizations, Institutions and Culture*. London: Routledge.

Bourdieu, P. and Wacquant, L. 1992: *An Invitation to Reflexive Sociology*. Chicago, IL: University of Chicago Press.

Bouty, I. 2000: Interpersonal and interaction influences on informal resource exchanges between R&D researchers across organizational boundaries. *Academy of Management Journal*, 43 (1), 50–66.

Brass, D. J. and Burkhardt, M. E. 1992: Centrality and power in organizations. In N. Nohria and R. G. Eccles (eds.), *Networks and Organizations: Structure, Form and Action*. Boston: Harvard Business School Press, 191–215.

Brown, J. S. and Duguid, P. 1991: Organizational learning and communities of practice: Towards a unified view of working, learning and innovation. *Organization Science*, 2 (1), 40–57.

Burt, R. S. 1992: *Structural Holes: The Social Structure of Competition*. Cambridge: Harvard University Press.

Burt, R. S. 1997: The contingent value of social capital. *Administrative Science Quarterly*, 42 (2), 339–66.

Burt, R. S., Hogarth, R. M., and Michaud, C. 2000: The social capital of French and American managers. *Organization Science*, 11 (2), 123–47.

Chung, S., Singh, H., and Lee, K. 2000: Complementarity, status similarity, and social capital as drivers of alliance formation. *Strategic Management Journal*, 21 (1), 1–22.

Coleman, J. S. 1988: Social capital and the creation of human capital. *American Journal of Sociology*, 94, S94–S120.

Coleman, J. S. 1990: *Foundations of Social Theory*, Cambridge: Harvard University Press.

DeFilippi, R. and Arthur, M. 1998: Paradox in project-based enterprises: the case of filmmaking. *California Management Review*, 40 (2), 125–40.

Eisenhardt, K. 1989: Building theories from case study research. *Academy of Management Review*, 14, 532–50.

Erlandson, D., Harris, E., Skipper, B., and Allen, S. 1993: *Doing Naturalistic Inquiry*. Newbury Park. CA: Sage Publications.

Gargiulo, M. and Benassi, M. 2000: Trapped in your own net? Network cohesion, structural holes and the adaptation of social capital. *Organization Science*, 11 (2), 183–96.

Granovetter, M. 1992: Problem of explanation in economic sociology. In N. Nohria and R. G. Eccles (eds.), *Networks and Organizations: Structure, Form and Action*. Boston: Harvard Business School Press, 25–56.

Hansen, M. T. 1999: The search transfer problem: the role of weak ties in sharing knowledge across organizational sub-units. *Administrative Science Quarterly*, 44, 82–111.

James, E. H. 2000: Race-related differences in promotions and support: underlying effects of human and social capital. *Organization Science*, 11 (5), 493–508.

Katz, R. and Allen, T. J. 1982: Investigating the Not Invented Here (NIH) syndrome: a look at the performance, tenure, and communication patterns of 50 R&D project groups. *R&D Management*, 12 (1), 7–19.

Kogut, B. and Zander, U. 1992: Knowledge of the firm, combinative capabilities and the replication of technology. *Organization Science*, 3 (3), 383–97.

Kogut, B. and Zander, U. 1996: What firms do? Coordination, identity, and learning. *Organization Science*, 3 (3), 502–18.

Leana, C. R. and Van Buren, H. J. 1999: Organizational social capital and employment practices. *Academy of Management Review*, 24 (3), 538–55.

Leonard-Barton, D. 1990: The intraorganizational environment: point-to-point versus diffusion. In F. Williams and D. V. Gibson (eds.), *Technology Transfer: A Communication Perspective*. Thousand Oaks, CA: Sage Publications, 43–62.

Levitt, B. and March, J. G. 1988: Organizational learning. *Annual Review of Sociology*, 14, 319–40.

Liebeskind, J. P., Oliver, A. L., Zucker, L., and Brewer, M. 1996: Social networks, learning and flexibility: sourcing scientific knowledge in new biotechnology firms. *Organization Science*, 7, 428–42.

Lincoln, J. R. and Miller, J. 1979: Work and friendship ties in organizations: a comparative analysis of relational networks. *Administrative Science Quarterly*, 24, 181–99.

Locke, E. A. 1999: Some reservations about social capital. *Academy of Management Review*, 24 (1), 8–9.

Marsden, P. V. 1988: Homogeneity in confiding relations. *Social Networks*, 10, 57–76.

McPherson, J. M. and Smith-Lovin, L. 1987: Homophily in voluntary organizations. *American Journal of Sociology*, 52, 370–9.

Moran, P. and Ghoshal, S. 1996: Value creation by firms. In J. B. Keys and L. N. Dosier (eds.), *Academy of Management Best Papers Proceedings*. Atlanta, GA: Georgia Tech University, 41–5.

Nahapiet, J. and Ghoshal, S. 1998: Social capital, intellectual capital, and the organizational advantage. *Academy of Management Review*, 23 (2), 242–66.

Newell, S., Edelman, L. F., Scarborough, H., Swann, J., and Bresnen, M. 2003: 'Best practice' development and transfer in the NHS: the importance of process as well as product knowledge. *Health Services Management Research*, 16, 1–12.

Portes, A. 1998: Social capital: its origins and applications in modern sociology. *Annual Review of Sociology*, 23, 1–24.

Putnam, R. D. 1993: The prosperous community: social capital and public life. *American Prospect*, 13, 35–42.

Putnam, R. D. 1995: Bowling alone: America's declining social capital. *Journal of Democracy*, 6, 65–78.

Szulanski, G. 1996: Exploring internal stickiness: impediments to the transfer of best practices within the firm. *Strategic Management Journal*, 17, 27–43.

Tsai, W. and Ghoshal, S. 1998: Social capital and value creation: the role of interfirm networks. *Academy of Management Journal*, 41 (4), 464–76.

Tsai, W. 2000: Social capital, strategic relatedness, and the formation of intraorganizational linkages. *Strategic Management Journal*, 21 (9), 925–40.

Walker, G., Kogut, B., and Shan, 1997: Social capital, structural holes, and the formation of an industry network. *Organization Science*, 8 (2), 109–28.

Williamson, O. 1975: *Markets and Hierarchies: Analysis and Antitrust Implications*. New York: The Free Press.

Winter, S. G. 1987: Knowledge and competence as strategic assets. In D. Teece (ed.), *The Competitive Challenge*. Cambridge, MA: Ballinger, 159–84.

Yin, R. K. 1989: *Case Study Research: Design and Methods*. Newbury Park, CA: Sage Publications.

Appendix C Interview protocol

Beginning of interview:
Can you tell me a little about the project on which you are working?

Type of project – Duration of project – Origins/history of project? (e.g., politics, organizational goals)

How were the project participants selected? (i.e., recruitment, analysis of skills, based on social capital contacts?)

Is this the way project participants are normally selected?

What is your role in the project?

Project Members:

How many project members are involved in project? Core team? Other?
Do the project members come from inside or outside of the company?
From which function(s) do the project members come?
How much do project members change over the life cycle of the project?
Do these members stay exclusively with the project or do they simultaneously work on a
     number of projects?
How are new project members trained and provided with an orientation to the work of the
     project team? (i.e., mentoring, working with other project members)

Tacit/Explicit Knowledge:

What type of documentation is created about the project?
Where is this documentation stored?
Do other projects have access to project documentation?
What is the primary mode of communication among project members?
Is most of the knowledge that you need to get from another project able to be written down?
Do you need to work closely with other project members to understand what they are doing?

Knowledge Redundancy:

Do careers in this company tend to develop within one specialist function or do people move
     between functions over time?
Is there a set of training that is the same for every new recruit?
How similar are the projects on which the project members are working?

Situated Learning:

When you need knowledge from another project, would you generally go to where that
     project is located, or would they come to you?
How important is situation in which the knowledge was created?

Absorptive Capacity:

When transferring knowledge or information, do you find that the recipient does not
     have the *a priori* skills necessary to understand what you are transferring? Can you give
     an example?
When receiving knowledge or information do you find that your project does not have
     the appropriate skills necessary to understand what is being transferred? Can you give an
     example?
What effect does language barriers have on effective transfer (i.e., either different language or
     unfamiliar word usage)? Is this a problem for your project?

Time:

Is there ample time to transfer knowledge in your firm?
Is transfer a planned organizational activity, as would be other important organizational
     activities?

Transfer Tools:

What IT systems do you have in place to facilitate transfer?
Any other transfer facilitators? (e.g., newsgroups, company-wide yellow pages, company news
letters, message boards)
What effort is made at your company to capture and catalog organizational knowledge? How
has this been done? Has this been successful?

Organizational Structure:

What does your organizational structure look like? (i.e., flat, hierarchy, matrix)
When you have an issue do you need to go through the proper channels or generally can you
go to the person who may have the knowledge to solve your problem directly? (i.e.,
decentralization of decision making or to informality of communication)

Organizational Culture:

Would you consider your organization to be open and conducive to sharing knowledge?
(an example here would be helpful)
If you have a problem would you tend to go inside the organization to solve it or to go
outside to consultants, for example?

Management Support:

Is management supportive of capturing and supporting learning? How do they show their
support?
Do they provide time for effective transfer?
Has there ever been formal or informal discussions about the importance of inter-group
knowledge transfer? Have they worked/been effective?

Social Capital:

What kind of networks are typically used during project work to bring outside knowledge
into the project (i.e., interpersonal, consultants, electronic, organizational,
interorganizational)?
What other sources of knowledge are drawn upon to facilitate project-based work?
Why are the networks that are used, chosen over other available alternatives?
Do you use your non-work contacts or friends to gain (or spread) work-related information?
When you receive knowledge from an individual outside the project, is there an expectation
that they will receive knowledge from you at an unspecified future date?
When you are gathering knowledge from outside the project team, who are you most likely
to go to first (i.e., colleague from your functional area, colleague with whom you have
previously worked)?

Organizational Incentives:

Are there any specific tangible or intangible incentives for sharing knowledge? (i.e., bonuses,
promotion, organizational recognition)?
Does the firm have in place an employee share ownership program?

Does the firm have in place a profit-sharing program (i.e., profit-related pay)?
Does the pressure of achieving your personal/project targets encourage or discourage the sharing of knowledge and information?

Motivational Issues:

If you were the source of the knowledge transfer, did you find any resistance to its adoption from the recipient? Can you provide an example?
If you were the recipient of the knowledge transfer, did you find any resistance to its being shared from the source? Can you provide an example?

Transfer Outcomes:

How would you define success for this project?
How would you know if a cross project knowledge transfer in your organization has been a success?
Does your company have any way of measuring transfer success?
Does your company discuss learning as a project outcome?
Does your company have any way to measure learning as an outcome?

Examples of Transfer:

Can you provide me with an example of when individuals in your project were the recipient of learning (knowledge) from other projects?
With respect to this transfer, did the transfer process meet your a priori expectations?
Were you satisfied with the project-based outcomes of the transfer? (e.g., on-time, on-budget)
Were you satisfied with the learning outcomes of the project?

Can you provide me with an example of when individuals in your project were the source of learning (knowledge) for other projects
With respect to this transfer, did the transfer process meet your a priori expectations?
Were you satisfied with the project-based outcomes of the transfer? (e.g., on-time, on-budget)
Were you satisfied with the learning outcomes of the project?

Can you provide me with an example where learning across projects was not successful? What do you think are the reasons for the lack of success?

Wrap-up:

What else do I need to know to understand the dynamics of inter-project learning at your company?

Thank you very much for your time.

CHAPTER 7

Strategic Innovation in Financial Services

Laura A. Costanzo

Keywords: Innovation, market positioning, strategic management.

Abstract

It has been suggested that innovation in the services industries is unsystematic and tends to coincide with organizational learning (Sundbo, 1997, 1998). This paper offers an alternative perspective by applying an exploratory approach to the study of innovation in the UK financial services. For this purpose the strategic and organizational profiles of two groups of companies defined respectively as the "fast innovators" and the "non-innovators" are compared and contrasted. It is revealed that the "fast innovators" innovate systematically by adopting a strategy of "continuous innovation." This is a "broad" market-oriented strategy. In such context, innovation is systematic and tends to coincide with both types of exploitation and exploration learning. The paper suggests that within the group of "fast innovators" innovation is not the exclusive domain of a particular functional area or individual. Instead, it is a broad collective business function that embraces the whole organization.

Introduction

In the past decade, the UK financial services sector has undergone a period of rapid change mainly driven by rapid technological development. Especially in the past five years, the Internet has caused increasing competition, with new players coming also from non-financial services. Both new and established players are aware of the need to innovate in order to secure competitive advantage in the future. Particularly, the executives of financial incumbents – the so-called dinosaurs – seem to be overwhelmed by the need to rejuvenate their companies. However, when asking the question "how can financial institutions innovate?" or "how does innovation actually occur within the single individual institution?" business executives can hardly find an answer. In the UK financial services industry, particularly among established players, there is common opinion that it is very difficult to innovate. The reasons generally addressed for this are that the financial products can be easily copied because of lack of a patent system, that early high-tech developments can be too risky and costly, and therefore affect negatively the City's views. In the financial services industry, it is perceived that innovation is not really useful for the purposes of differentiation. Is this a result of the way that these companies innovate?

A look at the existing literature on innovation reveals that the existing studies have not yet addressed this issue adequately. While studies on innovation are quite extensively developed in the manufacturing industry, existing studies on innovation in financial services seem to be quite limiting, as they seem to be mainly concerned about macroeconomic aspects of financial innovation. To date, they do not seem to have adequately addressed the issue of how financial services organizations innovate. Given this background and the new challenges that the new economy poses to both new and established financial players, the purpose of this paper is to provide a greater understanding of how financial services organizations innovate, and whether innovation effectively enables them to achieve differentiation from the competition. As a strategic perspective is applied to the study of this phenomenon, this paper also aims at contributing further theoretical insights into the existing literature of strategic innovation. In order to achieve these objectives, the paper is structured in the following way.

First, we will provide a review of the main theoretical contributions developed in the field of innovation, with a particular emphasis given to the emerging strategic theory of innovation. This will provide a background to research to date. Then we will briefly outline the research methodology adopted for the data collection and analysis. Particularly, given the exploratory nature of the study, it emphasizes that an inductive methodology has been applied. Next, we will outline the research findings. This is followed by a discussion of their implications for the theory of strategic management. Finally, the major conclusions are drawn.

Literature Review

The general theory of innovation

The general theory of innovation has its foundation in Schumpeter (1934) who defines innovation as the "carrying out of new combinations." Early studies of this phenomenon mainly focused on the industrial sector, with a particular regard to the relationships between the degree of innovation and factors such as firm size, intensity of R&D, and market power. A distinction between product and process innovation is also present in the general literature developed on innovation. Process innovations generally involve all those changes affecting the methods whereby outputs are produced, while product innovations are just aimed at the outputs themselves (Schmookler, 1966; Scherer, 1980; Mansfield, 1968). Also, distinctions are made between incremental and radical innovations in relation to the degree of newness. Hence, while an incremental innovation is a cumulative series of minor changes, a radical innovation is a major change (Tidd et al., 2001).

When looking at the topic of innovation in financial services, there seems to be a lack of research. A literature in the field of financial innovation started to develop in the early 1980s, when deregulation appeared to be one of the drivers of changes in the UK financial services. Therefore, some of the studies were mainly concerned with the macroeconomic aspects of financial innovation, for example how new financial instruments (options, swaps, etc.) could affect the efficiency of the UK financial system (Llewellyn, 1992). In this regard Llewellyn's (1992) approach follows the early approach of Silber (1975) who argues that financial innovation responds to economic forces in order to remove or lessen the financial constraints imposed on financial institutions. However, to date the limited literature does not clarify how innovation occurs in the financial services sector from a microeconomic point of view. Recent research conducted by Sundbo (1997) in the field of innovation in the services industry may represent a good references' source to refer to. In his studies, Sundbo concluded that innovation takes place in service firms, although it may empirically be difficult to distinguish it from organizational learning. He also stated that innovation theories developed from studies of the manufacturing sector are applicable to the services industry. Based on his studies of Danish service firms, Sundbo (1997) identified three main theoretical paradigms for the study of innovation: the entrepreneur paradigm, according to which the entrepreneur is the only person responsible for innovation; the techno-economic paradigm, which implies a shift from an individually based innovative activity centered on the figure of the entrepreneur to an innovation process characterized by the formalization of the innovation activity within firm-based R&D laboratories; and the basic strategic theory, which explains innovation as determined by the company's relationship with the customers and the market (Sundbo, 1998). For Sundbo, the basic strategic paradigm is not still defined clearly, but it represents the most adequate paradigm to explain innovation in today's business environment. In the present study, the most recent and significant developments of this strategic paradigm are presented in the next section.

The potential strategic theory of innovation

The strategic paradigm of innovation is a development of the early marketing theories and later theories of service management and quality management (Sundbo, 1998). During the 1980s, there was a growing emphasis on the market as the main determinant of innovation, whereby the marketing function assumed a prominent role over the R&D function. In this regard, an early contribution to the marketing theories derives from Levitt (1960) who argued that companies needed to develop the capability of market analysis rather than R&D capabilities, if they wanted to achieve higher growth. In the context of the marketing literature (Baker, 1985; Kotler, 1983, 1984), innovation is mainly concerned with new product development and it seems to play a defensive role as a reaction to the market conditions and solution to the problem of decreasing sales (Sundbo, 1998). In the marketing theories, the innovation activity is generally restricted to the sales activity – what can be currently sold in the market.

In the strategy theory, the innovative activity goes beyond the concept of new product development. It is a broader process entirely controlled by the top management (Sundbo, 1998). The external company situation, indeed, is not just concerned with the understanding of customers' needs, but also with competitors' strategies (Sundbo, 1998). However, as Sundbo (1998) observes, the strong analysis of the competition could lessen the ability of the company to develop its internal capabilities for innovation.

Over the past 20 years, the dilemma of whether strategies should focus on the development and leveraging of unique core competencies and capabilities (resource-based theories) or should just focus on the market (market-based theories) has represented a constant theme of debate in the strategic management literature. Different schools of thought have developed within these extreme positions. Mintzberg et al. (1998) have identified two main categories of schools developed during the period 1965–95, the "prescriptive schools" as contrasted with the "describing schools." The prescriptive schools are generally "concerned with how strategies should be formulated rather than with how they necessarily do form" (1998: 5). Belonging to this category are the design (Andrews, 1971), planning (Ansoff, 1965), and positioning (Porter, 1980) schools that were dominant during the 1960–80 period. The "describing schools" are concerned with "describing how strategies do get made" (Mintzberg et al., 1998: 6). Within this group, six schools of thought – entrepreneurial, cognitive, learning, power, cultural, and environmental – developed. These six schools started to be dominant during the 1980s and towards the 1990s, a period of rapid economic change. Particularly, the entrepreneurial school (Pinchot, 1985; Drucker, 1970) conceives the strategy process as a process of vision creation by the entrepreneur, while the cognitive school (Reger and Huff, 1993; Bogner and Thomas, 1993; Duhaime and Schwenk, 1985) is aimed at understanding strategic vision, and how this forms in the mind of the strategist. The other describing schools take into consideration other factors influencing the process of strategy, beyond the individual. For example, for the learning school (Lindblom, 1959; Wrapp, 1967; Quinn, 1980; Mintzberg, 1972, 1978; Weick, 1979) the world is too complex, such that it is not possible to formulate strategies through a clear plan. In this case strategies emerge as an organization learns. Similarly, the power school perspective (MacMillan, 1978; Sarrazin, 1975, 1977–78; Pettigrew, 1977; Bower and Doz, 1979) regards strategy as a process of negotiation between conflicting groups within an organization, or between different organizations as they try to react to the constraints coming from the external environment. In contrast with this view, the culture school (Feldman, 1986; Barney, 1986; Schwartz and Davis, 1981; Wernerfelt, 1984; Prahalad and Hamel, 1990; Prahalad and Bettis, 1986) considers strategy formation rooted in the culture of the organization, whereby the process is seen as a collective process. The environmental school (Miller, 1979; Hannan and Freeman, 1977; Meyer and Rowan, 1977) perspective sees strategy formation as a reactive process to the conditions in the external environment. The final group of Mintzberg et al. is the configuration school (Mintzberg, 1979; Kotter, 1995; Miller, 1976), which tries to combine all the best elements of each school into distinct stages or episodes, which reflect the life of organizations.

Recent theoretical developments recognize the need for more of a balance between the extreme positions of the market-based and resource-based theories (Mintzberg et al., 1998). There are different reasons for this observed need:

1   On the one hand, it is observed that the economic environment has become unstable, and the traditional processes used for strategy formulation are not flexible enough to cope with the rapid changes in the economic and business environment mainly driven by technological change (Mintzberg et al., 1998). Although Porter's concepts of value chain and competitive advantage are still influential, the massive growth of e-commerce has made value chains more open and dynamic than they used to be in the past (Donlan, 1999), when they were just regarded as simple static sequences of physical activities.

2   On the other hand, the excessive focus on the internal resources of the firm seems to be inadequate in a period of rapid change in the economic and business environment. It is a one-sided approach, which is in danger of over-emphasizing internal analysis to the extent of neglecting the problems that the accelerating rate of change in the environment poses for organizations. While fully accepting that organizations continually need to adapt to change by developing their core competencies, it is also evident that the analysis of the environment is now even more vital in order to know how to adapt (*Management Accounting*, 1998).

During the last decade, the topic of innovation has also been addressed more specifically within the context of the general literature of strategic management. Particularly, recent studies have increasingly emphasized the need for companies to shift their strategic focus from incremental innovation to radical innovation. Incremental strategic innovation is the logic behind many programs of continuous improvement aimed at achieving the objectives of cost reduction and efficiency. These programs have dominated the boardrooms' decisions to the extent that incremental strategic innovation became the strategic imperative of the last decade. However, in the new age, companies are increasingly exhorted to move beyond incrementalism and embrace radical strategic innovation or "non-linear innovation" (Hamel, 1998a). For example, Hamel (1998a) argues that in the new economy "... *innovation is concerned with the strategy to reconceive the existing industry model in ways that create new value for customers, wrong foot competitors, and produce new wealth for all the stakeholders.*" Markides (1999) argues in a similar way by introducing the concept of "*breakthrough strategy.*" As Markides (1999) clarifies, this is the kind of strategy that successful companies pursued to innovate in their industry. A "*breakthrough strategy*" involves making tough decisions on the "*who-what-how*" dimensions that define the company's unique strategic position in the industry (Markides, 1999: 3). Similarly, Kim and Mauborgne (1997) use the expression "*value-innovation logic,*" to indicate a strategy aimed at shaping the industry's conditions by bringing to it a "*quantum leap in value buyer,*" whereby the analysis of the competition becomes irrelevant. The *"value-innovation logic"* brings managers to look across different industries' boundaries, and in doing so they can find unoccupied market space (Kim and Mauborgne, 1999). Significant is the point made by Geroski (1998) in relation to the ability of a company to create a new market space. His argument is that new market space is not created by new technology, yet it is born from strategic innovation, that is, the capability of a firm to redefine its customer base and its market position vis-à-vis its rivals. For this purpose, Hamel (1998a) argues that companies will have to make radical strategic innovation a systemic capability, like the quality revolution of 20 years ago. Loewe et al. (2001) propose that companies should harness their resources and energy to the cause of continuous innovation. Particularly, in fast-changing industries – i.e., computing and telecommunications – with short product cycles and rapidly shifting competitive landscapes, the ability to engage in continuous change becomes a crucial capability for survival (Eisenhardt, 1989a).

However, current practice of successful companies still shows that soon after the introduction of an innovative strategy, successful companies become committed to incremental change – making the existing strategy better. In other words, little or no effort is devoted to discovering new ways of playing the game. This attitude is typical of the established companies whose vast experience accumulated over the years acts as a barrier to unlearn received wisdom (Kumar et al., 2000). Organizational learning and successful performance may, in fact, prevent firms from spotting opportunities that would lead them beyond programmatic practices to occupy the domain of breakthrough innovation (Colarelli O'Connor and Rice, 2001). Also, the commitment of top management to the pursuit of efficiency inadvertently drives out the unconventional strategic thinking that is the very fuel of radical innovation (Hamel, 2001). Therefore, it is argued that radical innovations generally come from outsiders whose strategic thinking is not unencumbered by conventional business assumptions (Campbell and Collins, 2001). Also, it is suggested that radical innovations will have to be developed in separate business units – "greenfield" operations. This would avoid the problem of tension in the resource allocation between the new and the old business (Christensen, 1997).

However, "greenfield" operations carry with them a strategic risk – to split the company's commitment to radical innovation, which is core to the future business strategy, from the commitment to incremental innovation, which is core to the current business strategy (Abell, 1999). The separation of the business unit devoted to radical innovation from the rest of the organization is actually in conflict with some recent studies (Hamel, 2000) that underline a need of continuity for radical strategic innovations. Indeed, the "greenfield" organizational approach is criticized because *"dedicated innovation units have a purpose and are not a substitute for an innovation pipeline overflowing with ideas for revitalizing the core busines"* (Hamel, 2001). Also, the "ghetto-incubator" – the dedicated innovation unit – bears the danger of causing a drain of brain and entrepreneurial talent from the core business: *"anybody who has any passion can migrate out of the core business to the ghetto-incubator!"* (Hamel, 2001).

The reality is that, in practice, established companies already have too many obligations and face too many risks to justify their commitment to radical strategic innovation. Bearing this in mind, it is suggested that these companies, while devoting the overwhelming majority of their innovation efforts to incremental innovation and traditional market research, should also devote some of their time to radical business innovation (Kumar et al., 2000). In other words, established companies would need to have a balanced portfolio of incremental innovation and radical innovation. For this purpose, firms need to become ambidextrous, capable of simultaneously managing incremental, as well as radical innovation (Tushman and O'Reilly, 1996). Similarly, Abell (1999) proposed that companies should always have in place dual strategies, a "today for today strategy" and a "today for tomorrow strategy." While recognizing that in periods of rapid change the future component might receive more attention, Abell (1999) admits that, in any situation, the two components should be addressed simultaneously.

Taking into consideration all these recent contributions to the theory of innovation, this paper applies a strategic approach to the study of innovation in the UK financial services by applying an inductive methodology. An overview of this is provided in the next section.

Methodology, Data Collection and Analysis

This study applies the logic of inductive inquiry. The main reason for this is that exploratory fieldwork is appropriate to research a phenomenon-which has been poorly researched before or shows a lack of extant theory and data (Glaser and Strauss, 1967). Also, such a method is appropriate when the phenomenon-under investigation – the strategic and organizational mechanisms leading to innovation – cannot be easily measured from a quantitative point of view (Strauss and Corbin, 1990; Yin, 1993). Moreover, exploratory fieldwork enables the researcher to formulate theoretical propositions that might become the object of further research (Noda and Bower, 1996). Particularly, the underlying logic of the research presented here is "grounded-theory" building through the case study method. There is no uniformity of perspectives regarding the number of case studies that researchers should consider. For example, on one extreme, Dyer and Wilkins (1991) contend that a "deep single case study" is the optimum form of case study research, as this is more suitable to reveal the more tacit and less obvious aspects of the social context under investigation. In contrast, Eisenhardt (1989b) contends that researchers should use multi-case studies and search for underlining constructs and that they should stop adding cases when their understanding is saturated. Eisenhardt suggests that a number of cases between four and ten generally works well. Here, the research design is based on multiple case studies and the application of the replication logic (Yin, 1984). This logic enables the researcher to look for common emerging insights across a number of case studies. In other words, each case is analyzed in depth in order to confirm or not theoretical insights emerging from case studies analyzed previously. The dataset includes twelve UK-based financial services organizations, generally identified in four main categories: established banks, building societies, insurance companies, and a mixed category of new entrants into the market. Table 7.1 outlines the companies[1] included in the study.

Table 7.1 Company details

Company name	Strategic profile	Employee nos
Fast innovators		
Money First (MF)	Credit card company (new entrant)	16,000
The Helping Bank (HB)	Established bank	4,010
Building Society B (BSB)	Building society	3,045
Sunshine (SS)	Internet bank (new entrant)	1,876
Focused innovators		
Phone Banking (PB)	Telephone bank (new entrant)	4,000
Star Insurance (SI)	Insurance company	3,000
Non-innovators		
STA Bank (STAB)	Established bank	31,268
Sassoon Bank (SSB)	Established bank	85,847
Building Society A (BSA)	Building society	8,924
Building Society C (BSC)	Building society	2,717
Global Insurance (GI)	Insurance company	3,511
Golden Insurance (GoI)	Insurance company	72,749

Data collection

Altogether 59 in-depth face-to-face interviews, five for each company, were carried out at the level of top-middle management. For Star Insurance only four informants took part in the interview process. On some occasions, two senior informants working in two distinguished key roles in the same department were interviewed. The initial informal contact with key people within the organizations facilitated the identification of potential informants for the interviews. The informal contact with key people within the organizations was useful in terms of gaining support from the top and getting the informants on board. An average of five years' experience at a senior level as head of a department was the main criterion for the selection of the informants. A list of interviewees is presented in Appendix D.

Primary data were collected through semi-structured interviews with open-ended questions, while secondary data from magazines and newspapers were collected prior to the visit to the sites. Secondary information gathered from internal documentation was collected during the visit or supplied by the informants soon after the visit. The interviews were conducted on the basis of several trips to the sites. Once collected, they were tape-transcribed. Each interview lasted on average one or two hours. In order to ensure that the same themes were covered in each interview the protocol interview reported in Appendix E was put in place.

Data analysis

The data were analyzed by first building case studies for each of the companies included in the sample. Each case was written on the basis of the information provided by the informants during the interview process and secondary data collected on the companies. Once the case studies were completed, a cross-case analysis (Miles and Huberman, 1984) was applied in order to identify common insights across all the cases. The aim of this cross-case analysis was to identify unique characteristics for each of the case studies included in the sample, but also common features so that clusters of companies presenting a similar strategic behavior could be built. Each time that a pass was completed in the identification of an emerging insight, it was necessary to go back to the interviews and additional secondary information to make sure that the emerging insights were consistent with the data. This was a lengthy process that lasted over 12 months. The reason for this was that during the analysis several breaks were taken in order to refresh the process of ideas development.

Research Findings

Three distinguished patterns of strategic behavior emerged from the data analysis. The first pattern is related to the group of companies (MF, HB, BSB, SS) who made "continuous innovation" central to their strategy of competitive advantage. While these companies addressed both the resource and the market perspectives in their strategic approach to innovation, the focus on the market appeared to be more critical than the focus on the internal resources. The companies included in this group tended to be fast moving and react quickly to changes in the external environment. Hence, these companies will be identified with the terminology of "fast innovators" in the rest of the paper.

The second pattern is related to a group of companies (PB, SI) who made "radical innovation" central to their strategy of renewal and differentiation from the competition. While these companies were initially focused on the external environment, they increasingly shifted their strategic focus to the internal resources. A consequence of this was that "continuous improvement" of the existing products and services became their main strategic focus. In a sense, continuous improvement took the place of radical innovation. As these companies were highly focused on the improvement of their existing business, they will be identified with the terminology of "focused innovators."

Finally, the third pattern is related to a group of companies who did not conceive innovation as central to their strategy of competitive advantage. They generally regarded innovation only as a way of improving existing products and services and/or internal processes. They were aware that this type of innovation, mainly incremental, could be easily imitated from the competition and, therefore, was not useful to their competitive advantage. Business consolidation rather than business innovation was the main focus of these companies, whose strategic behavior was mainly adaptive to the evolution of the external environment. These companies will be identified with the terminology of "non-innovators" in the rest of the paper.

The two remaining companies, STA Bank and Golden Insurance, did not present a distinguished strategic profile. Their strategic behavior appeared to be a mix of some of the strategic features of each of the three groups described above. After careful revision of the interviews, the decision to omit the case studies of STA Bank and Golden Insurance was taken. A reason for this was that ten case studies were regarded as sufficient to the understanding of the emerging knowledge. In other words, following Eisenhardt's (1989b) recommendation, the understanding of the phenomenon-under investigation (innovation) was already saturated with the use of ten case studies. Given this, the consideration of the additional two case studies would not have generated further insights.

A summary of the research findings is outlined in Appendix F. For this paper, the group of "focused innovators" is eliminated from the analysis. A reason for this is that this "middle" group of companies does not present unique strategic features. These are partly similar to some of the features of the "fast innovators" and the "non-innovators." Thus, this elimination will enable us to distinguish more clearly the key processes of innovation by only considering the strategic behavior of two opposite groups – the "fast innovators" who considered innovation central to their strategy of competitive advantage and the "non-innovators" who did not. Hence, in the following section the strategic profile of the "fast innovators" will be contrasted with the strategic profile of the "non-innovators" in relation to the following emerging theoretical constructs: a market-driven approach to innovation, organizational learning, and organizational structures.

A market-driven approach to innovation

Fast innovators

The "fast innovators" made continuous innovation central to their strategy of competitive advantage. A "continuous innovation" based strategy constantly answers the question "what is next?" In the sample of companies studied, the "fast innovators" were constantly wondering and thinking of the next development or idea to bring to the market. One informant at BSB, for example stated,

> "You have to have the next idea out. It is leap frog ideas all the time; it is a constant process . . . we don't . . . do something and sit back and think 'wow, we can rest for two years now that we have launched that product.' It may be for the rest of the two weeks of the month, but then we have to think the next idea."

This constant thinking about the next development was regardless of whether the new development was a new product, service, process, or completely a new business proposition outside the core business of the company. For example, Money First did not just offer consumers credit cards, but also other offerings outside the main scope of the company. There, one informant told me, "We do not consider ourselves to be just a credit card company or a financial services company. We are looking at opportunities outside. One example is our telecommunication business . . . where we are selling a telecommunication service to the customer. We are buying time from a telephone communication company, packaging it . . . and offering it to the customers at a price that is more attractive . . ."

The "fast innovators" were never complacent or satisfied with the status quo of their business, whereby they were always searching for the next market opportunity to jump on. At SS, there was a strong belief that without continuous innovation, the company would not have survived: "The almost internal conversation is that the day everything stays steady is the day we are not surviving, and we are not winning."

This constant search for new market opportunities led the "fast innovators" to carefully monitor the developments occurring in the external environment. However, by doing so, the "fast innovators" did not aim to align the company's strategy to the evolution of the external environment. Instead, they tended to break any established rule of the industry by creating market gaps that competitors were not able to anticipate. For example, the launch of the ethical positioning by the HB was the result of a market opportunity identified in the early 1990s, when there was a growing concern in the society about the issue of environmental pollution, nuclear weapons, social exclusion, etc. Basically, this constant focus on the market developments represents a source for ideas generation.

Moreover, although market-oriented, the "fast innovators" never locked themselves into a myopic view of their own business industry – financial services. Instead, they also used to watch constantly the changes occurring in adjacent industries or other geographical markets. For instances, at SS there was a constant focus on the developments occurring in the US market and the UK retail industry.

The "fast innovators" also used to watch carefully competitors' strategic moves. However, their strategy was not one of "wait and see." They used to monitor competitors in order to understand how the strategic moves of these would have affected their strategy. By monitoring the competition, they pursued the aim of being different. For example, one informant commented:

> "...In 1996, since we built our capital again, we started to go on our own way. Although we looked at what the market was doing, we did not necessarily follow the market, because...if we had followed the market, we would have never introduced free unemployment insurance, we would have never introduced base rate track mortgages. You know, it is an easy option to follow the market...if you do not find and fill gaps, then you are just fighting on prices all the time..."

Customers also were a constant focus of a strategy of "continuous innovation." Particularly, this is a strategy which is based on the adoption of a customer-centric perspective of the entire business. One informant, for example, stated: "I would not say that we are excessively focused on the external market. I would say that we are very focused on customers and what our customers want and what our future customers would want, and I think we need to be, because that's what the company is about. It is about providing value to people..."

Non-innovators
In contrast to the "fast innovators," the "non-innovators" did not consider innovation as important to the company's strategy of competitive advantage. The informants showed concerns about the increasing competition in the industry, whereby they felt that their companies had to become more innovative. However, unlike the "fast innovators" who made "continuous innovation" central to their strategy of competitive advantage, the "non-innovators" were only considering innovation as a strategic response to the external challenges. This strategic behavior led them to

align their strategies to the evolution of the external environment, rather than shaping the industry conditions. Hence, in contrast to the "fast innovators" who were constantly monitoring the external environment to spot market opportunities and actually drive changes in the industry, the "non-innovators" were simply seeking a fit between the company's internal situation and the changes in the external environment.

In contrast to the "fast innovators" who tended to jump quickly on new and untested developments before any of their competitors or potential new entrants did so, the "non-innovators" tended to adopt more a "wait and see" approach, with the consequence that their developments tended to be slow and similar to most new developments occurring in the industry. In other words, being fast and quick to the market was not a priority of the "non-innovators." The risk of failure was an inhibitor for them to jump quickly on new market developments. One informant commented,

> "Doing the wrong things before somebody else does the wrong things is not helpful . . . What matters is that you do the right thing rather than the wrong thing, even if it means that it takes a few years longer to work out what the right thing is . . . The value that you destroy by doing the wrong things is tremendous. So, avoidance of doing the wrong things is probably more important than jumping in and doing something, which turns up to be the wrong thing."

Although the "non-innovators" watched the developments of the external market, their strategic focus remained too myopic. Unlike the "fast innovators" who had a broader look at the external developments, including those occurring outside the boundaries of their own industry, the "non-innovators" simply kept their strategic monitoring close to the developments occurring within their own industry.

Unlike the "fast innovators" who made the customer central to their strategy for radical innovation, the "non-innovators" did not take a customer perspective in their approach to innovation. The main concern of these players was the creation of shareholder value rather than customer value. Therefore, their strategies were about strengthening their financial resources, achieving big scale, rather then directing their resources toward innovation in products and/or services for the consumer. For instance, informants respectively at SSB and GI commented,

> "Particularly Sassoon Bank has been innovative in the overall strategy and in the approach to create value for shareholders. Where it is a follower rather than a leader is in the introduction of new products for the customers . . ."

> "I think the greater focus has been on what is happening outside . . . there has been a tremendous amount of legislation going on over the last ten years so it has been looking after that. A lot of consolidation of the companies and a lot of effort has gone into that. Not so much change over the last five/six years in the design of the products. They have stayed pretty much the same – new things, new features brought in but the actual basic products are pretty much the same."

The "non-innovators" also used to monitor carefully the strategic moves of their rivals. However, in contrast to the "fast innovators," this monitoring was not aimed at anticipating competitors' strategic moves and/or adopting strategies of differentiation from them. The closer look at competitors simply enabled the "non-innovators" to monitor their strategic actions and, if necessary, to adopt reactive strategies. This type of strategic focus on competitors easily led the "non-innovators" to follow the dominant logic of their industry, whereby their strategies were not innovative. For example, one informant at GI commented that,

> "...I would say if you took the company as a whole we are a follower...The insurance companies are very good at following what other people are doing, it needs somebody to take the lead and then others will follow..."

Innovation and organizational learning

Fast innovators
The fieldwork revealed that the two strategic approaches to innovation by the "fast innovators" and the "non-innovators" could be mapped to some common patterns of organizational behavior. Particularly, it appeared that the "fast innovators" behaved as learning organizations (Senge, 1991). These companies showed the ability to learn from their own experience, simply through exploitation of existing knowledge in some key areas of the business and reapplication of it into other business contexts. Sources of learning were not just internal, but also external, i.e., customers. The "fast innovators" usually learned from their customers via the use of surveys and various types of forums, where customers' feedback enabled them to retrospectively evaluate the efficacy of their past behavior and the subsequent strategic actions to be taken. For instance, informants respectively at SS and MF commented that,

> "...[Creativity] comes from our customers in terms of feedback. How are our customers using our services? What do they like? What don't they like? How satisfied are they?"

> "We will be testing very carefully a lot of different product variations and see which one the customer likes. So, basically doing a lot of testing (philosophy of MF) and learning from that testing."

In addition to exploitation learning, the "fast innovators" were also engaged in exploration learning. While exploitation learning was more based on leveraging existing knowledge by reapplying it to different organizational contexts, exploratory learning was more about the ability to learn proactively and, on the back of that, to build new knowledge. Again, in this context, sources of learning were either internal or external. SS, for example, had a technology knowledge, which was built on the basis of the company's early experience and mistakes (internal learning) in dealing with Internet-based technologies. One informant said:

"... There is definitely a technology competency. So, we spent the last two years with systems falling over and doing stuff with systems that have never been asked to do that sort of thing before, building platforms, making mistakes ... So, now we have technical competence that is hard to match ... Internet knowledge you can find here."

The "fast innovators" were also able to learn proactively from external sources, such as customers, competitors, and consultants. Learning proactively from customers, for example, mainly meant developing an understanding about the future and latent customers' needs. Here, the process of idea development and building knowledge around the customers was mainly put in place through forums or focused groups where customers actively contributed inputs around a theme of interest. At SS, one informant commented that,

"... The customer knowledge is critical in terms of understanding what products and services customers want in the future, how they want them, how you can offer it to them ... you know all that learning."

These companies learned proactively from consultants specialized in a particular area of the business, i.e., new technology. For example, it was common practice to actively involve external consultants, which would have helped senior managers to sense the future developments occurring in their own market and other industries. For instance, external market research companies used to provide HB with insightful information about the future market developments in terms of new technology and changes in consumers' behavior and expectations. At SS, managers used to learn from external technologists who usually run training courses in order to help managers develop skills in the arena of visioning the future. Here, proactive learning, which was very much forward-looking, was considered to be more important than exploitation learning, in terms of leveraging existing knowledge and vast experience built over the years. For instance, one informant commented,

"We have a lot of management development and a lot of internal training, developing senior management in those sorts of skills, particularly transformation technology from the US. We run courses with one of the experts ... that have written a number of transformation technology books. We have actually run courses and we use a number of other sources ... which is all about visioning the future ... rather than working on vast experiences."

In addition to the previous forms of exploration learning, the ability to continuously challenge established norms or rules frequently emerged as another common pattern of learning, enabling the generation of new knowledge. This was like a process of "unlearning the past," which involved a continuous reinvention of existing products, services, or business models. For instance at MF, one informant commented,

"It is hard to sustain competitive advantage through just one product, one action. So, one needs to consistently be reinventing ..."

Similarly, at HB proactive learning derived from the ability to think out of the "normal box." One informant commented,

> "Essentially I think it [strengths of the company] . . . is about the ability to step away from . . . things that are sacrosanct, the status quo . . . I do not think that we have any[thing] sacrosanct. If there is a better way of doing something, we are always prepared to consider it."

Challenging the established norms, in some cases, also meant a complete reinvention of the business model. For example, since SS was launched, its business model had changed many times:

> ". . . We really pushed hard on our business model . . . I think since then [when we launched] that [the business model] has evolved and we are now on a slight different business model again where we sell our own products and other people's own products through our investment supermarket and our insurance supermarket. So, that's a slight different business model again, not relying on our own products to sell, but selling other people's stuff. Now, we are just entering into another phase, different completely again, which is the multi-channel model, where the business model looks substantially different again . . . Yes, it [the business model] is all changing all the time."

This ability of learning proactively and building new knowledge by challenging existing norms, or thinking outside the "normal box," or consequently reinventing products, services, or processes may recall some Schumpeterian concepts of entrepreneurship and radical change. However, from the informants' comments, what really emerged was that the type of innovation resulting from this type of proactive learning was never too dramatic or radical. Simply said, it was not the type of revolutionary or breakthrough innovation which is generally the cause of major dramatic shifts within an existing business or industry. However, neither was it the sort of incremental innovation based on small changes brought to existing products, services, or processes. Although this might have occurred on an occasional basis, it never represented the strategic focus of the "fast innovators." As these were like "shape shifters," they were all the time shifting their focus from a current customer segment, product, business model, etc., to another. So, the innovation process was continuous, although the resulting innovation itself was always at the edge between the two extreme situations of incremental and radical innovation. Moreover, the evolutionary nature of the innovation process involved here was not the sort of process causing a slow and steady type of change. It was mainly a type of "dynamic" evolution from one situation to another without causing any dramatic shift.

Non-innovators
Organizational learning did not emerge as a systematic feature of the "non-innovators." If, sometimes, some aspects of organizational learning emerged across a few companies, these were mainly the type of adaptive or exploitation learning. Within this group, it appeared that organizational learning only occurred as a response to a specific stimulus, generally a business need. In contrast with the "fast-innovators" which used to learn proactively from small and cheap experiments, the

"non-innovators" did not engage in exploratory learning through experimentation. Instead, they tried to avoid experimentation. They simply learned from competitors' experimentation. In most of the cases, this learning did not generally cause any sort of breakthrough innovation; it was mainly adaptive, leading the "non-innovators" to simply follow competitors' moves. At SSB, one informant commented,

> "If we want to be innovative and come first with something, we have to spend a lot of money on it, and a lot of research, and a lot of testing it out, before going to the market. Very much we avoided that, because if somebody else comes up with the product first, we can learn from them and it would be much cheaper for us to come with it very closely behind."

Unlike the "fast innovators," whose exploitation learning derived from an ability to leverage existing knowledge across different organizational boundaries, among some of the "non-innovators" there was an inability to reapply the ideas generated in one part of the organization to other areas:

> "I think ideas come from all over the organization. But, I do actually think one of our weaknesses is that we do not harness and utilize that creativity across the organization."

Moreover, within this group of companies, there was a complete lack of exploration learning through challenging established norms. At SSB, for example, one informant commented:

> ". . . I believe we never look at it [innovation] from a different perspective. We have always done it this way, and therefore, we will do just something a little bit different, and we think that's innovative, but it is not! I think it is coming up with a complete different idea."

Organizational mechanisms enabling innovation

A third theoretical construct emerging from the field research was about the nature of the organizational mechanisms affecting the innovation process. It appeared that while organic systems were the dominant organizational paradigm among the "fast innovators," the "non-innovators" had in place more mechanistic systems. One reason for this might be due to the different market orientation of the two groups of companies. As the "fast innovators" tended to quickly react to changing market conditions, organic structures provided them with some internal flexibility in terms of moving quickly to the market and, therefore, innovating faster than their competitors. For instance, one informant commented,

> ". . . We have a more informal management structure. You do not have to have a full business plan and committees and all the rest of it to get things moving."

Teamwork and cross-functional teams were common features of the organic structures existing among the "fast innovators." In most of the cases, the informants referred to an organizational environment as open, with no boundaries, and where most of the new business initiatives were undertaken through teamwork and/or cross-functional teams. One reason why the informants stressed teamwork as a useful mechanism for enhancing innovation was that it tended to favor an adequate climate where new ideas could develop. For example, some informants commented that,

> "I think team working enhances creativity, develops initiatives . . . We encourage people to work in teams at all levels, to solve problems or to look at new ideas, new ways of working."

Another reason why teamwork and/or cross-functional teams were regarded as important mechanisms for innovation was that they constituted an ideal place where experimentation could take place – in a sense they were regarded as self-contained, protected environments. For instance, some informants commented that,

> "If we do believe that it is such an idea, then we put our resources behind that, we recommend it to the executive board, and get funding for it, which tries to be a team, an area where ideas can grow and are visible as well . . ."

> "If you have got an idea, go to set up a cross-functional team and do it, whether you would be an associate or an MD."

Also, teamwork, particularly in the form of cross-functional teams, would have impacted on innovation by bridging diversity of perspectives across the different organizational boundaries. This would have favored an environment where learning could take place. One informant commented,

> "Because we are reasonably unstructured, most of the projects that go on within the organization are cross-functional teams . . . It encourages innovation by bringing diversity of [the] different departments you have, and particularly not putting a rigid framework around it . . ."

Non-innovators

In contrast with the "fast innovators," the organizational structures of the "non-innovators" tended to be hierarchical, although at the time of the field research, some of them were moving toward the adoption of flatter structures. For example, some comments were,

> ". . . At the moment it is traditionally hierarchical, but we are moving away from that into something which will be flatter."

One reason for this move toward flatter organizational structures was that highly hierarchical structures with a lot of grades tended to slow down the decision-making process, particularly in presence of risk-averse attitudes. In other words, flatter structures would have given managers more flexibility and autonomy in the decision-making process, with great potential for innovation. One informant at BSC, for example, explained,

"... My view is that, [the] less structure you have, [the] more innovation you are capable of ... Now, in the banking sector to do things that you have not done before, there is quite an element of risk ... That's one aspect. But, I think in terms of structures, the more structure you have in doing something actually reduces the scope of the individuals to feel comfortable in making decisions because there is a referral tree ... [the] less structure you have, the wider the gap or the wider the ceiling is for these people to make more decisions."

However, it also appeared that flatter organizational structures did not always have a positive effect on the capability of the company to be more innovative. A common view among the informants was that flatter organizational structures had to be complemented by an adequate culture, supportive toward risks taken, empowerment, and therefore innovation. Also, these organizational structures had to be complemented by the right processes, enhancing an environment where new idea development and experimentation could easily take place. For instance, an informant at SSB commented,

"I think that one of the keys to me is first of all a culture that encourages risk taking, making mistakes, experimentation ... Also, I think, process. I think you can probably get as much innovation, research, clever ideas, if you have the correct process, where you stimulate thinking, changing people's environment, giving them circumstances where [they can] experiment, etc."

In contrast with the "fast innovators," where teamwork and cross-functional teams were key to promoting experimentation and/or to generating new knowledge across the organizational boundaries, there was not a similar pattern in the group of "non-innovators." Particularly, in the case of the "non-innovators," evidence from the data suggests that knowledge creation occurred within "fixed" organizational boundaries, with adaptive learning outcomes. Here, the top-middle management was convinced that specialist skills rather than a broad range of knowledge had a crucial role in terms of inputs to the innovation process. For example, one informant stated,

"... I suppose it will depend on what you describe as lower levels ... but if we are talking about people in administrative functions, whether they are capable of producing an idea which has a major impact on the kind of offering that goes to the marketplace – I think the chances of that are quite small. Can they come up with ideas, which influence the way they do their own work? – Yes they can and very often do. They are very often accepted and adopted, and that's fine. But, I think for the bigger issues, life in the industry is now so complex that you need specialist people looking at that kind of activity."

Also, within BSA, there was the conviction that people at the top of the organizational hierarchy had a better knowledge than the people at the lower levels. Hence, in contrast with the "fast innovators," the innovation process did not systematically embrace all the organizational members. Only accidentally might some innovative outcomes have been generated at the lower levels of the organization. One inform-ant, for example, commented,

> "No, if something [innovation] came through it was chance, and I suspect that it is still the case today. There is still a feeling that people at the top know best and do not genuinely encourage creativity."

Discussion: Strategic Implications

The research findings presented in this paper offer some useful insights into the study of innovation in financial services. Particularly, the data analysis suggests that financial services organizations can innovate on a regular basis by having in place a strategy of continuous innovation. In the following sub-sections, this paper addresses the major implications of this finding for the strategic theory of innovation.

Organizational learning and innovation

Previous research (Sundbo, 1997, 1998) suggested that the services industries, including the financial services sector, innovate like their counterparts in the manu-facturing industry. However, the phenomenon of innovation is generally unsys-tematic and, when it occurs, it tends to be close to organizational learning (Sundbo, 1998). Evidence from the present study suggests a contrasting perspective. It appears, here, that innovation is unsystematic only when it is not a core component of the company's strategy of competitive advantage. Instead, when innovation is central to the strategy of competitive advantage, it is a systematic process that tends to coincide with organizational learning. Moreover, it appears that this process is the result of a market approach to innovation. Indeed, the innovative companies learned proactively from a wide variety of external sources, including customers, but also competitors, consultants, and partners. These results are not completely new. Previous research (Day, 1994: 43) already suggested that "a market orientation emphasizes the ability of the firm to learn about customers, competitors, and channel members in order to continuously sense and act on events and trends in present and prospective markets." Such a broad construction of the market will expand the "learning zone" of the innovative firm by taking it out of the scope of its current business. In contrast, a narrow market perspective will generally lead to overlooking the broad range of opportunities that managers might pursue. Hamel and Prahalad (1991: 83) argued that a consequence of the narrow construction of the market is that "the resulting learning boundary constrains organizational learning to the adaptive variety, which is usually sequential, incremental, and focused on issues or opportunities that are within the traditional scope of the organiza-tion's activities." Also, a narrow construction of the market will lead to overlooking possible threats from non-traditional competitors (Slater and Narver, 1995: 68). Hence, it is proposed that when innovation is central to the company's strategy of competitive advantage and is market-led, then it is systematic and tends to be close to exploration learning.

An entrepreneurship paradigm to the study of innovation in financial services

A second emerging field insight is that the companies who have in place a strategy of "continuous innovation" are able to learn proactively and generate new knowledge (exploratory learning) by mainly continuously understanding not only the customers' current needs, but also their latent needs, and challenging established norms. Researchers Slater and Narver (1995: 68) have considered such aspects within the type of the market-oriented organization. They concluded that when market-oriented organizations focus on the understanding of the customers' latent needs, such organizations are inherently entrepreneurial, too. According to Slater and Narver (1995: 68), entrepreneurial organizations are able to generate new knowledge through exploration, for example, by challenging established assumptions, avoiding bureaucracy, by being proactive, and so on. In previous research, it is indicated that resistance to bureaucracy (Kanter, 1989; Mintzberg, 1991; Quinn, 1985) and proactiveness (Naman and Slevin, 1993) are some of the cultural values of entrepreneurship. Slater and Narver (1995: 65) concluded that a market-orientation has to be complemented by entrepreneurship if a company wants to learn faster than its competitors and secure competitive advantage in very dynamic and turbulent business environments. On the basis of these arguments and results from the previous data analysis, it is argued that the market and entrepreneurial orientations of the firm influence exploration learning and, ultimately, the behavior of innovative companies. Such a conclusion is clearly in contrast to the results of prior research by Sundbo (1997) who stated that the strategic paradigm is the most adequate paradigm to explain innovation in the services industry, while, for this purpose, the entrepreneurship paradigm has a limited application generally restricted to the case of funding a new firm. Hence, this study leads us to propose that the entrepreneurship paradigm is adequate to explain innovation in financial services and that, jointly with the strategic paradigm, contributes to enhancing the understanding of the phenomenon of innovation in the services industry.

Innovation and the boundary of knowledge creation

The data analysis presented in this paper also suggests that in today's knowledge economy, the innovation process is not the exclusive domain of a few people or organizational divisions. It appears to be a highly participative process, involving organizational members at any level of the organizational hierarchy. Within the companies that innovate continuously, there was, indeed, an ongoing innovation process, whose scope spanned different organizational boundaries, requiring access to a broad range of expertise. Instead, when innovation is not institutionalized, there is no need for such type of knowledge. In this circumstance, the innovation process is only aimed at improving the existing products and services, whereby it tends to occur within the domain of the existing knowledge. This often resides in a specialized area, i.e., the marketing or IT department, which is highly involved in the improvement of existing products or services. Hence, when the scope of innovation is narrow, the innovation process occurs within well-defined organizational boundaries.

In other words, when the innovation process is broad and continuous, companies tend to act on the basis of emerging market opportunities, whereby they need easy access to new knowledge, wherever this might be. This finding supports previous research by Hamel and Prahalad (1996: 240), who concluded that "as industry boundaries are increasingly melding, industry specialization might become a handicap for competing in today's networked economy." On the basis of these arguments, it is proposed that, when innovation is central to the company's competitive strategy, the innovation process cannot be identified with the marketing function of many financial services organizations. In contrast, it is a collective business function whose scope embraces the whole organization.

The institutionalization of creativity

Another emerging insight, linked to the previous one, is concerned with the role of creativity for the purpose of innovation. In the extant theory of innovation and creativity, there has been an increasing emphasis on the important role of creativity (Amabile et al., 1996; Amabile, 1998; Robinson and Stern, 1998; Markides, 1999; Hamel, 1998a) and consequently the need for hiring creative people as a way to energize the innovation process. In most of these theoretical contributions, attention is on a single person as a highly creative individual capable of influencing the innovation process, either at the lower level of the organization (Amabile et al., 1996) or at the higher level (Hamel, 1998a). Hamel (1998a) calls these people "activists" who are generally outsiders. The present study shows a contrasting result. First of all, it appears that these "activists" or "champions" are not necessarily outsiders. Instead, they have been within the organizations for a long time. They are highly creative people who learn by doing, and strongly feel that they belong to their organization. These champions are also willing to accept risk and therefore unlearn the organization by breaking through any current rule that represents a barrier to radical innovation. Secondly, the field research seems to suggest that it is not the creativity of the single individual that matters. If innovation is central to the company's competitive strategy, creativity has to be institutionalized as a core component of the innovation process. For this purpose, appropriate processes have to be in place in order to create an environment where creativity and innovation can take place. Particularly, it is proposed that a combination of formal mechanisms, such as suggestion schemes, informal procedures, and structured processes, is more useful than the creativity of a single individual to create such a type of environment.

Escaping the trap of continuous improvement

In the extant literature of innovation, it is argued that soon after the introduction of a breakthrough innovation or a radical business concept (Hamel, 2000), there is a tendency for companies to fall into the trap of continuous innovation. In this study, for example, this has been the case for PB and SI. Soon after the introduction of the radical idea – telephone banking – PB fell into the trap of continuous improvement. Similarly, SI became committed to the improvement of its product range soon after the launch of its radical business idea – the offering of "peace of mind" to consumers. Levinthal and March (1993) argued that this is a situation common to many companies where exploitation learning tends to drive out exploration learning. In the extant theory, it is suggested that one solution to this problem is to have in place dual strategies, a today strategy for today and a tomorrow strategy for tomorrow (Abell, 1999). Tushman and O'Reilly (1996) have stressed the difficulty of dealing with such dual strategies and suggested that "ambidextrous managers" must be prepared to cannibalize their own business at times of industry transitions (Tushman and O'Reilly, 1996). In order to overcome resistance to change and internal change, Christensen (1997) and Markides (1998) have proposed setting up a separate company to support the new strategic innovation.

The research presented here offers an alternative perspective to those presented in the extant theory (Christensen, 1997; Tushman and O'Reilly, 1996; Abell, 1999). The "fast innovators" were committed not just to the improvement of the existing innovations, but also to the development of new business ideas. They achieved this by having in place a strategy of "continuous innovation." This is a strategy that constantly answers the question "what is next?" In the sample of companies studied, the "fast innovators" were constantly wondering and thinking of the next development or idea to bring to the market. As these innovators are constantly preoccupied with the future, they address the strategy for the future continuously. This kind of strategy evolves all the time to the extent that it requires continuous strategic focus and a continuous redefinition of what the business is all about.

This study also offers a new insight into the concept of "continuous innovation." This is different from the concept of innovation generally used in the extant literature of innovation (Christensen, 1997; Tushman and O'Reilly, 1996; Abell, 1999). The kind of innovation involved in the study of the strategic behavior of "fast innovators" is not incremental innovation. This type of innovation was never the strategic focus of the "fast innovators" as these players tended to shift their focus quite rapidly. However, neither is it the type of radical innovation which is generally described in the punctuated equilibrium model (Gersick, 1991). In this model, long periods of incremental change are interrupted by periods of brutal change that alter the structure of an industry (Abernathy and Utterback, 1978; Tushman and Anderson, 1986; Rosenkopf and Tushman, 1995). It appears that the type of innovation described in the study of the "fast innovators" is closer to the concept of "continuous change" described by Brown and Eisenhardt (1997) in the study of firms operating in high-velocity environments, i.e., computing and communication. "For these firms, change is not the rare, episodic phenomenon described by the punctuated equilibrium model, but, rather, it is endemic to the way these organizations compete" (Brown and Eisenhardt, 1997: 1). As argued by Brown and Eisenhardt, the punctuated equilibrium model has become the "foreground of academic interest." The two authors proposed that in today's business environment, many companies compete by changing continuously, whereby researchers should devote their efforts to researching the mechanisms by which firms achieve continuous innovation. The research findings presented in this paper not only reinforce Brown and Eisenhardt's arguments, but also show that the concept of "continuous innovation" is increasingly useful to explain the strategic behavior of firms not only competing in a high-velocity industry – i.e., computing – but also in other industries, such as financial services, characterized by a slower pace of change.

Conclusions

This exploratory study offers rich insights into the strategic and organizational aspects of innovation among UK financial services organizations. Particularly, the study shows that although innovation is a top priority of many executives, it is not fully incorporated in the strategy of competitive advantage. Thus, most innovations tend to be incremental, simply representing a modification of existing products and services, with little or no differentiation from the competition. In this case, innovation is not systematically organized and tends to coincide with exploitation learning. This confirms prior research by Sundbo (1997) who investigated the phenomenon of innovation in the broad Danish services industries. However, the research presented in this study also offers an alternative perspective. Particularly, it shows that when companies make "continuous innovation" central to their strategy of competitive advantage, the occurrence of innovation is a systematic process and presents aspects of both exploitation and exploration learning. In this circumstance, innovation is not the prerogative of particular functional areas, such as marketing and IT. Instead, it appears to be a collective business function that embraces the whole organization. An implication of this is that the area of knowledge creation through experimentation and exploration learning is not the exclusive domain of a few individuals or departments. It is an area that spans different organizational and industry boundaries.

The aspect of creativity also assumes a different relevance. As innovation is a broad business function, creativity has to be institutionalized within the company through the combination of formal and informal procedures and more structured processes. In a sense, it is not the creativity of the single individual that matters anymore, but the fact that within the organization there are appropriate mechanisms that enable any member at any level to have an input into the innovation process. Given this portrait of how financial services organizations innovate continuously, it appears that the innovation phenomenon is based on the firm's capability to bridge internal and external diversity of knowledge.

This study also suggests that a strategy of "continuous innovation" requires constant strategic focus. This offers an alternative to many studies that have recommended different solutions to the problem of dealing with dual strategies – a strategy for today mainly leading to incremental innovation and a strategy for tomorrow mainly leading to radical innovation. Here, it appears that such problems become redundant when companies have in place a strategy of continuous innovation. As already suggested by Brown and Eisenhardt (1997), research should devote more effort to the study of continuous innovation, which is the way through which companies compete not just in high-velocity environments – i.e., computing and electronics – but also in industries, such as financial services, characterized by a slower pace of change.

However, given the exploratory nature of this research and the kind of methodology adopted, this study presents some limitations. First of all, the research findings rely heavily on the informants' descriptions of how they conceive innovation within the context of their companies. Secondly, the data are highly descriptive and only represent the views of the companies' top-middle management. If the dataset also included the views of less senior informants, a better understanding of the organizational mechanisms underlying the process of innovation would be gained. A third limitation of the study is that the dataset does not allow establishing whether and how top management controls the innovation process. Further research should be carried out in this direction. Particularly, if companies are going to compete in fast-changing business environments through continuous innovation, it would be useful to understand the kind of strategic and organizational routines that enable them to retain the unity of strategic direction, while changing all the time.

Acknowledgements

The author would like to thank Kevin Keasey for helpful comments on an earlier draft of the paper and Richard Bettis for his invaluable comments for the full redevelopment of the paper.

Note

1   For confidentiality reasons, the companies' names are pseudonyms.

References

Abell, D. F. 1999: Competing today while preparing for tomorrow. *Sloan Management Review*, 40 (3), 73–81.
Abernathy, W. J. and Utterback, J. M. 1978: Patterns of industrial innovation. *Technology Review*, 80, 40–7.
Amabile, T. M. 1998: How to kill creativity. *Harvard Business Review*, 76 (5), 76–87.
Amabile, T. M., Conti, R., Coon, H., Lazenby, J. and Herron, M. 1996: Assessing the work environment for creativity. *Academy of Management Journal*, 39 (5), 1154–84.
Andrews, K. R. 1971: *The Concept of Corporate Strategy*. Homewood, IL: Irwin.
Ansoff, H. I. 1965: *Corporate Strategy*. New York: McGraw-Hill.
Baker, M. 1985: *Marketing Strategy and Management*. London: Palgrave Macmillan.
Barney, J. B. 1986: Organizational culture: can it be a source of sustained competitive advantage? *Academy of Management Review*, 11 (3), 656–65.
Brown, S. L. and Eisenhardt, K. M. 1997: The art of continuous change: linking complexity theory and time-paced evolution in relentlessly shifting organizations. *Administrative Science Quarterly*, 42 (1), 1–34.
Bogner, W. C. and Thomas, H. 1993: The role of competitive groups in strategy formulation: a dynamic integration of two competing models. *Journal of Management Studies*, 30 (1), 51–67.
Bower, J. L. and Doz, Y. 1979: Strategy formulation: a social and political process. In D. E. Schendel and C. W. Hofer (eds.), *Strategic Management*. Boston, MA: Little Brown, 152–66.
Campbell, M. and Collins, A. 2001: In search of innovation. *The CPA Journal*, 71 (4), 26–35.
Christensen, C. M. 1997: *The Innovator's Dilemma. When New Technologies Cause Great Firms to Fail*. Boston, MA: Harvard Business School Press.

Colarelli O'Connor, G. and Rice, M. P. 2001: Opportunity recognitions and breakthrough innovation in established firms. *California Management Review*, 43 (2), 95–116.

Day, G. S. 1994: The capabilities of market-driven organizations. *Journal of Marketing*, 58, October, 37–52.

Donlan, J. P. 1999: Who owns the customer? *Chief Executive*, 147, September, 66–73.

Drucker, P. F. 1970: Entrepreneurship in business enterprise. *Journal of Business Policy*, 1 (1), 3–12.

Duhaime, I. M. and Schwenk, C. R. 1985: Conjectures on cognitive simplification in acquisition and divestment decision making. *Academy of Management Review*, 10 (2), 287–95.

Dyer, W. G. and Wilkins, A. L. 1991: Better stories, not better constructs, to generate better theories: a rejoinder to Eisenhardt. *Academy of Management Review*, 16 (3), 613–19.

Eisenhardt, K. M. 1989a: Making fast strategic decisions in high-velocity environments. *Academy of Management Journal*, 32, 543–76.

Eisenhardt, K. M. 1989b: Building theories from case study research. *Academy of Management Review*, 14 (4), 488–511.

Feldman, S. P. 1986: Management in context: an essay on the relevance of culture to the understanding of organizational change. *Journal of Management Studies*, 23 (6), 587–607.

Geroski, P. 1998: Thinking creatively about your market: crisps, perfume and business strategy. *Business Strategy Review*, 9 (2), 1–10.

Gersick, C. J. G. 1991: Revolutionary change theories: a multilevel exploration of the punctuated equilibrium paradigm. *Academy of Management Review*, 32, 274–309.

Glaser, B. G. and Strauss, A. L. 1967: *The Discovery of Grounded Theory: Strategies for Qualitative Research.* London: Weidenfeld and Nicholson.

Hamel, G. 1998a: The challenge today: changing the rules of the game. *Business Strategy Review*, 9 (2), 19–26.

Hamel, G. 1998b: Strategy innovation and the quest for value. *Sloan Management Review*, 39 (2), 7–14.

Hamel, G. 2000: *Leading the Revolution.* Boston, MA: Harvard Business School Press.

Hamel, G. 2001: Innovation's new math. *Fortune*, July 9, 130–2.

Hamel, G. and Prahalad, C. K. 1991: Corporate imagination and expeditionary marketing. *Harvard Business Review*, 69 (4), 81–91.

Hamel, G. and Prahalad, C. K. 1996: Competing in the new economy: managing out of bounds. *Strategic Management Journal*, 17, 232–42.

Hannan, M. T. and Freeman, J. 1977: The population ecology of organizations. *American Journal of Sociology*, 82 (5), 929–64.

Kanter, R. M. 1989: *When Giants Learn to Dance.* New York: Touchstone.

Kim, W. C. and Mauborgne, R. 1997: Value innovation: the strategic logic of high growth. *Harvard Business Review*, 66 (1), 102–12.

Kim, W. C. and Mauborgne, R. 1999: Creating new market space. *Harvard Business Review*, 77 (1), 83–93.

Kotler, P. 1983: *Principles of Marketing.* Englewood Cliffs, NJ: Prentice Hall.

Kotler, P. 1984: *Marketing Management.* Englewood Cliffs, NJ: Prentice Hall.

Kotter, J. P. 1995: Leading change: why transformation efforts fail. *Harvard Business Review*, March–April, 59–67.

Kumar, N., Scheer, L., and Kotler, P. 2000: From market driven to market driving. *European Management Journal*, 18 (2), 129–42.

Levinthal, D. A. and March, J. G. 1993: The myopia of learning. *Strategic Management Journal*, 14, 95–112.

Levitt, T. 1960: Marketing myopia. *Harvard Business Review*, 38 (4), 45–56.

Lindblom, C. E. 1959: The science of muddling through. *Public Administration Review*, 19 (2), 79–88.

Llewellyn, D. 1992: Financial innovation: a basic analysis. In H. Cavanna (ed.), *Financial Innovation*. New York: Routledge, ch. 1, 15–51.

Loewe, P., Williamson, P., and Chapman, R. 2001: Five styles of strategy innovation and how to use them. *European Management Journal*, 19 (2), 115–25.

Macmillan, I. C. 1978: *Strategy Formulation: Political Concepts*. St. Paul, MO: West.

Management Accounting 1998: Strategic management: which way to competitive advantage? *Management Accounting*, 76 (1), 32–7.

Mansfield, E. 1968: *Industrial Research and Technological Innovation*. New York: W.W. Norton.

Markides, C. 1998: Strategic innovation in established companies. *Sloan Management Review*, 39 (3), 31–42.

Markides, C. 1999: Six principles of breakthrough strategy. *Business Strategy Review*, 10 (2), 1–10.

Meyer, J. W. and Rowan, B. 1977: Institutionalized organizations: formal structure as myth and ceremony. *American Journal of Sociology*, 83, 340–63.

Miles, M. B. and Huberman, A. M. 1984: *Qualitative Data Analysis*. Beverly Hills, CA: Sage.

Miller. D. 1976: *Strategy making in context: ten empirical archetypes*. PhD Thesis, Faculty of Management, McGill University, Montreal.

Miller. D. 1979: Strategy, structure, and environment: context influences upon some bivariate associations. *Journal of Management Studies*, 16, October, 294–316.

Mintzberg, H. 1972: Research on strategy-making. *Proceedings of the 32an Annual Meeting of the Academy of Management*. Minneapolis.

Mintzberg, H. 1978: Patterns in strategy formation. *Management Science*, 24 (9), 934–48.

Mintzberg, H. 1979: *The Structuring of Organizations: A Synthesis of the Research*. Englewood Cliffs, NJ: Prentice Hall.

Mintzberg, H. 1991: The innovative organization. In H. Mintzberg and J. B. Quinn (eds.), *The Strategy Process, Concepts, Contexts, Cases*, 2nd edn. Englewood Cliffs, NJ: Prentice Hall, 731–46.

Mintzberg, H., Ahlstrand, B., and Lampel, J. 1998: *Strategy Safari. A Guided Tour through the Wilds of Strategic Management*. Hemel Hempstead: Prentice Hall.

Naman, J. L. and Slevin, D. P. 1993: Entrepreneurship and the concept of fit: a model and empirical tests. *Strategic Management Journal*, 14 (2), 137–53.

Noda, T. and Bower, J. L. 1996: Strategy making is iterated processes of resource allocation. *Strategic Management Journal*, 17 (Summer Special Issue), 159–92.

Pettigrew, A. M. 1977: Strategy formulation as a political process. *International Studies of Management and Organization*, Summer, 78–87.

Pinchot, G., III 1985: *Entrepreneuring*. New York: Harper and Row.

Porter, M. E. 1980: *Competitive Strategy: Techniques for Analyzing Industries and Competitors*. New York: Free Press.

Prahalad, C. K. and Bettis, R. A. 1986: The dominant logic: a new linkage between diversity and performance. *Strategic Management Journal*, 7, 485–501.

Prahalad, C. K. and Hamel, G. 1990: The core competence of the corporation. *Harvard Business Review*, 68, May–June, 79–91.

Quinn, J. B. 1980: *Strategies for Change: Logical Incrementalism*. Homewood, IL: Irwin.

Quinn, J. B. 1985: Managing innovation: controlled chaos. *Harvard Business Review*, 63, May/June, 73–84.

Reger, R. K. and Huff, A. S. 1993: Strategic groups: a cognitive perspective. *Strategic Management Journal*, 14, 103–24.

Robinson, A. G. and Stern, S. 1998: *Corporate Creativity. How Innovation and Improvement Actually Happen*. San Francisco. CA: Berrett-Koehler Publishers Inc.

Rosenkopf, L. and Tushman, M. L. 1995: *Network evolution over the technology cycle: lessons from the flight simulation community*. Working Paper, Department of Management, University of Pennsylvania.

Sarrazin, J. 1975: *Le role des processus de planification dans les grandes entreprises francaises: un essai d'interpretation*. These 3ieme cycle, Université de Droit, d'Economie et des Sciences d'Aix-Marseille.

Sarrazin, J. 1977/1978: Decentralized planning in a large French company: an interpretative study. *International Studies of Management and Organization*, Fall/Winter, 37–59.

Scherer, F. M. 1980: *Industrial Market Structure and Economic Performance*. Boston, MA: Houghton Mifflin.

Schmookler, J. 1966: *Invention and Economic Growth*. Cambridge, MA: Harvard University Press.

Schumpeter, J. 1934: *The Theory of Economic Development*. London: Oxford University Press.

Schwartz, H. and Davis, S. M. 1981: Matching corporate culture and business strategy. *Organizational Dynamics*, Summer, 30–48.

Senge, P. 1991: *The Fifth Discipline: The Art and Practice of the Learning Organization*. New York: Doubleday.

Silber, W. L. 1975: Towards a theory of financial innovation. In W. L. Silber (ed.), *Financial Innovation*. Lexington, MA: D. C. Heath & Co, 53–85.

Slater, S. and Narver, J. 1995: Market Orientation and the Learning Organization. *Journal of Marketing*, 59, July, 63–74.

Strauss, A. and Corbin, J. 1990: *Basics of Qualitative Research*. Newbury Park: CA: Sage.

Sundbo, J. 1997: Management of innovation in services. *The Service Industries Journal*, 17 (3), 432–55.

Sundbo, J. 1998: *The Theory of Innovation. Entrepreneurs, Technology and Strategy*. Northampton, MA: Edward Elgar.

Tidd, J., Bessant, J., and Pavitt, K. 2001: *Managing Innovation. Integrating Technological, Market and Organizational Change*. Chichester: Wiley.

Tushman, M. L. and Anderson, P. 1986: Technological discontinuities and organizational environments. *Administrative Science Quarterly*, 31, 439–65.

Tushman, M. L. and O'Reilly, C. A. 1996: Ambidextrous organizations: managing evolutionary and revolutionary change. *California Management Review*, 38 (4), 8–30.

Yin, R. 1984: *Case Study Research: Design and Methods*. Beverly Hills, CA: Sage.

Yin, R. 1993: *A Review of Case Study: Research: Design and Methods*. Newbury Park, CA: Sage.

Weick, K. E. 1979: *The Social Psychology of Organizing*. Reading, MA: Addison-Wesley.

Wernerfelt, B. 1984: A resource-based view of the firm. *Strategic Management Journal*, 5, 171–80.

Wrapp, H. E. 1967: Good managers don't make policy decisions. *Harvard Business Review*, September–October, 91–9.

Appendix D Details of interviewees

Company name	Total interviews – departments involved in the interviews
Fast innovators	
Money First (MF)	Head of HR Recruitment
	Head of Marketing Analysis
	Head of Information Systems
	Head of E-commerce
	Head of External Communication
The Helping Bank (HB)	Head of HR
	Head of Marketing
	Director of IT
	Senior manager of Information Systems
	Director of Communication
Building Society B (BSB)	Head of HR
	Head of Marketing and Product Development
	Director of Customer Service
	Senior manager of E-commerce
	Senior manager of Business Strategy Development
Sunshine (SS)	Head of HR
	Director of IT
	Director of Customer Experience
	Director of Marketing
	Director of Multi-delivery Channels
Focused innovators	
Phone Banking (PB)	Head of HR
	Senior manager of IT
	Head of Marketing
	Head of E-commerce
	Manager of E-commerce
Star Insurance (SI)	Director of HR
	Head of IT
	Head of Marketing
	Head of Strategy
	Senior manager of E-commerce
Non-innovators	
STA Bank (STAB)	Director of HR
	Director of IT
	Head of Product Development
	Director of E-commerce
	Head of Marketing
Sassoon Bank (SSB)	Head of HR Development
	Director of IT
	Director of Customer Service
	Head of Business Strategy
	Senior manager of E-commerce

Appendix D (*cont'd*)

Company name	Total interviews – departments involved in the interviews
Building Society A (BSA)	Head of HR Development
	Senior manager of HR recruitment
	Senior manager of IT
	Head of Business Strategy
	Senior manager of marketing
Building Society C (BSC)	Head of HR development
	Director of IT
	Head of Mortgage business
	Head of Investment business
	Director of E-commerce
Global Insurance (GI)	Director of HR
	Senior manager of IT
	Head of sales
	Head of Product Development
	Head of Marketing
Golden Insurance (GoI)	Senior manager of HR development
	Head of IT
	Head of E-commerce
	Head of Product Development
	Head of Customer Service

Appendix E Interview themes

- Background of the interviewee
- The role of the interviewee within the company
- Interviewee's perception of the importance of innovation for competitive advantage
- Interviewee's perception of the drivers and barriers of innovation
- Comment on the major strategic change of the company in the last five years
- Interviewee's perception of what the company's strategic focus has been in the past
- Interviewee's views about competitors and changes in the external environment
- Interviewee's perceptions of where he/she will see the company in the future
- The nature of the organizational structures
- Interviewee's perception of the importance of creativity for innovation
- Interviewee's perceptions of the processes in place to generate and develop new ideas
- Interviewee's perceptions of the importance of the customer in the process of innovation
- Interviewee's perception of the importance of new technology for innovation

Appendix F Summary of key findings

Market driven approach

Fast innovators
Innovation answers the question of
"what's next?"
Innovation aims at shaping industry
conditions
Innovation creates gaps in the market

Innovation aims at being different from
competitors
Focus on adjacent industries
Create value for current and future
customer needs

Non-innovators
Innovation as a strategic response to the
external challenges
Innovation aims at aligning strategy to the
industry conditions
Innovation seeks a fit between the company
situation and changes in the market
Innovation adjusts to the dominant logic

Myopic focus on financial services industry
Create shareholder value

Innovation and organizational learning

Fast innovators
Exploitation learning
Exploration learning
Broad sources of learning: specialist and
broad range of skills

Non-innovators
Exploitation learning

Limited sources of learning: internal
specialist skills

Innovation and organizational structures

Fast innovators
Organic structures
Experimentation through cross-functional
teams
Institutionalization of creativity through
a combination of formal and informal
procedures and structured processes

Non-innovators
Mechanistic structures

Focus on individual creative skills

The Role of Organizational Culture in the Corporate Branding Process at Silicon Valley firms

Stanley J. Kowalczyk

Keywords: Corporate branding, reputation, organizational culture.

Abstract

This paper suggests that there may be a relationship between the external perception of organizational culture and corporate branding as measured by reputation. Using an instrument based on the Organizational Culture Profile, 179 industry professionals evaluated eight culture dimensions in seven well-known Silicon Valley firms (Apple Computer, Cisco Systems, Hewlett Packard, National Semiconductor, Oracle, Sun Microsystems, and 3 Com). Corporate branding was measured by utilizing a reputation measuring instrument. Reputations of the firms were obtained by using the eight dimension results of Fortune's 1999 Most Admired Company Survey. Regression analysis was performed on 64 possible pair-by-pair reputation/culture dimension relationships. A total of 36 correlations were found significant. The findings suggest that in these seven firms, the strategic resource of corporate brand (as measured by reputation) may partially reflect external perceptions of culture.

Introduction

Business firms are increasingly moving from the branding of products towards the branding of the firm. Hatch and Schultz (2000) suggest that the locus of competitive advantage is shifting from products to organizations and that a shift from product branding to corporate branding is under way. They state: "Differentiation requires positioning not products but the whole corporation. Accordingly, the values and emotions symbolized by the organization become key elements of differentiation strategies, and the corporation itself moves to center stage . . . corporate branding brings to marketing the ability to use the vision and culture of the company explicitly as a part of its unique selling proposition" (2000: 3).

The names Apple Computer, Cisco Systems, Hewlett Packard, National Semiconductor, Oracle, Sun Microsystems and 3 Com generate images not only of products but also of the work environments of the firms. For example, Cisco conveys an image of its egalitarian nature (signaled by the granting of stock options to all employees). Hewlett Packard is known for its strong consensus-driven culture ("The HP Way"). Oracle is known as a company reflecting the maverick style of its founder and chairman, Larry Ellison. Sun Microsystems has a similar maverick image that reflects the style of its CEO, Scott McNealy. Apple Computer is known for its informality. Each firm projects a corporate image that is dependent not only on its economic performance but also on how it conducts business or its unique culture. According to Keller (2000), "a corporate image can be thought of as the associations in the consumer's memory to the company or corporation making the product or providing the service as a whole" (118). Keller (2000) goes on to say:

> . . . some marketing experts believe that a factor increasing in importance in consumer purchasing decisions is consumer perceptions of a firm's whole role in society, including how a firm treats its employees, shareholders, local neighbors, and others. As Procter & Gamble's one time CEO Ed Artz remarked, 'Consumers now want to know about the company, not just the products.' (118)

Keller (2000) identifies four types of corporate image associations: (1) common product attributes (for example, 3M is perceived by many outsiders as having a strong culture that emphasizes innovation); (2) people and relationships (Nordstrom is legendary for having sales people with a personalized touch and a willingness to go to extraordinary lengths in their relationships with customers); (3) values (McDonald's image as a socially responsible company may be enhanced by their operation of the Ronald McDonald Houses for sick children); (4) corporate credibility (Enron and Arthur Andersen are examples of organizations that have had severe issues of trustworthiness that have impacted on their images).

Hatch and Schultz (2001) recognize the important role that organizational culture may play in generating an image to outside stakeholders. One of the implications of their model is that it may be important to examine how external observers perceive the internal phenomena of culture. We are suggesting that external perceptions of organizational culture may be related to the corporate brand. We are proposing that corporate branding is a multidimensional construct and that two of those dimensions are the recognition of the corporate brand and the quality of the corporate brand. Recognition is the measure of how widely known the brand is. Quality refers to the positive or negative image of the brand. Because the study of corporate branding is a relatively new field, the availability of measuring instruments is problematic. Our study did not attempt to measure the recognition of the corporate brand. Rather, our focus was on the measurement of the quality of the corporate brand. To measure quality, we turned to the concept of reputation.

Much of the research on corporate reputation suggests that reputation is dependent on prior economic performance (Vergin and Qoronfleh, 1998; Brown and Perry, 1994; Fryxell and Wang, 1994). However, there is a stream of research that suggests that there is more to an organization's reputational status than its economic standing (see Fombrun, 1996; Fombrun and Shanley, 1990; Brown and Perry, 1994). Fombrun and Shanley conclude ". . . that publics appear to construct reputations from a mix of signals derived from accounting and market information, media reports, and other non-economic cues" (Fombrun and Shanley, 1990: 252). Using the Hatch and Schultz (2000) Corporate Branding Model as a basis, we suggest that one of the non-economic cues – external perception of organizational culture – may be used by outsiders to construct corporate reputations.

Background

Corporate reputation

Reputation is a perceptual judgment of a company's past actions that is developed over time. Fombrun defines corporate reputation as "the overall estimation in which a company is held by its constituents" (Fombrun, 1996: 37). Like Fombrun (1996), we suggest that developing this intangible asset – corporate reputation – will become increasingly important in the years ahead. We believe this is so because the recent cases of damage to the credibility of corporations have made it more likely than ever that organizations will be competing based on their ability to express who they are and what they stand for.

A positive reputation creates a strategic advantage or what Fombrun (1996) refers to as reputational capital. Fombrun adds: "Ultimately, reputations have economic value to companies because they are difficult to imitate. Rivals simply cannot replicate the unique features and intricate processes that produced those reputations. Reputations are therefore a source of competitive advantage. To sustain that relative advantage requires a commitment to the ongoing management of a company's reputation – that is, the extent to which the images a company projects coincide with and reinforce its identity" (Fombrun, 1996: 387). A good reputation permits a company to command premium prices for its products, pay lower prices for purchases through its ability to leverage in negotiations, recruit the top candidates to its company, enhance employee morale and loyalty, have greater stability in stock prices, and reduce its risks during a crisis (Vergin and Qoronfleh, 1998; Fombrun, 1996). Strategically, reputation offers a firm greater value, rarity, inimitability, and sustained competitive advantage (Boyd et al., 1995).

The importance of reputation as a strategic resource is best summarized by Barney (2002: 285): "Of all the bases of differentiation . . . perhaps none is more difficult to duplicate than a firm's reputation . . . Reputations are not built quickly, nor can they be bought and sold . . . A firm with a positive reputation can enjoy a significant competitive advantage, whereas a firm with a negative reputation, or no reputation, may have to invest significant amounts over long periods of time to match the differentiated firm." While reputations cannot be built quickly, however, they can be degraded or destroyed in a very short time span. For example, in the 2001 Fortune survey of the Most Admired Companies in America, Enron was ranked in the top 5 percent (out of approximately 500 corporations). One year later, in the 2002 Fortune survey, they were ranked in the bottom 5 percent. A further example of the fragile nature of reputation may be found in the Hewlett Packard acquisition of Compaq Computer. The extremely contentious and personal battle over this acquisition may have degraded the sterling reputation that Hewlett Packard had built up over a more than 50-year period.

From a resource dependence perspective, reputation offers a firm a sustainable competitive advantage because it is difficult to duplicate and/or because it offers unique capabilities or competencies. According to Collis and Montgomery (1995), the resource-based view of the firm is particularly useful to business practitioners because "it derives its strength from its ability to explain in clear managerial terms why some competitors are more profitable than others" (119). Rao (1994) summarizes this perspective aptly: "The resource-based perspective defines resources as inputs into the production process and depicts capabilities as capacities to coordinate and deploy resources to perform tasks. Resources may be tangible (e.g., equipment, finance) or intangible (e.g., brand name, trade secrets) and capabilities may consist of subroutines and master routines (e.g., product development, distribution) that integrate subroutines into performance" (Rao, 1994: 29). Hall (1992) asked CEOs to identify the most important intangible resource (from a list of 13 intangible resources) and to rank its replacement period. He found that a company's reputation was the most important intangible resource as well as the one requiring the longest replacement period.

Reputations reflect the general esteem in which a firm is held by its multiple stakeholders (Fombrun, 1996). Given that the stakeholders represent economic and non-economic sectors, the resultant reputation of a company reflects both of these sectors. Although Fombrun and Shanley (1990) find a stronger economic contribution towards reputation, they also acknowledge that a significant portion of a company's reputation can be attributed to its "softer" side or institutional record. Our paper extends the research by Fombrun (1996) and Fombrun and Shanley (1990) by focusing on one of the "institutional" components of corporate reputation, external perception of its culture.

Organizational culture

The research on organizational culture is extensive (e.g., Deal and Kennedy, 1982; Schein, 1985; Kotter and Heskett, 1992; Collins and Porras, 1994) and culture is generally defined as a set of values, beliefs, and norms shared by members of an organization. Schein (1985) characterizes corporate culture as consisting of symbols, rites, and ceremonies; these characteristics then are artifacts of the underlying values, beliefs, assumptions, and feelings shared by the members of the organization. Schein (1996) defines culture as "a set of basic tacit assumptions about how the world is and ought to be that a group of people share and that determines their perceptions, thoughts, feelings, and, to some degree, their overt behavior." He goes on further:

> To discover the basic elements of a culture, one must either observe behavior for a very long time or get directly at the underlying values and assumptions that drive the perceptions and thoughts of the group members. . . . Cultures arise within organizations based on their own histories and experiences. Starting with the founders, those members of an organization who have shared in its successful growth have developed assumptions about the world and how to succeed in it, and have taught those assumptions to new members of the organization. Thus, IBM, Hewlett Packard, Ford, and any other company that has had several decades of success will have an organizational culture that drives how its members think, feel, and act. (1996: 11–12)

Kotter and Heskett (1992) conceptualize organizational culture as having two levels. "At the deeper and less visible level, culture refers to values that are shared by the people in a group and that tend to persist over time even when group member-ship changes. These notions about what is important in life can vary greatly in different companies; in some settings people care deeply about money, in others about technological innovation or employee well being" (1992: 4). They go on to identify a second more visible level in which "culture represents the behavior patterns or style of an organization that new employees are automatically encour-aged to follow by their fellow employees" (1992: 4). Kotter and Heskett (1992) conducted an empirical study to determine whether a relationship exists between corporate culture and long-term economic performance. They identify three cat-egories of culture performance theories. The first theory examines the connection between "strong" cultures and economic performance. The second examines the notion that the culture of a firm must be strategically appropriate for its industry. The third theory is that only cultures that are adaptive will be associated with superior performance over long periods of time. Their study concludes that the third theory is the most predictive of long-term economic performance.

In a similar fashion, Collins and Porras (1994) conducted an empirical study to identify the factors that distinguished companies that had consistent outstanding economic performance. They concluded that having a cult-like culture was a signi-ficant factor common to many of the successful firms that they studied. According to Collins and Porras (1994), organizations with cult-like cultures are excellent places to work for those who are in agreement with the core ideology of the firm.

Organizational culture is an important element in developing and implementing the best strategy for competitive advantage (Barney, 2002; Peteraf, 1993). Daft (1998) presents a model to show how the fit between an organization's environ-ment and its strategic focus identifies specific types of culture, such as adaptability/entrepreneurial, mission, clan, and bureaucratic. Tushman and O'Reilly (1997) argue that an organization's culture serves as a social control mechanism that enhances or impedes organizational change or innovation.

Prior research has identified various cultural dimensions (e.g., Sackmann, 1992; Hofstede et al., 1990; Rousseau, 1990). The 'Organizational Culture Profile' (OCP) is an instrument developed by O'Reilly et al. (1991) to identify cultural dimensions. The OCP approach focuses on the central values of the organization, which requires identifying the relevant range of values, their intensity and their consensus among the members of the organization (O'Reilly et al., 1991; Saffold, 1988). Using the

OCP, O'Reilly et al. (1991) reduced 54 culture values to eight underlying factor dimensions. As is the case with any factor analysis, O'Reilly et al. (1991) named their factors based on subjective interpretations ("innovation," "attention to detail," "outcome orientation," "aggressiveness," "supportiveness," "emphasis on rewards," "team orientation," "decisiveness"). Further research by Chatman and Jehn (1994) validate O'Reilly et al. (1991) and they note that the OCP finding of eight factors is similar to categories identified by Hofstede et al. (1990) and Sackman (1992).

The link between reputation and the external perception of culture

An example of how a firm's reputation can be related to the external perception of its culture may be found in Southwest Airlines. The firm has been one of the most consistently profitable companies in America for many years. In addition, Southwest Airlines has earned one of the highest scores in the Annual Fortune Most Admired Companies Survey since the inception of the survey in the early 1980s. Its underlying philosophy "which emphasizes the importance of people, having fun, and recognition of employees is largely unchanged since the company was founded" (Pfeffer, 1994: 61). A similar phenomenon occurs at Nordstrom. Like Southwest Airlines, Nordstrom has had financial success and an excellence in reputation for many years. Collins and Porras (1994) characterize Nordstrom's culture as being very strong and cult-like. Through this culture Nordstrom has "created a zealous and fanatical reverence for its core values, shaping a powerful mythology about the customer service heroics of its employees" (1994: 135). In both companies culture becomes a marketing tool because it spills over onto the public and results in a positive customer service oriented experience. Pfeffer (1994) suggests that these cultures that stress customer service result in a sustainable competitive advantage because they are very difficult to imitate.

Our paper is suggesting that the reputation of firms can be linked to such non-economic factors as culture. Exactly what comprises the non-economic portion of reputation is open to debate, but cultural aspects are suggested by Fombrun (1996). He states that such values as credibility, reliability, trustworthiness, and responsibility are at the core of the perceptual representation of a company's reputation. "A company's reputation sits on the bedrock of its identity – the core values that shape its communications, its culture, and its decisions" (Fombrun, 1996: 268). He adds: "Identity is therefore closely aligned with notions of corporate character, personality, and culture" (1996: 277). Hatch and Schultz (1997) claim that identity is grounded in organizational culture and that identity is constituted within cultural contexts. Our paper accepts their grounding argument and we have therefore elected to examine the external perception of culture rather than the external perception of identity.

On a conceptual basis, the study of corporate branding recognizes the importance of the external perceptions of internal phenomena such as culture. Hatch and Schultz (2000) state: "Corporate branding exposes corporations to far greater scrutiny. This means that organizational behavior, even at the level of everyday employee interactions, becomes visible (and sometimes newsworthy) so that, for example, the organization becomes more transparent than ever before. This in turn, elevates the importance of a healthy (i.e., non cynical, non repressive) organizational culture" (2000: 5).

On an empirical basis Flatt and Kowalczyk (2000) have conducted a study that examines the link between perceived culture and reputation. Their study made use of the Organizational Culture Profile Instrument and Fortune's Most Admired Corporations Survey Instrument. Using a "clustering technique," they developed a probabilistic methodology and suggest that external observers may factor perceptions of firm culture into their assessments of a company's reputation. By using an instrument directly based on the OCP, we hope to show that cultural dimensions are related to dimensions of reputations.

Method

Seven firms headquartered in Silicon Valley were chosen as subjects for this project. They are: Apple Computer, Cisco Systems, Hewlett Packard, National Semiconductor, Oracle, Sun Microsystems, and 3 Com. These companies are well known not only in Silicon Valley but also nationally and globally.

Instruments and respondents

Reputation
The most commonly used (e.g., Fombrun and Shanley, 1990; Dollinger et al., 1997; Vergin and Qoronfleh, 1998) reputation measure is the annual Fortune's Most Admired Corporations (MAC) survey, where either the overall rating is used or one of the eight question areas is used. The seven firms are included in the 1999 Fortune MAC survey which measures the following eight key attributes or dimensions of reputation: (1) Innovativeness, (2) Quality of management, (3) Ability to attract and develop talented people, (4) Quality of product and services, (5) Long-term investment value, (6) Soundness of financial position, (7) Social responsibility, and (8) Wise use of corporate assets.

Each dimension uses an 11-point scale with 0 = poor and 10 = excellent. This survey has been conducted by Fortune since 1982. The strengths of the survey are a sample size of approximately 8,000 and a response rate of about 50 per cent. A further strength is the knowledge base of the respondents. They consist of executives, directors, and analysts all of whom rate only firms in their industries. Although there is debate about the use of the Fortune MAC survey to measure reputation, it is the only comprehensive corporate reputation survey available (Fombrun, 1996: 397) and it is widely used and influential.

External perception of organizational culture
Perceived culture was measured by using an instrument based on the Organizational Culture Profile developed by O'Reilly et al. (1991). Their eight factors were used as a basis to construct an eight-item organizational culture questionnaire. For example, the O'Reilly et al. factor #2 consists of the following three culture values: analytical, attention to detail, precise. Using this as a basis, our culture question #2 is "This company pays attention to details, strives for precision and stresses the importance of

analytical skills." Using a five-point scale ranging from 1, "Highly Disagree" to 5, "Highly Agree" respondents were asked to evaluate the culture of the seven firms by answering eight questions about each firm (each question corresponding to one of the O'Reilly et al. factors or dimensions). Respondents were also given a "Cannot Judge" option (see Appendix G for the complete questionnaire). The five-point scale and the "Cannot Judge" option are similar to those used by Turban and Greening (1997) in their examination of the relationship between a firm's reputation and its attractiveness as an employer.

Respondents
The respondents in this study were 273 individuals who were employed as professionals in various high technology firms. They were attendees of the WESTECH Career Fairs held in Santa Clara, CA, in April and December 1999. The Santa Clara WESTECH Career Expo is a major two-day forum held several times a year. Over 300 high technology companies were represented and several thousand people attended the event at which admission was free. Each respondent was approached individually and asked to complete the 56-item questionnaire (8 questions about 7 firms). Using criteria similar to those developed by Turban and Greening (1997), respondents who used "Cannot Judge" more than one-third of the time were eliminated from the study (1/3 × 56 questions = 18.67, rounded to 19). These criteria reduced the sample size to 179. Eighty-three percent of the respondents had at least one to three years of work experience in high technology industries and 85 percent had at least a four-year college degree.

Results

Using the means of the 179 respondents, each of the seven firms received a score (1 through 5) on each of the eight dimensions of culture (Table 8.1). The names of our eight culture dimensions were changed slightly from the O'Reilly et al. (1991) names in order to more accurately reflect our questionnaire: (1) "risk taking and innovative," (2) "detail and precision oriented," (3) "achievement oriented," (4) "aggressive and opportunistic," (5) "supportive and acknowledges performance," (6) "high pay and opportunities for growth," (7) "team oriented," and (8) "decisiveness."

The Fortune MAC survey uses simple averaging to obtain a score (1 through 10) on eight dimensions of reputation (Table 8.2). Pair-by-pair regression analysis was used to determine if there was a significant relationship between a given Fortune reputation dimension and a given culture dimension. Since there are eight reputation dimensions and eight culture dimensions, there are 64 possible pairs. For example, in looking at the relationship between Fortune reputation dimension #1 (innovativeness) and Culture dimension #1 (risk taking and innovative), the seven points to be examined are: (1) National Semiconductor; Fortune score 5.10 on innovativeness and Culture score 3.04 on risk taking and innovative. (2) Cisco Systems; Fortune score 7.63 on innovativeness and Culture score 4.21 on risk taking and innovative . . . and so on for the other five firms. A regression analysis was done

Table 8.1 Culture survey summary results

		Risk Taking & Innovative	Detail & Precision Oriented	Achievement Oriented	Aggressive & Opportunistic	Supportive & Acknowledge Performance	High Pay & Opportunities for Growth	Team Oriented	Decisiveness
Oracle	Mean	3.83	3.83	4.11	4.01	3.65	3.9	3.82	3.51
	N =	162	156	162	165	133	162	151	152
Hewlett-Packard	Mean	3.6	4.06	3.95	3.7	4.06	3.83	4.07	3.59
	N =	173	170	173	174	153	172	164	165
Cisco Systems	Mean	4.21	4.03	4.23	4.25	3.87	4.05	3.99	3.67
	N =	170	168	172	172	151	169	153	163
3Com	Mean	3.84	3.85	3.89	3.84	3.6	3.68	3.74	3.45
	N =	163	159	150	168	143	154	153	152
National Semi.	Mean	3.04	3.42	3.63	3.29	3.33	3.37	3.42	3.23
	N =	151	149	152	155	133	139	136	132
Apple Computers	Mean	4.21	3.69	3.88	3.63	3.84	3.8	3.97	3.24
	N =	169	167	168	175	155	164	163	163
Sun Micro.	Mean	4.11	4.06	4.21	4.24	3.97	4.07	4.12	3.71
	N =	169	169	168	170	152	173	163	159

Table 8.2   Fortune reputation survey results March 1, 1999

	Innovativeness	Quality of Management	Employee Talent	Quality of Products	Investment Value	Financial Soundness	Social Responsibility	Use of Assets
Oracle	7.12	7.24	7.00	7.27	6.98	7.80	5.44	6.76
Hewlett-Packard	7.18	7.34	7.36	8.00	7.12	7.91	7.43	7.00
Cisco Systems	7.63	8.89	8.58	7.53	8.63	9.42	7.26	8.63
3Com	6.84	7.05	6.63	7.42	6.37	7.16	6.63	6.16
National Semi.	5.10	4.87	4.33	5.63	3.90	4.10	5.13	4.40
Apple Computers	7.34	5.14	4.75	6.57	3.93	3.98	5.83	5.40
Sun Micro.	7.51	7.00	7.21	7.57	6.67	6.95	6.55	7.05

on these seven points and the correlation coefficient was found to be .915 at a significance level of $p < .01$. This suggests that there is a positive near linear relationship between dimension #1 of Reputation and dimension #1 of Culture. Similar analysis was done on the other 63 pairs and in 35 additional cases a significant relationship was found at a level of $p < .05$ (see Table 8.3).

Reputation dimension #1, innovativeness, is most linked with culture – showing a significant correlation with seven of the eight dimensions of culture. Reputation dimension #7, social responsibility, is least linked with culture – showing a significant relationship with two dimensions of culture. The other six reputation dimensions show a significant relationship with either four or five culture dimensions.

Below are shown the seven strongest correlations:

Reputation dimension	Culture dimension	Correlation
Quality of product/services	Detail and precision oriented	0.964
Ability to attract and develop talented people	Decisiveness	0.934
Innovativeness	High pay and opportunities for growth	0.932
Innovativeness	Team oriented	0.920
Innovativeness	Risk taking and innovative	0.915
Long-term investment value	Decisiveness	0.902
Wise use of corporate assets	Achievement oriented	0.900

Of the seven firms in the study, two specific cases are particularly noteworthy – Cisco Systems and National Semiconductor. In eight of eight culture dimensions, Cisco Systems has a higher score than National Semiconductor (See Figures 8.1a and 8.1b). In eight of eight reputation dimensions, Cisco Systems has a higher score than National Semiconductor. This result is consistent with our suggestion that there is a significant pattern between dimensions of organizational culture and dimensions of reputation.

Table 8.3 Correlation table

	Risk Taking & Innovative	Detail & Precision Oriented	Achievement Oriented	Aggressive & Opportunistic	Supportive & Acknowledge Performance	High Pay & Opportunities for Growth	Team Oriented	Decisiveness
Innovativeness	.915**	.839*	.845*	.801*	.842*	.932**	.920**	.664
Quality of management	.470	.831*	.808*	.817*	.503	.725	.530	.871*
Employee talent	.489	.899**	.852*	.848*	.599	.786*	.623	.934**
Quality of products/services	.529	.964**	.736	.717	.786*	.761*	.789*	.847*
Long-term investment value	.414	.845*	.812*	.811*	.514	.725	.534	.902**
Financial soundness	.366	.817*	.766*	.768*	.469	.672	.483	.868*
Social responsibility	.417	.848*	.510	.505	.757*	.564	.690	.717
Use of corporate assets	.619	.883**	.900**	.875**	.663	.861*	.698	.895**

** $p \le .01$.
* $p \le .05$.

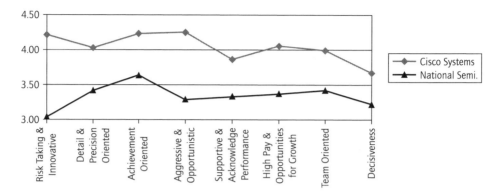

Figure 8.1a   Culture survey summary results

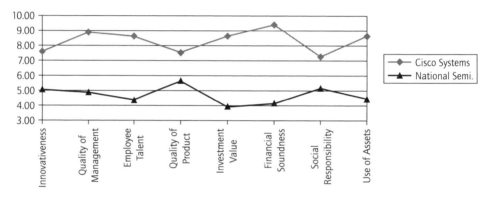

Figure 8.1b   Fortune reputation summary results

Discussion and Conclusion

We recognize that culture is conceptualized and traditionally measured as a set of internal perceptions of employees of an organization. However, like Flatt and Kowalczyk (2000), we propose that external observers may rely on perceptions of culture to construct corporate reputations. Our proposal is consistent with the study of corporate branding, which also suggests that external perceptions of culture may be related to overall corporate image. Hatch and Schultz (2000) depict branding as a process that draws simultaneously on organizational culture, strategic vision, and corporate images. They add: "Branding efforts are generally an attempt to project the distinctiveness of the company to external audiences who are encouraged to perceive and judge them as attractive and desirable. We argue that successfully managing the corporate brand also involves reaching inside the corporation to better project and communicate practiced organizational values to external stakeholders" (2000: 11). One of the implications of their model is that it is important to examine how external observers perceive the internal phenomena of culture. We believe our paper is a step in that direction.

There is much debate about the interconnectedness among identity, culture, image, and reputation. Identity and culture are generally conceptualized as internal constructs while image and reputation are seen as external. The concept of identity is often posed as the internal question – who are we? In a seminal piece on organizational identity, Albert and Whetten (1985) define identity as that which is central, enduring, and distinct about an organization. They note that core features of identity are presumed to be resistant to ephemeral or faddish attempts at alteration, because of their ties to the organization's history. Hatch and Schultz (1997) point to the importance of organizational culture as a framework for the expression of identity. They "view organizational identity as grounded in local meanings and organizational symbols and thus embedded in organizational culture, which (they) see as the internal symbolic context for the development and maintenance of organization identity" (1997: 358). Figure 8.2 shows the Hatch and Schulz's (1997) model of the relationship between culture, identity, and image. The left half of Figure 8.2 represents the internal context while the right half represents the evaluation of external stakeholders. Their model proposes that there is an increasing breakdown of the internal/external boundary of organizations. Consequently, there is a heightened visibility of inside phenomena to outsiders. We suggest that one of the implications is that managers need to be aware of how outsiders may interpret the culture of their organizations.

While we accept the Hatch and Schultz (1997) model of the identity/culture relationship, we propose a slight modification to their concept of external context. Specifically, we envision the external context of image to be identical to reputation. We agree with Berg (1985) who takes an external approach when he defines image as the public's perception or impression of an organization. "This definition is similar to Fombrun's (1996) definition of reputation as the collective judgments (by outsiders) of an organization's actions and achievements" (Gioia et al., 2000: 66). In their Corporate Branding model, Hatch and Schultz (2000) also take an external approach when they define image as "views of the organization developed by its

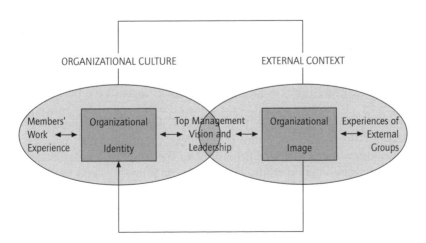

Figure 8.2 Hatch and Scultz's (1997) model of the relationship between culture, identity, and image

stakeholders; the outside world's overall impression of the company including the views of customers, shareholders, the media, the general public, and so on" (p. 9). Using Berg (1985), Fombrun (1996), and Hatch and Schultz (2000) as a basis, we suggest that the external context of the Hatch and Schultz (1997) and (2000) models fit the definition of reputation. We therefore propose the model below:

IDENTITY/CULTURE<----------------------------->IMAGE/REPUTATION

Although causality cannot be demonstrated, our empirical results suggest that a consistent pattern exists between the external phenomena of reputation (as measured by the Fortune MAC Survey) and externally perceived culture. Our results offer some empirical support to Hatch and Schultz's (2000) suggestion that there is a connection between how outsiders perceive a firm's culture and that firm's reputation.

Recent events involving firms such as Enron, Tyco, and Arthur Andersen have shown how quickly a firm's reputation can be degraded or destroyed. The bitter personal battle between the Hewlett and Packard families and the Hewlett Packard Company over the acquisition of Compaq may have eroded the high quality reputation that the firm had developed over many years. Furthermore, the recent economic downturn in the US economy makes the issue of reputation even more critical. If our study were to be conducted today, the results might be quite different. Each of the seven firms that we studied had poorer results in the 2002 Fortune MAC Survey than they had in the 1999 Survey. If our Culture Survey were taken today, its results might also result in poorer scores for each of the seven firms.

Our study has several limitations. In doing their empirical work examining organizational reputation and organizational attractiveness as an employer, Turban and Greening (1997) used senior level students ($n = 75$ and $n = 34$) to evaluate companies. On the other hand, we have used 179 respondents with fairly extensive industry experience to evaluate companies. We believe that one of the strengths of our study is the knowledge base of our respondents. One of the strengths of Turban and Greening (1997) is the number of firms they evaluated ($n = 189$). Our study examined seven firms. Because we felt that our respondents at the Santa Clara WESTECH Career Fairs would be far more familiar with the culture of local firms, we included only corporations that were headquartered in Silicon Valley. Furthermore, the firms had to be a part of the Fortune's Most Admired Corporations Survey. The only other firm we could have included would have been Intel. Future research might also factor in the halo effect that financial performance has on reputation as well as on culture. Finally, while this study has examined outside perception of culture, perhaps future research can use insider evaluation of culture.

Our study offers some useful contributions to the study of organizational culture, reputation, and corporate branding. First, while the primary effects of organizational culture are internal, in that the beliefs, knowledge, customs, and values are what bind organizations together, there may be a secondary effect of culture. This effect may be external, in that the outside perception of culture may influence the corporate brand as measured by firm reputation. Practitioners should be aware of this possibility and recognize that having a positive culture and signaling a favorable image of that culture to the public may have a positive impact on reputation. In fact, corporate branding models suggest that communicating a "good" image of organizational culture may be a critical element in the relatively new corporate function of reputation management. Second, while the research on reputation is influenced by the "halo effect" of financial performance, the literature acknowledges other non-economic factors as important in affecting reputation. We have brought the study of organization culture into the domain of the reputation research by examining seven of the largest firms in Silicon Valley. We see this as an important inclusion and propose that further study of the culture–reputation linkage would be useful.

Acknowledgment

Special thanks go to Michael Pawlish of Lund University in Sweden. This project would not have been possible without his help.

References

Albert, S. and Whetten, D. 1985: Organizational identity. In L. L. Cummings and B. M. Staw (eds.), *Research in Organizational Behavior*, vol. 7. Greenwich, CT: JAI Press, 263–95.

Barney, J. B. 2002: *Gaining and Sustaining Competitive Advantage*. Reading, MA: Addison-Wesley.

Berg, P. O. 1985: Organization change as a symbolic transformation process. In P. Frost, L. Moore, M. R. Louis, C. Lundberg, and J. Martin (eds.), *Organizational Culture*. Beverly Hills, CA: Sage, 281–99.

Boyd, B. K., Carroll, W. O., and Dess, G. G. 1995: *Determining the strategic value of firm reputation: a resource-based view*. Presented at the 15th annual Conference of the Strategic Management Society, October 1995.

Brown, B. and Perry, S. 1994: Removing the financial performance halo from Fortune's "most admired" companies. *Academy of Management*, 37, 1347–59.

Chatman, J. A. and Jehn, K. A. 1994: Assessing the relationship between industry characteristics and organizational culture: How different can you be? *Academy of Management Journal*, 37 (3), 522.

Collins, J. and Porras, J. 1994: *Built to Last*. New York: HarperCollins Business.

Collis, D. and Montgomery, C. 1995: Competing on resources: strategy in the 1990s. *Harvard Business Review*, 73 (4) 118–28.

Daft, R. L. 1998: *Organization Theory and Design*, 6th edn. Cincinnati, OH: South-Western College Publishing.

Deal, T. E. and Kennedy, A. A. 1982: *Corporate Cultures*. Reading, MA. Addison-Wesley.

Dollinger, M. J., Golden, P. A., and Saxton, T. 1997a: The effect of reputation on the decision to joint venture. *Strategic Management Journal*, 18 (2), 127–40.

Flatt, S. J. and Kowalczyk, S. J. 2000: Do corporate reputations partly reflect external perceptions of organizational culture? *Corporate Reputation Review*, 3 (4), 334–42.

Fombrun, C. J. 1996: *Reputation: Realizing Value from the Corporate Image*. Boston, MA: Harvard Business School Press.

Fombrun, C. and Shanley, M. 1990: What's in a name? Reputation building and corporate strategy. *Academy of Management Journal*, 33(2), 233–58.

Fortune. 1999: *America's Most Admired Companies*, 139 (4), 68–73.

Fryxell, G. E. and Wang, J. 1994: The Fortune corporate "Reputation" Index: reputation for what? *Journal of Management*, 20 (1), 1–14.

Gioia, D. A., Schultz, M., and Corley, K. G. 2000: Organizational identity, image and adaptive instability. *Academy of Management Review*, 25 (1), 63–81.

Hall, R. 1992: The strategic analysis of intangible resources. *Strategic Management Journal*, 13, 135–44.

Hatch, M. J. and Schultz, M. 1997: Relations between organizational culture, identity and image. *European Journal of Marketing*, 31, 356–65.

Hatch, M. J. and Schultz, M. 2000: *Bringing the corporation into corporate branding*. Paper presented at the 2000 Academy of Management Meeting. Toronto, Canada, August, 2000.

Hatch, M. J. and Schultz, M. 2001: Are the strategic stars aligned for your corporate brand? *Harvard Business Review*, 79 (2), 128–34.

Hofstede, G., Neuijen, B., Ohay, D., and Sanders, G. 1990: Measuring organizational cultures: A qualitative and quantitative study across twenty cases. *Administrative Science Quarterly*, 35, 286–316.

Keller, K. L. 2000: Building and managing brand equity. In Schultz et al. (eds.), *The Expressive Organization: Linking Identity, Reputation, and the Corporate Brand*. New York: Oxford University Press, 115–37.

Kotter, J. P. and Heskett, J. L. 1992: *Corporate Culture and Performance*. New York: Free Press.

O'Reilly, C., Chatman, J., and Caldwell, D. 1991: People and organizational culture: A Q-sort approach to assessing person–organization fit. *Academy of Management Journal*, 34, 487–516.

Peteraf, M. A. 1993: The cornerstones of competitive advantage: a resource based view. *Strategic Management Journal*, 14, 179–91.

Pfeffer, J. 1994: *Competitive Advantage Through People*. Boston, MA: Harvard Business School Press.

Rao, H. 1994: The social construction of reputation: certification contests, legitimation, and the survival of organizations in the American industry: 1895–1912. *Strategic Management Journal*, 15, 29–44.

Rousseau, D. 1990: Quantitative assessment of organizational culture: the case for multiple measures. In B. Schneider (ed.), *Frontiers in Industrial and Organizational Psychology*, vol. 3. San Francisco: Jossey-Bass, 153–92.

Sackmann, S. 1992: Culture and subcultures: an analysis of organizational knowledge. *Administrative Science Quarterly*, 37, 140–61.

Saffold, G. 1988: Culture traits, strength, and organizational performance: moving beyond "strong" culture. *Academy of Management Review*, 13, 546–58.

Schein, E. 1985: *Organizational Culture and Leadership*. San Francisco: Jossey-Bass.

Schein, E. 1996: Three cultures of management: the key to organizational learning. *Sloan Management Review*, 38 (1), 9–20.

Turban, D, and Greening, D. 1997: Corporate social performance and organizational attractiveness to prospective employees. *Academy of Management Journal*, 40, 658–72.

Tushman, M. and O'Reilly, C. 1997: *Winning Through Innovation*. Boston, MA: Harvard Business School Press.

Vergin, R. and Qoronfleh, M. 1998: Corporate reputation and the stock market. *Business Horizons*, Jan–Feb, 19–26.

Appendix G

Circle 1 = Highly Disagree (HD) Circle 2 = Disagree (D) Circle 3 = Neither Agree or Disagree (N) Circle 4 = Agree (A) Circle 5 = Highly Agree (HA) Circle 6 = Cannot Judge (CJ)

	This company takes risks, is innovative, and is open to experimenting with different ways of doing things. HD D N A HA CJ 1 2 3 4 5 CJ Please Circle # Choice	This company pays attention to details, strives for precision, and stresses the importance of analytical skills. HD D N A HA CJ 1 2 3 4 5 CJ Please Circle # Choice	This company is achievement-oriented, has high expectations and demands results from its employees. HD D N A HA CJ 1 2 3 4 5 CJ Please Circle # Choice	This company is an aggressive competitor and takes advantage of opportunities. HD D N A HA CJ 1 2 3 4 5 CJ Please Circle # Choice
Oracle	1 2 3 4 5 CJ	1 2 3 4 5 CJ	1 2 3 4 5 CJ	1 2 3 4 5 CJ
Hewlett Packard	1 2 3 4 5 CJ	1 2 3 4 5 CJ	1 2 3 4 5 CJ	1 2 3 4 5 CJ
Cisco Systems	1 2 3 4 5 CJ	1 2 3 4 5 CJ	1 2 3 4 5 CJ	1 2 3 4 5 CJ
3Com	1 2 3 4 5 CJ	1 2 3 4 5 CJ	1 2 3 4 5 CJ	1 2 3 4 5 CJ
National Semi-conductor	1 2 3 4 5 CJ	1 2 3 4 5 CJ	1 2 3 4 5 CJ	1 2 3 4 5 CJ
Apple Computers	1 2 3 4 5 CJ	1 2 3 4 5 CJ	1 2 3 4 5 CJ	1 2 3 4 5 CJ
Sun Microsystems	1 2 3 4 5 CJ	1 2 3 4 5 CJ	1 2 3 4 5 CJ	1 2 3 4 5 CJ

Circle 1 = Highly Disagree (HD) Circle 2 = Disagree (D) Circle 3 = Neither Agree or Disagree (N) Circle 4 = Agree (A) Circle 5 = Highly Agree (HA) Circle 6 = Cannot Judge (CJ)

	This company is supportive of its employees, shares information with them and praises their performance.	This company is noted for its high pay for performance and offers opportunities for professional growth.	This company has a team-oriented work environment and encourages collaboration.	This company's decision-making process is decisive, and entails little conflict.
	HD D N A HA CJ 1 2 3 4 5 CJ Please Circle # Choice	HD D N A HA CJ 1 2 3 4 5 CJ Please Circle # Choice	HD D N A HA CJ 1 2 3 4 5 CJ Please Circle # Choice	HD D N A HA CJ 1 2 3 4 5 CJ Please Circle # Choice
Oracle	1 2 3 4 5 CJ	1 2 3 4 5 CJ	1 2 3 4 5 CJ	1 2 3 4 5 CJ
Hewlett Packard	1 2 3 4 5 CJ	1 2 3 4 5 CJ	1 2 3 4 5 CJ	1 2 3 4 5 CJ
Cisco Systems	1 2 3 4 5 CJ	1 2 3 4 5 CJ	1 2 3 4 5 CJ	1 2 3 4 5 CJ
3Com	1 2 3 4 5 CJ	1 2 3 4 5 CJ	1 2 3 4 5 CJ	1 2 3 4 5 CJ
National Semi-conductor	1 2 3 4 5 CJ	1 2 3 4 5 CJ	1 2 3 4 5 CJ	1 2 3 4 5 CJ
Apple Computers	1 2 3 4 5 CJ	1 2 3 4 5 CJ	1 2 3 4 5 CJ	1 2 3 4 5 CJ
Sun Microsystems	1 2 3 4 5 CJ	1 2 3 4 5 CJ	1 2 3 4 5 CJ	1 2 3 4 5 CJ

Intangible Capital in Industrial Research: Effects of Network Position on Individual Inventive Productivity

Jukka-Pekka Salmenkaita

Keywords: Industrial research, social network analysis, inventive productivity.

Abstract

The paper investigates the effects of collaboration networks on inventive productivity within an industrial research environment. A distinction is made between two kinds of network structures: structural holes offering information brokerage opportunities to individuals and network closures supporting co-specialization of individuals. Hypotheses regarding the effects of network positions on the development of technological know-how are tested based on longitudinal individual-level network data. The analysis provides partial support of both the structural hole and the network closure argument. However, contrary to literature emphasizing innovation via inter-organizational collaboration, the positive effects of ties between the research center and business units are highlighted. The interpretation of these results seems to call for more refined models of firm boundaries to better explain how the research activities are organized within firms.

Introduction

Innovation, along with the associated "creative destruction" that occurs as successful new products, production methods, and ways of organizing economic activity replace old ones, forms the core of competition in the capitalist process (Schumpeter, 1942). Start-ups innovate in order to identify, pioneer, and capture key positions in emerging markets. Incumbents innovate in order to leverage their existing assets, match changing customer preferences, and expand their markets. Thus, understanding the sources of innovation is a high priority for all managers. Indeed, theories of knowledge-based competition and intangible capital are receiving increasing attention (e.g., Grant, 1996; Nahapiet and Ghoshal, 1998).

This paper focuses on one major antecedent of successful innovation – inventive activity – and the related form of intangible capital – technological know-how. The development of technological know-how is investigated within the context of the industrial research laboratory, a form of organizing inventive activities that dates back to the early twentieth century (Mowery, 1990). A stream of research inspired by Schumpeter (1934, 1942) has studied the factors that affect inventive productivity in industrial research. Specifically, Schumpeter surmised that "technological progress is increasingly becoming the business of trained specialists who turn out what is required and make it work in predictable ways" and, as a result, "innovation itself is being reduced to routine" (1942: 132). The stylized context for this routinization of innovation was "the perfectly bureaucratized giant industrial unit" (1942: 134), although in his earlier work he emphasized that inventions are economically irrelevant unless put into practice (1934: 88), and presumed that the new combinations (i.e., innovations) do not generally arise from old firms but from new ones beside them (1934: 66). Thus, some words of caution are in order regarding the present study. First, regularities in inventive activities may not be directly associated with regularities in innovative activities. In extreme cases, some factors that benefit inventive activities may reduce the chance of fully translating the inventions into practical use, i.e., hinder innovation. Second, innovations may arise from different contexts via distinct mechanisms, and as such the dynamics related to industrial research are only a very specific lens to the issue. Nevertheless, investigation of the regularities of inventive activities within the context of industrial research is an important intermediary stage for understanding which, if any, aspects of the innovation process can be routinized, and for identifying boundary conditions for this routinization.

This study aims to contribute to this research agenda by examining the role of collaboration networks as a form of intangible capital in inventive activities. A competence-based approach is used to develop measures for technological know-how internal to the organization (Henderson and Cockburn, 1994, 1996). The innovation network approach provides the rationale to extend the analysis beyond the focal organization (Powell et al., 1996). Finally, the social capital approach is used to integrate the external and internal perspectives (Burt, 2000).

The measures of factors contributing to inventive productivity are based on archival data that is readily available to managers of R&D organizations. Thus, the methods applied in this paper can be adapted for practical research portfolio management purposes. The present analysis quantifies the relative effects of internal and external networks on inventive productivity, providing managers with insight on how to monitor and fine-tune their research organizations.

The research contributes to the existing competence-based strategy literature by applying measures that address the role of individual-level collaboration networks in the development of technological know-how. The network measures are one approach to investigating aggregate firm-specific differences in inventive productivity, i.e., competence, in detail (cf. Henderson and Cockburn, 1994). The empirical results are a direct test, within the specific context of industrial R&D, of the relative importance of competing hypotheses developed in the social capital literature regarding the role of network closures and structural holes as mechanisms of social capital creation.

Theoretical Background and Hypotheses Development

This section develops the hypotheses regarding the sources of inventive productivity drawing on three streams of previous research. First, competence-based perspective emphasizes hard-to-imitate resources, e.g., technological know-how embedded within firm-specific routines and collaboration structure, as sources of competitive advantage. Second, the innovation networks perspective elaborates on access to and absorption of external knowledge as factors critical to inventive performance. Third, social capital literature examines the relative merits of tight collaboration closures and brokerage across closures as mechanisms supporting inventive activities.

The argumentation is based on the following network terminology. *Internal collaboration network* refers to the collaborative ties between members of the case organization, i.e., the personnel of the research center. *Boundary-spanning* refers to network nodes that have connections, i.e., collaborative ties, outside the case organization. *Network closures* are parts of the internal network in which there are tight connections between the nodes. There are a number of different network analysis definitions that define what is "tight enough" for the nodes to be considered to form a closure. *Structural holes* are connections between otherwise separate parts of the network. These separate parts can be network closures, and often the structural hole is bridged by only one node that has connections to each closure. Detailed discussion of network terminology and methods to operationalize the concepts are presented by Wasserman and Faust (1994).

Competence-based perspective on within-firm sources of inventive productivity

The resource-based view of the firm proposes that firms are essentially pools of heterogeneous resources, such as technological know-how (e.g., Wernerfelt, 1984). If the resources are hard to imitate or replicate, the firm's unique resource combinations may provide a source of temporarily sustainable competitive advantage (Amit and Schoemaker, 1993). The firm's ability to develop and apply new resources has been referred to as "competence" (Prahalad and Hamel, 1990), "capability" (Leonard-Barton, 1992) or "dynamic capability" (Teece et al., 1997). In the context of industrial research, research scientists and engineers apply their skills, such as cryptographic expertise or knowledge of data-mining methods, to various R&D projects. The projects yield novel ways to apply technological know-how to produce practical results, i.e., inventions. Inventions, some of which the firms patent, are thus a measure of the technological know-how accumulated.

The resource-based framework can be used to conceptualize firm-level processes of invention and innovation. Schumpeter defined innovation as an activity in which an entrepreneur "carries out new combinations" in the economy (1934: 132). In the realm of industrial research, inventions are the result of R&D personnel carrying out new combinations of technological know-how. Firms and individuals gain experience of those entrepreneurial opportunities they undertake, and the increased knowledge opens up new areas of entrepreneurial activity (Penrose, 1959). The new knowledge, a part of which is tacit, can be embodied in the skills of employees (Polanyi, 1958), or become routinized in the firm's way of operating (Nelson and Winter, 1982). Neither tacit knowledge nor firm-specific routines are easily imitated or replicated by competitors. Technological progress often follows paradigms as the R&D activities cumulatively increase certain performance characteristics of a technology, thus creating technological trajectories (Dosi, 1982, 1988). Given the cumulative, and partially tacit, nature of technological know-how, we posit:

Hypothesis 9.1. Technological competence is cumulative at the individual level in the sense that previous contribution to inventive accomplishments positively affects the likelihood of producing new inventions.

To a degree, the above hypothesis is confounded by several different mechanisms. First, it is conceivable that the individuals have semi-permanent characteristics that affect their inventiveness (e.g., Amabile, 1988). These characteristics could include specific skills (e.g., ability to communicate ideas clearly), norms (e.g., likes to challenge status quo), psychological features (e.g., creativity), as well as personal history (e.g., highest educational degree completed). Second, there could be differences in "technological opportunity" that affect the fertility of R&D activities between technological fields (e.g., Cohen, 1995). This effect could be either "global," in the sense that all the R&D efforts by various organizations in a given field result in a particularly high or low inventive productivity, or "local," in the sense that a specific organization provides extensive management attention (including, e.g., special incentives) and support (e.g., priority in patenting process) to R&D activities in select fields. Although it is not within the scope of the present study to extensively examine the specific effects of individual characteristics, the hypothesized combined cumulative effect will provide a baseline for investigating the effects of network position on inventive productivity. To a degree, variation in technological opportunity can be controlled by dummy variables of broad technological fields.

R&D activities confront uncertainty regarding both emerging business needs and development of competing technological trajectories. As a result of bounded rationality considerations, scientists and engineers resort to selective communication channels and information filters, as well as tacit problem-solving strategies (Henderson and Clark, 1990). Communication channels are formed between groups and individuals with interacting tasks. Individuals focus on information which their previous experience suggests as relevant. Successful solutions to old problems are adapted to new ones with relatively little conscious effort. Henderson and Cockburn hypothesized that R&D scientists and engineers "embedded within particular firms develop deeply embedded, taken for granted knowledge or unique modes of working together that make the group particularly effective" (1994: 65). In terms of social network analysis, a group of individuals with direct connections to each other forms a closure within the broader network (cf. Burt, 2001). For example, all the members of a research laboratory form one kind of network, and within that network there can be several project teams. The members of those teams, in so far as they work closely together, form closures within the broader network of the research laboratory. Another example of network could be researchers working on some next-generation technology. This network could consist of personnel from several laboratories. Within that network there can be numerous small groups active, for example, in standardization of specific aspects of the technology. Those groups form closures within the broader network. Based on these arguments we propose:

> Hypothesis 9.2. Membership in a closure within the collaboration network positively affects the inventive productivity of the associated individuals.

Close collaboration with other scientists and engineers can facilitate inventive productivity by several mechanisms. First, it provides access to partially tacit knowledge, thus enabling more effective transfer of knowledge between individuals than would be possible otherwise (cf. Nonaka, 1994; Nonaka and Konno, 1998). Second, collaboration is likely to proxy some amount of trust between the collaborators. The trust can be either a prerequisite or byproduct of collaboration; nevertheless, it may support an exchange of ideas, especially in the early phases of the inventive process. Third, the collaborating individuals may develop co-specialized skills and knowledge that as a combination provide fertile ground for inventive activities.

Innovation networks perspective on distributed sources of inventive productivity

Cohen and Levinthal (1989, 1990) note that internal R&D contributes to the firm's ability to evaluate and utilize innovations external to the firm, that is, it provides them with "absorptive capacity." Powell et al. (1996) expand this view by arguing that "the locus of innovation is found within the networks of inter-organizational relationships" (1996: 142) if the knowledge base of an industry is complex and expanding. Informal and formal collaborative relationships, especially in R&D, enable knowledge transfer across organizational boundaries. Thus, individuals and organizations in central network positions have timely access to information of new breakthroughs or obstacles, and are thus better able to leverage their own R&D capabilities.

The innovation networks contribute to inventive productivity via several mechanisms. First, knowledge of the R&D capabilities of potential partners is often hard to acquire without direct ties. Some collaboration, perhaps informal, can act as a prerequisite for the formation of more complex joint arrangements that aim to combine complementary capabilities, reduce risks, or seek synergies in R&D efforts

(Teece, 1992, 1998). Collaboration supports strategic structuring of R&D activities into the most effective make-or-buy variants (Pisano, 1990, 1991). Second, knowledge transfer via collaborative ties may provide insights that lead to new inventions. Inventive productivity is increased almost immediately due to better flow of knowledge (cf. Zucker et al., 1998). Third, the diverse sources of knowledge increase the participants' awareness of future R&D opportunities. Inventive productivity is increased with some delay, as the R&D portfolio starts to reflect the new insights from external sources. Of these arguments, the second and third can be examined at the individual level:

Hypothesis 9.3a. Boundary-spanning network connections positively affect the inventive productivity of the associated individuals.

However, collaboration across organizational boundaries may entail significant communication and coordination costs for the "gatekeepers" (Allen, 1977). The gatekeepers have to resolve differences due to conflicting organizational values, priorities, working practices, and so on. Thus, although the boundary-spanning connections may be beneficial for research groups, the gate-keeping activities may be burdensome at the individual level.

Hypothesis 9.3b. Boundary-spanning network connections positively affect the inventive productivity of research groups.

In addition to direct boundary-spanning network connections, scientific contributions can also be interpreted as "currency of exchange" (Pake, 1986) in the communities of practice furthering technological development (cf. Rappa and Debackere, 1992; Brown and Duguid, 1998). Thus, a less strict operationalization of the boundary-spanning activities could be based on all academic contributions, whether or not those outputs involve cross-organizational co-authorship. In this study, internal networks refer to the collaborative relations between the researchers of the case organization. For example, three researchers jointly producing an invention are considered to be connected to each other in the internal network. This internal network may span organizational boundaries if the act of jointly producing also involves members from other organizations. For example, a researcher can make an invention with a marketing manager from some business unit of the company, or a researcher can produce an academic publication in collaboration with university researchers. Thus, although the connections in the internal network as well as the connections from internal to external network are formed by the same mechanisms, i.e., joint production of inventive or academic outputs, it is still analytically feasible to distinguish nodes of internal network that are tied to the external network, i.e., boundary-spanning network connections.

Social capital perspective on integration of internal and external sources of inventive productivity

Nahapiet and Ghoshal define social capital as "the sum of actual and potential resources embedded within, available through, and derived from the networks of relationships possessed by an individual or social unit" (1998: 243). Whereas the competence-based perspective examines the processes through which new resources are generated and existing resources are utilized in ways that provide competitive advantage, the social capital perspective emphasizes that the exchange and combination of information is embedded in a network of relationships. Burt (2000, 2001, 2002) distinguishes between two different network mechanisms that make somewhat contradictory predictions about how social capital can facilitate the creation of competitive advantage. *Network closures*, i.e., parts of networks in which the nodes are closely connected to each other, provide access to information and support the development of trust, common norms, and shared language. *Structural holes* are connections between otherwise separate parts of the network. In so far as different information flows in the different parts of the network, the structural holes offer opportunities for information brokerage. Burt argues that as the "structural holes are gaps between non-redundant sources of information" (2000: 10), contact networks rich in structural holes are the ones that provide entrepreneurial opportunities. For example, within an industrial research center there can be one research group specializing in data-mining methods, and another in user interface design. In so far as these groups do not typically interact, a researcher interested in applying data-mining techniques for user interface optimization could be in a brokerage position that spans this structural hole. Hargadon and Sutton (1997) present a detailed analysis of how a product design company has organized itself to leverage the brokering possibilities.

The closure argument at the individual level of analysis underlies hypothesis 9.2. At the social unit level of analysis, all organizations are network closures to some degree. Thus, hypothesis 9.3 is a structural hole argument. Structural holes, however, can also be examined in the within-firm network structure. Especially in expert organizations, including research laboratories, the individuals possess highly specialized bodies of knowledge. Both the awareness of "who-knows-what" and the appreciation of the kind of problems to which the knowledge could be applied are likely to correspond to the collaboration network structure within the organization. That is, individuals with collaborative relationships tend to be more familiar with each other's areas of expertise than unconnected individuals. An individual with collaborative ties to otherwise unconnected experts thus spans a structural hole in the within-firm network structure. In terms of network analysis, positions of high betweenness centrality provide brokerage opportunities between otherwise disconnected parts of the network (Burt, 2000). Thus we propose:

Hypothesis 9.4a. Information brokerage, in the sense of individual's betweenness centrality within the collaboration network, positively affects the inventive output of the individual.

Table 9.1   Summary of internal and external perspectives on network benefits

	Competence-based internal perspective	Innovation networks external perspective
Closure benefits	Co-specialization; sharing of tacit knowledge	Flow of information is facilitated by co-membership in communities-of-practice
Brokerage benefits	Unique resource combinations	Access to non-redundant sources of information

Both internal information brokerage and external boundary-spanning activities require translation between diverse perspectives and specialized terms, as well as unique routines and working practices. Although it is reasonable to presume that these communication and coordination costs are less notable in within-firm brokerage situations than in boundary-spanning collaboration, information brokerage may also be beneficial at the level of research groups rather than that of individual R&D personnel.

Hypothesis 9.4b. Opportunities for information brokerage positively affect the inventive productivity of research groups.

To summarize, technological competence is hypothesized to be cumulative, partly tacit and embedded in routines. Network closures facilitate the use of tacit knowledge and the formation of routines. Both external boundary-spanning activities and internal information brokerage offer opportunities to create novel combinations of diverse sources of expertise, i.e., inventions. However, whether the boundary-spanning and information brokerage activities are beneficial at the individual or group level of analysis is an open empirical question. The synthesized framework is presented in Table 9.1.

The hypotheses are examined at the individual level of analysis. At this level, the collaboration networks are clearly defined, i.e., each individual is a node in the network and ties between nodes indicate collaborative relationships; thus, the use of network measures is feasible. Some of the measures can also be aggregated to the level of technological programs. At the program level of analysis, a number of control variables regarding the research portfolio structure can be introduced. These effects are investigated in Salmenkaita (2001).

Data Sources, Measures and Analysis

This section describes the data sources, operationalization of the measures, and analysis methods.

Case organization and data sources

The main network data comprises the industrial research activities of one major communications equipment corporation in 1995–2000. In 2000, the corporation had sales of 30.376 billion euros (6.191 billion euros in 1995), total R&D expenditures of 2.584 billion euros (425.7 million euros in 1995), and employed some 60,000 people (32,000 in 1995) (Nokia Corporation, 1996, 2001). The company has one corporate research center that serves the business units. The business units have their own R&D activities, mainly at the product development end of the research–development continuum, which are not included in the study. It should be noted that only a small portion of all the R&D effort, slightly over 5 percent in 2000 as measured by person-years, is conducted at the corporate research center. The research center is divided into seven laboratories based on broad technological fields (e.g., software, electronics), and the center operates at several sites in Europe, Asia, and the US. The laboratories are divided into research groups, based on more specific technological disciplines (e.g., software architectures, data mining). The research center explores new technological opportunities and develops technological know-how for both current and future business areas of the corporation.

The research activities are divided into projects, that is, the organization is a matrix of research groups and projects. Funding for the research activities is negotiated on a per project basis, with the majority of funds coming directly from the business units. Following common research management practices, the corporation has also "earmarked" some funding for research that is beyond the current interests of business units (cf. Buderi, 2000). Participation in research collaboration is partly funded by external sources, e.g., European Union Framework Programmes[1] and Finnish National Technology Agency.[2]

The years 1995–2000 were a period of significant growth both for the company and for the research center. The growth was primarily internally generated, i.e., there were relatively few acquisitions and no mergers during the period. The research center grew from fewer than 500 to slightly over 1,000 employees during the period. Technological change during the period was rapid. Mobile terminals became increasingly miniaturized, incorporated new features and supported new radio transmission technologies. Communication networks evolved from circuit-switched voice networks to packet-switched data networks with integrated support for Internet protocols. With regard to emphasis on internal growth within a technologically turbulent environment, the case setting is representative of the kind of incumbent firms Schumpeter (1942) and Penrose (1959) envisioned in their theories of economic change and growth of firms.

Corporate research centers play a dual role in internal inventive activities, as well as in monitoring developments in the external environment (Mowery, 1983). Promising inventive outputs are first documented in invention reports, in which the employees disclose their findings to the employer. Thereafter, the employer has an opportunity to evaluate the importance of the findings and seek patent protection for the invention, if appropriate. The patenting process is costly; therefore, a decision to seek patent protection for an invention is a measure of the perceived quality of the invention (Patel and Pavitt, 1995). For the purposes of this study, the author had access to the company's internal database of inventions created by the research center's personnel. Compared to publicly available patent data, this arrangement had several benefits. First, it made it possible to distinguish between inventors from different organizational sub-units (e.g., corporate research center, business units), which is not feasible based on information available in patents. Second, it provided a richer view of the collaborative structure, since inventions that eventually were not patented were also accounted for in examining the collaboration network. Third, it removed the artificial collaborative relationships that are formed when a corporation merges several invention reports into one patent application.

The monitoring of the external environment is often active in the sense that the R&D personnel collaborate with people external to the company, such as in joint projects with university researchers (Debackere and Rappa, 1994). The outputs of the boundary-spanning collaboration include publications and conference presentations, the co-authors of which are from different institutions. This data is available from specialized databases, including Science Citation Index by the Institution for Scientific Information and INSPEC by IEEE, which are used in this research. Science Citation Index covers a broad range of major publications in a multitude of scientific fields, but lacks information about conference activities. INSPEC focuses on electrical engineering and related disciplines and also has coverage of key conferences in these areas. The results of author affiliation-based searches from the Science Citation Index and INSPEC databases were combined, and duplicate entries were manually removed.

The journals and conferences covered by the external publication databases have selective inclusion processes, typically based on peer review. Thus, by examining outputs that have passed the review, a standard of quality regarding the academic and scientific contributions is created. Also, the associated collaborative relationships are likely to entail a significant investment of time and effort between the parties (Cockburn and Henderson, 1998). However, there are some caveats. First, the author–affiliation relationships are not fully recorded on a one-to-one basis in the databases. The standard bibliometric approach to mitigating this problem is to first identify the affiliations of sole authors, and to then apply those author affiliations to database items with several authors in which the same authors also participate. In this study, an internal company phonebook was used to complement this method. Second, some authors can have multiple affiliations and may exercise discretion regarding which affiliations they report to each publication. Thus, an affiliation-based search can miss some relevant items. Unfortunately, complementary individual-based searches, besides being costly compared to a relatively simple affiliation-based search, would create new problems with the data. Specifically, different individuals can have similarly abbreviated names in the databases, or one individual can have differently abbreviated names in the databases.

In total, the databases comprise 2,427 records of inventions and 443 records of academic outputs for years 1995–2000. Of the inventions, 360 (14.8 percent) involved collaboration across research center boundaries, mainly with personnel in business units (331 items). Of the academic outputs, 169 (38.1 percent) involved collaboration across research center boundaries, mainly with researchers at universities

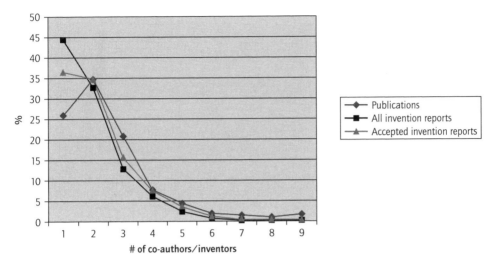

Figure 9.1   Distribution of number of co-authors / inventors; 443 publications, 2,427 invention reports, of which 1,135 accepted for patenting

(145 items). Figure 9.1 shows the distribution of the number of co-authors and co-inventors for publications, inventions and inventions accepted for patenting.

For the network analysis, *an internal collaboration matrix* was constructed for each year from 1995 to 2000 by combining the co-inventor relationships from the internal invention report database and the co-author relationships from the external database. The internal collaboration matrix is a square matrix in which the nodes represent collaboration relationships among those R&D personnel that have participated in at least one invention report or publication in a given year. The size of the internal collaboration matrix increased from 137 in 1995 to 445 in 2000. In Figures 9.2 and 9.3, illustrations of the structure of the internal collaboration matrix are presented for years 1997 and 1998, respectively. The matrices are hierarchically clustered using Johnson's hierarchical clustering by applying the single link (minimum) method on the similarity data. The algorithm finds nested partitions of the nodes of the matrix, starting from all nodes in different clusters, and then joins together those nodes that are most similar. The figures illustrate that there are some areas in the collaboration network where dozens of researchers have overlapping collaborative relationships. In addition, there are many smaller collaborative clusters, and a number of the researchers remain unconnected in terms of collaborative ties.

The databases were also used to construct *an external collaboration matrix* for each year from 1995 to 2000. The external collaboration matrix is an affiliation matrix in which the nodes represent collaborative relationships between the R&D personnel and external institutions. Specifically, the number of collaborative ties for each researcher with both business units and other organizations was recorded separately.

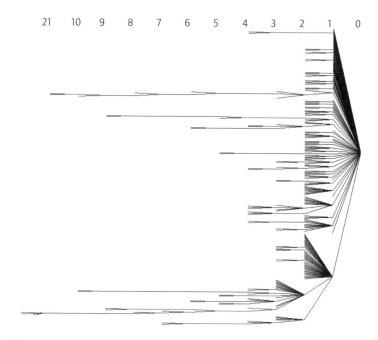

Figure 9.2   Cluster presentation of the internal collaboration matrix; year 1997

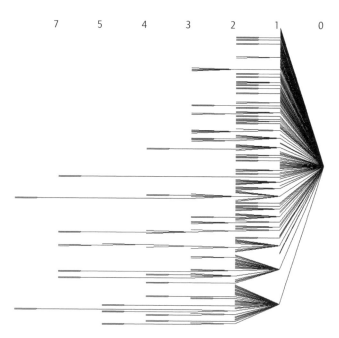

Figure 9.3   Cluster presentation of the internal collaboration matrix; year 1998

In 1995–2000, some 4,000 man-years of research were conducted at the research center. However, a significant share of this effort produced deliverables other than inventions or publications, and was therefore not accounted for in the data sources used in this study. Specifically, the data sources involved 1,717 observations of 881 individuals. The observations were arranged in longitudinal panel format, and all one-year "holes" in the panel were identified. Thus appended, the data set under analysis consists of 1,820 researcher/year observations.

Some technology development tasks, especially in the field of software technologies, result in relatively few potentially patentable outputs. Industrial research often involves a great deal of consulting (Kline and Rosenberg, 1986). Consulting can be critical to the firm's ability to appropriate the technological expertise, but results in few academic or inventive outputs. Moreover, industrial research is not devoid of managerial responsibilities. Even though senior technological experts might prefer to spend their time on inventive activities and scientific research, practical project and personnel management duties may consume a significant share of their attention (cf. Katz and Allen, 1997). Lastly, previous studies have observed that an "induction period" in tasks requiring highly specialized expertise can take up to one year or more before the person assigned to the task is able to fully contribute to new knowledge creation (cf. Katz, 1997). The effects of this induction period are more noticeable in fast-growing organizations, and in those that frequently recruit new graduate students instead of head-hunting seasoned experts. Both conditions apply to the case organization during the observation period.

Measures

Individual technological productivity is measured based on invention reports patented by the company. Inventive productivity INV_P is the sum of contributions to invention reports in a given year by a researcher, with each qualified invention report providing a contribution of one to the sum. For inventions produced by several collaborators, equal contribution by each individual was presumed. Thus, collaboration as such does not increase inventive productivity, i.e., two researchers producing a total of two inventions are allocated an inventive productivity of one regardless of whether each of the two inventions is the product of a single inventor or of the two inventors collaborating on both inventions.

Based on INV_P, an inventive stock variable, INV_S, of individual technological know-how is constructed by adding INV_P to past stock INV_S on a yearly basis. To allow for gradual obsolescence of technological know-how, the invention stock measure is depreciated yearly using a depreciation rate of 0.25 (cf. Henderson and Cockburn, 1996). The values of INV_S for the first year of observation (1995) in this study were based on archival data that covers the whole history of the research center from 1987. Therefore, it was not necessary to estimate starting stock values (cf. Henderson and Cockburn, 1996: 58).

Analogous to inventive productivity and stock measures, academic productivity ACA_P and academic stock ACA_S variables were constructed based on individual contributions to publications and conferences. Academic outputs are indicators of the individual's ability to contribute to the frontiers of knowledge in their fields.

Production of academic outputs involves both formal (e.g., peer review) and informal interaction with the scientific community. The interactions, and especially the informal ones, may involve transfer of knowledge regarding the latest developments in the field. Because of these factors, publication counts can be used as indicators of investment in absorptive capacity (Cockburn and Henderson, 1998). Many academic outputs involve co-authorship of individuals from different institutions. In these instances, the flow of knowledge across organizational borders is likely to be more intense, although not all co-authorships reflect joint research and problem-solving but may, instead, indicate sharing of data or research instruments, for example. The distributions of inventive productivity and stock, as well as academic productivity and stock, are presented in Figure 9.4.

A number of definitions for identifying cohesive subgroups, or closures, within networks have been developed in social network analysis literature (for review, see Wasserman and Faust, 1994: 249–90). The most common definitions for cohesive subgroups within symmetric networks include cliques, n-cliques, n-clans, and n-clubs (Mokken, 1979). Cliques are maximal complete sub-graphs of three or more nodes. The definition is very strict in that the absence of a single tie between network nodes prevents the sub-graph from being a clique. N-cliques, n-clans and n-clubs are definitions that aim to capture the "clique-like" structures that frequently appear in empirical network data. An n-clique is a maximal sub-graph in which the largest geodesic distance between any two nodes is no greater than n. This definition is somewhat loose in terms of identifying cohesive subgroups. For example, two nodes belonging to the same n-clique may have no path connecting them that includes only n-clique members. N-clans and n-clubs are n-cliques that have restrictions that make them more cohesive (Wasserman and Faust, 1994: 260–2).

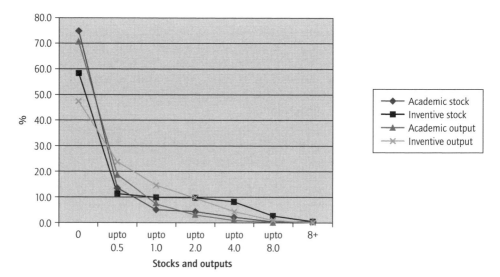

Figure 9.4 Distribution of inventive and academic stocks and outputs; n = 1,820 person-year observations

Specifically, n-clans are n-cliques with a diameter no greater than n. In this study, 2-clans are used as the network definition of cohesive subgroups.

Although n-clan is a reasonably robust definition for a cohesive subgroup within network in terms of connections among nodes, the definition does not account for different strengths of the ties. For example, the definition does not distinguish between three collaborators with one joint output and three collaborators with numerous joint outputs. In practice, the definition allows rather indiscriminate cohesive subgroups to be identified based on single, perhaps ad hoc, acts of collaboration. To better capture the deeply embedded ways of working together, the internal collaboration matrices were dichotomized using a tie-strength of two as cut-off value before the cohesive subgroups were identified. Thus, individuals were considered to be collaborating closely if they were involved in two or more joint outputs in a given year. Individual membership in one or more n-clans is indicated in the binary variable *SCLANB* for each year.

In order to measure opportunities for information brokerage, a measure which identifies those individuals who have connections to otherwise unconnected individuals is required. Betweenness centrality is based on calculating the shortest paths and geodesics among all the nodes in the network (Wasserman and Faust, 1994: 188–92; Freeman, 1979). In the case of several equally short paths between two nodes, the paths are presumed equally likely to be used. In this case, the betweenness index for a network node is the sum of estimated probabilities that the network node is between other nodes. Specifically, let g_{jk} be the number of geodesics connecting nodes j and k, and $g_{jk}(n_i)$ the number of geodesics that contain node i. In this case, node betweenness index for node i is

$$C_B(n_i) = \Sigma_{j<k}\, g_{jk}(n_i)/g_{jk}.$$

The betweenness index is normalized using the number of potential geodesics of the network, thus limiting the value of normalized betweenness centrality *NBETW* to between 0 and 1. That is, normalized network betweenness index for node i is

$$C'_B(n_i) = C_B(n_i)/[(g-1)(g-2)/2].$$

Boundary-spanning across organizational boundaries is measured based on individuals' participation in collaborative outputs in which at least one collaborator was from an external organization. *TIE_EXT* is the number of boundary-spanning opportunities in which the external organization was a university, research institute, or company. *TIE_BU* is the number of opportunities in which the external organization was a business unit of the same company.

The overlapping n-clan membership indicator matrix (n-clan * individual) is used to identify individuals with at least one common n-clan membership (individual * individual). Based on shared n-clan memberships, potential spill-over effects of *NBETW*, *ACA_P*, *TIE_EXT*, and *TIE_BU* measures are studied. Specifically, measures *MNBETW*, *MACA_P*, *MTIE_EXT*, and *MTIE_BU* are calculated for each individual by taking the maximum values of respective individual measures over the individuals sharing n-clan membership.

To allow for the varying role of patent protection in different technological fields, dummy variables are defined based on broad technological areas (laboratories of the research center, e.g., software, electronics, mobile networks). These dummies may also capture laboratory-wide differences in management practices, as well as variance in technological opportunity between the fields. It should be noted that the administrative practices and specialized personnel for protecting intellectual property were the same for all the laboratories during the study period.

Routines provided in UCINET 5 are used for calculation of all the network measures (Borgatti et al., 1999). Descriptive statistics are presented in Table 9.2.

Analysis

The dependent variable, individual inventive productivity INV_P, can receive non-negative values. Unlike many patent count studies made at higher levels of aggregation, the dependent variable is not tied to integer values (cf. Hausman et al., 1984). The hypothesized inventive productivity function is of the form

$$y = f(\mathbf{x}, \mathbf{n}, \mathbf{c}, \mathbf{c} * \mathbf{n}')$$

where $\mathbf{x}$ is a vector of inputs to the inventive process that includes the individual's inventive stock, $\mathbf{n}$ is a vector of network measures, $\mathbf{c}$ is a binary measure of membership in a network closure, and $\mathbf{n}'$ is a vector of network measures related to the network closure. Previous studies have often assumed the patent counts to be generated by a Poisson process (e.g., Henderson and Cockburn, 1994, 1996). As for most patent count data, "the mean is equal to variance" property of the Poisson distribution is not followed by the present invention report data. To partially allow for the skewed distribution of the dependent variable, a logarithmic transformation is performed.

The regression analysis is divided into five models. Model 1a includes the stock variables INV_S and ACA_S, academic productivity ACA_P, as well as the technological area dummies. Models 1b and 1c add network variables ($SCLANB$, $NBETW$, TIE_BU, and TIE_EXT) and variables related to potential spill-over effects within closures ($MACA\_P$, $MNBETW$, $MTIE\_BU$, and $MTIE\_EXT$), respectively. In model 2, which is an alternative to model 1c, the independent variables INV_S, ACA_P, and ACA_S are also entered in logs, with appropriately coded dummy variables indicating zeros. Model 3 investigates the effects of academic stock by including the cross-terms $ACA\_S*SCLANB$, $ACA\_S*NBETW$, ACA_S*TIE_BU, and ACA_S*TIE_EXT in model 1b. Model 4 is otherwise comparable to model 1b, but the dependent variable is academic rather than inventive productivity. Lastly, in model 5 the network variables of model 1b are lagged by one year. Thus, this is the only model in which the dependent and independent variables do not interact indirectly via the collaboration network being constructed from the outputs that also constitute the dependent variable.

The hypothesized effects of the variables on inventive productivity are summarized in Table 9.3.

Table 9.2 Descriptive statistics

Variable	Minimum	Maximum	Mean	Std. Deviation	Correlations															
					1	2	3	4	5	6	7	8	9	10	11	12	13	14	15	
1 INV_P	0	12.58	0.549	0.939	1															
2 INV_S	0	12.67	0.678	1.343	0.29 0	1														
3 ACA_P	0	4.99	0.192	0.424	0.013 0.573	0.147 0	1													
4 ACA_S	0	6.26	0.208	0.559	0.067 0.004	0.381 0	0.331 0	1												
5 SCLANB	0	1	0.190	0.400	0.32 0	0.201 0	0.062 0.008	0.046 0.052	1											
6 NBETW	0	9.985	0.078	0.451	0.248 0	0.153 0	0.076 0.001	0.056 0.017	0.215 0	1										
7 TIE_BU	0	11	0.270	0.840	0.304 0	0.144 0	-0.022 0.338	0.029 0.224	0.139 0	0.168 0	1									
8 TIE_EXT	0	4	0.160	0.460	-0.075 0.001	0.024 0.314	0.4 0	0.181 0	0.002 0.922	0.023 0.336	0.002 0.926	1								
9 MACA_P	0	3.65	0.129	0.458	0.189 0	0.171 0	0.282 0	0.087 0	0.575 0	0.336 0	0.058 0.013	0.078 0.001	1							
10 MNBETW	0	9.99	0.176	0.850	0.222 0	0.091 0	0.037 0.119	0.018 0.454	0.422 0	0.52 0	0.119 0	-0.023 0.322	0.316 0	1						
11 MTIE_BU	0	11	0.360	1.340	0.309 0	0.147 0	0.041 0.08	0.02 0.384	0.544 0	0.206 0	0.317 0	-0.01 0.656	0.386 0	0.403 0	1					
12 MTIE_EXT	0	3	0.086	0.350	0.119 0	0.093 0	0.128 0	0.037 0.119	0.497 0	0.14 0	0.077 0.001	0.238 0	0.599 0	0.305 0	0.343 0	1				
13 SCLANB*ACA_S	0	3.97	0.051	0.291	0.252 0	0.429 0	0.209 0	0.471 0	0.354 0	0.149 0	0.078 0.001	0.052 0.027	0.321 0	0.146 0	0.184 0	0.203 0	1			
14 ACA_S*NBETW	0	5.59	0.030	0.277	0.159 0	0.223 0	0.108 0	0.226 0	0.161 0	0.524 0	0.14 0	0.08 0.001	0.15 0	0.347 0	0.176 0	0.124 0	0.377 0	1		
15 ACA_S*TIE_BU	0	8.25	0.069	0.472	0.174 0	0.217 0	0.124 0	0.29 0	0.102 0	0.166 0	0.493 0	0.046 0.05	0.116 0	0.118 0	0.21 0	0.116 0	0.273 0	0.368 0	1	
16 ACA_S*TIE_EXT	0	10.49	0.079	0.578	-0.012 0.616	0.108 0	0.344 0	0.472 0	-0.002 0.921	0.034 0.149	-0.001 0.971	0.525 0	0.069 0.003	-0.003 0.914	-0.004 0.849	0.117 0	0.19 0	0.135 0	0.11 0	

N = 1820
Pearson correlation
Sig. (2-tailed)

Table 9.3 Summary of variables and hypothesized effects on inventive productivity

Variable name	Proxy for	Definition	Hypothesized effects
INV_P	Output of new (technological) knowledge	# of invention reports accepted for patenting	Dependent variable
INV_S	Accumulated technological knowledge capital	Stock of INV_P calculated using a 25% annual depreciation rate	Positive (Hypothesis 9.1)
ACA_P	Output of new (scientific) knowledge; measure of absorptive capacity	# of journal and conference contributions	
ACA_S	Accumulated scientific knowledge capital	Stock of ACA_P calculated using a 25% annual depreciation rate	
SCLANB	Access to individuals with co-specialized skills and knowledge	Membership in a network closure (n-clan); binary variable	Positive (Hypothesis 9.2)
NBETW	Ability to broker information	Betweenness centrality in the internal collaboration matrix	Positive (Hypothesis 9.4a)
TIE_BU	Access to diverse sources of knowledge	# of network ties in the external collaboration matrix towards business units	
TIE_EXT	Access to diverse sources of knowledge; measure of absorptive capacity and gate-keeping	# of network ties in the external collaboration matrix towards other organizations	Positive (Hypothesis 9.3a)
MACA_P	Access to latest scientific knowledge; measure of absorptive capacity	Maximum of ACA_P of closure members	
MNBETW	Ability to broker information	Maximum of NBETW of closure members	Positive (Hypothesis 9.4b)
MTIE_BU	Access to diverse sources of knowledge	Maximum of TIE_BU of closure members	
MTIE_EXT	Access to diverse sources of knowledge; measure of absorptive capacity and gate-keeping	Maximum of TIE_EXT of closure members	Positive (Hypothesis 9.3b)
TECH_AREA DUMMIES	Cross-sectional variation in technological opportunity and importance of patenting	Dummy variables (8) for broad technological areas	

Results

The regression results for individual level data are presented in Table 9.4. In model 1a, inventive stock is a significant predictor of inventive productivity. Neither academic stock nor productivity has a significant effect on inventive productivity. In model 1b, the network measures are added. The inventive stock variable remains as a significant predictor of inventive productivity. Network positions of high betweenness centrality, which support structural hole arguments, are associated positively with inventive productivity. In addition, memberships in network closures are associated with inventive productivity. Ties to business unit personnel are associated positively with inventive productivity. However, ties to individuals outside the company are associated negatively with inventive productivity.

Model 1c includes the network measures related to potential spill-over effects within network closures. The coefficients for variables introduced in the previous models retain their signs and significance. Neither brokerage opportunities nor academic productivity seems to provide spillovers to close collaborators. Both internal and external ties are associated with spillovers with regard to inventive productivity – positive in case of internal business unit ties and negative in case of external ties.

Model 2 provides the same results as model 1c, with two exceptions. First, academic productivity has a positive effect on inventive productivity, but the dummy variable denoting an academic productivity of zero also has a significant positive coefficient. Second, the negative sign of the coefficient for external ties is no longer significant.

In model 3, academic stock in combination with membership in a network closure is associated positively with inventive productivity, whereas the other cross-terms do not have significant effects.

In model 4, academic stock has a strong positive effect, inventive stock a relatively weak but statistically significant positive effect, and inventive productivity has no effect on academic productivity. As with model 1b, membership in a network closure and brokerage positions have a positive effect. However, as can be expected based on the collaboration patterns related to inventive and academic outputs, external ties are positively and business unit ties negatively associated with academic productivity.

In model 5, inventive stock has a positive effect on inventive productivity, whereas academic stock and productivity do not. Of the lagged network variables, brokerage positions and ties to business units have positive effects. Interestingly, past membership in a network closure does not have a positive effect on inventive productivity.

Overall, the regression results support hypothesis 9.1 regarding the cumulative nature of technological knowledge capital. However, it should also be noted that of all the research personnel, only a subset actively produces outputs of the kinds measured in this study. The hypothesized cumulative nature of the knowledge capital can also be interpreted as a hypothesis of the membership dynamics of that active subset. That is, the stronger the positive association between technological knowledge capital stock and inventive productivity, the more stable the "inventive core" of research personnel. The core–periphery dynamics have been discussed in

Table 9.4   Regression results, OLS-model

	Equation						
	(1a)	(1b)	(1c)	(2)	(3)	(4)	(5)
Intercept	0.375**	0.285**	0.281**	0.0908	0.291**	0.0734**	0.348**
	(.022)	(.021)	(.021)	(.050)	(.021)	(.013)	(.035)
INV_S	0.0716**	0.0457**	0.0457**	0.132**	0.0398**	0.0144**	0.0609**
	(.008)	(.007)	(.007)	(.028)	(.007)	(.004)	(.010)
INV_P						0.00231	
						(.006)	
ACA_P	−0.00263	0.00405	0.00576	0.142*	−0.00101		0.03859
	(0.023)	(.023)	(.024)	(.058)	(.023)		(.030)
ACA_S	−0.0173	−0.00524	−0.00589	0.0551	−0.0307	0.100**	−0.0106
	(.018)	(.017)	(.017)	(.046)	(.020)	(.010)	(.023)
SCLANB		0.238**	.242**	0.242**	0.210**	0.0409**	0.0112
		(.022)	(.030)	(.029)	(.023)	(.014)	(.036)
NBETW		0.0990**	0.0941**	0.0977**	0.105**	0.0340**	0.0567*
		(.019)	(.022)	(.022)	(.022)	(.012)	(.024)
TIE_BU		0.114**	0.108**	0.105**	0.119**	−0.0167**	0.0263
		(.010)	(.011)	(.011)	(.012)	(.006)	(.015)
TIE_EXT		−0.0612**	−0.0463*	−0.0280	−0.0635**	0.181**	−0.0411
		(.020)	(.021)	(.021)	(.022)	(.011)	(.031)
MACA_P			−0.00462	0.00655			
			(.026)	(.026)			
MNBETW			0.00475	0.00395			
			(.013)	(.013)			
MTIE_BU			0.0177*	0.0188*			
			(.008)	(.008)			
MTIE_EXT			−0.0763*	−0.0740*			
			(.032)	(.032)			
SCLANB*ACA_S					0.141**		
					(.037)		
ACA_S*NBETW					−0.0273		
					(.039)		
ACA_S*TIE_BU					−0.0163		
					(.022)		
ACA_S*TIE_EXT					0.0116		
					(.019)		
N	1820	1820	1820	1820	1820	1820	913
R-sqr	0.168	0.304	0.308	0.313	0.309	0.357	0.186

* Significant at the 5% level.
** Significant at the 1% level.
Standard errors in parentheses.
All models include 8 dummy variables for technological areas.
In model 2, ACAS, INVS and ACAP are entered in logs with appropriate coded dummy variables.
In model 5, SCLANB, NBETW, TIE_BU and TIE_EXT are lagged by one year.

the learning literature regarding "communities of practice" (Brown and Duguid, 1991, 1998).

From the knowledge capital perspective, both academic outputs and inventions qualify as indicators of the ability to contribute to the frontiers of knowledge. Consequently, both academic and inventive stock measures should have similar positive effects on inventive productivity. Interestingly, the academic stock variable is insignificant as a predictor of inventive productivity in all the models. This suggests that the "inventive community" is distinct from the more general "community of individuals in the frontiers of knowledge". Model 1c also includes academic outputs as one of the potential sources of spillovers between individuals within a network closure. If this variable had positive effects on inventive productivity, we could interpret production of inventive and academic outputs as forms of co-specialization among research personnel. However, this interpretation is not supported by the data.

Brokerage opportunities do not seem to provide spillovers to close collaborators. Assuming a static network structure, brokerage opportunities could be interpreted as a form of intangible capital appropriable by the individual. In practice, however, the network structure is dynamic. The brokerage performed by an individual will increase the knowledge flow between the previously separated parts of the network, and in so far as joint efforts seem beneficial (as indicated by the positive effects on the individuals initially active in brokering), a closure encompassing both sides of the structural hole can emerge. Thus, brokerage can be an important mechanism in the evolution of network structures, even if the brokerage does not seem to offer spillover benefits in the static investigation of networks. In fact, in model 5 prior brokerage position has significant positive effects, whereas previous membership in a network closure does not. Thus, the results support an interpretation of brokerage positions as a form of intangible capital, while the closures are associated with the realization of the value of potentially complementary knowledge capital stocks (cf. Burt, 2001).

Both internal and external ties are associated with spillovers – positive for internal business unit ties and negative for external ties. This seems to reflect a relationship between inventive productivity in industrial research and the relatedness of research activities to the firm's main operations. Research activities in which business unit personnel participate seem to be especially productive, as measured by inventions qualified for patenting. Several factors may contribute to this finding. First, the patenting process is selective and inventions related to existing operations may be more likely to be perceived as important enough for patenting. Indeed, of the 2,096 inventions that did not involve co-inventors from business units, 43.5 percent were accepted for patenting, compared to 67.4 percent of the 331 inventions with at least one co-inventor from a business unit. Second, business unit personnel may choose to invest their time and effort in collaborating only with exceptionally capable researchers. The correlation between inventive stock and business unit ties is 0.144 (significant at the 0.001 level), whereas the correlation between academic stock and business unit ties is 0.029 (not significant). This suggests that individuals with proven, perhaps firm-specific, inventive capabilities are indeed sought-after collaborators within the firm. Third, inventions are not equal in importance (e.g., Scherer and Harhoff, 2000) or degree of novelty (e.g., Ahuja and Lampert, 2001). In so far as business units are more involved with technologies based on already established paradigms and follow-up inventions in those areas are less costly to produce (in terms of research and engineering effort or cognitive capabilities, including attention) than more radical breakthroughs, the results may reflect the limitations of using only non-weighted invention counts as a productivity measure. These limitations and related future research opportunities are further discussed in the conclusions.

Hypothesis 9.2 of the positive effects of network closures is supported by the analysis. The quantitative data used is not detailed enough to allow us to make causal interpretations regarding the way in which network closures contribute to inventive productivity. However, the insignificance of prior closure membership as a predictor of inventive productivity in model 5 should be noted. This would suggest that closure membership as such is not a form of intangible capital, but rather the individual's ability to identify potential collaborators and initiate joint efforts with them. In general, organizations facilitate joint activities by offering common experiences, shared norms, language, and objectives (Nahapiet and Ghoshal, 1998). The level of social capital generated within different organizations is likely to vary according to firm-specific differences in organizational routines. Moreover, differences in organizational routines may cause semi-permanent differences in the rate at which individuals form collaborative network closures. A study of these differences, however, would necessitate cross-organizational observations not available in this study. Rather, the present analysis should be considered as affirmation of the important role of internal collaboration networks regarding inventive activities in an industrial research environment. Further research should examine the formation dynamics of specific network closures (cf. Kreiner and Schultz, 1993), as well as factors in organizational design that facilitate or hinder the underlying mechanisms.

Hypotheses 9.3a and 9.3b are supported, but with a different interpretation than envisaged based on previous studies (e.g., Henderson and Cockburn, 1994; Liebeskind et al., 1996). Underlying the hypotheses is a presumption that inventive capacity is distributed among research personnel, and that boundary-spanning network connections would provide a rich flow of external knowledge resulting in increased inventive productivity. However, the inventive and academic outputs are produced by somewhat distinct groups of individuals, as only inventive, not academic, stock is a predictor of inventive productivity in the models. The regression results reflect corresponding differences in collaboration practices. Specifically, collaboration with business unit personnel is relatively common in inventive activities, and collaboration with external personnel is often associated with academic outputs. Thus, innovation in the corporate research environment under study is associated positively with boundary-spanning ties to the business units, not toward external innovation networks.

With regard to hypotheses 9.4a and 9.4b, the regression results support individual-level benefits of brokerage (9.4a), but the benefits do not seem to spill over to collaborators (9.4.b). However, as was already mentioned, these results refer only to a static view of network structure. In a dynamic network, a potential spillover effect could be the formation of new network closures. That is, brokerage offers benefits to the organization by contributing to the renewal of collaboration clusters. This potential benefit is of increased importance if collaboration closures including experts from multiple areas are especially effective in R&D (Ancona and Caldwell, 1992).

Conclusions

The quantitative results provide some points of departure with regard to the routinization of inventive activities and associated boundary conditions. The inventions are not isolated flashes of genius, but a rather systematic output of continuous work by R&D professionals. Although there are some peaks in the inventive output of a small subset of the research personnel, the inventive capability is spread over a large number of researchers. The majority of inventions are collaborative, involving two or more co-inventors. Moreover, the collaborative relationships overlap in some areas, with the overlapping collaboration clusters involving dozens of researchers. From these patterns two propositions emerge.

First, as inventive outputs involve collaboration and previous contributions to inventive outputs are positively associated with inventive productivity, organizations that aim to routinize inventive activities benefit from mechanisms that support the formation of collaboration clusters. Collaboration clusters are needed to realize the complementarities of technological know-how possessed by the individuals. Also, collaboration clusters are avenues by which new individuals are introduced to the tacit and firm-specific elements of the technology. In addition, the more systemic the underlying technological knowledge, the relatively more beneficial these mechanisms can be hypothesized to be.

Second, the knowledge required to identify beneficial collaboration opportunities may involve a significant tacit element. Consequently, internal brokerage, i.e., individuals who by their collaboration connect otherwise unconnected clusters within the organization, is valuable for the organization. Internal brokerage provides immediate benefits when previously unrecognized opportunities between complementary bodies of technological know-how are realized via new combinations. In addition, internal brokerage contributes to the renewal of the organization's internal collaborative structure. Thus, mechanisms that support internal brokerage are beneficial to both inventive productivity and the organization's internal adaptation aimed at better realizing inventive opportunities.

From the perspective of social network analysis, both membership in a collaboration network closure and a brokerage position are plausible candidates for intangible capital for an individual. The analysis, however, provides support only for brokerage as a form of intangible capital, whereas closures are instrumental in promoting individual-level technological know-how in various combinations. Although important for the overall inventive productivity, the benefits of closures seem to be fully captured in the individual-level knowledge capital measures. That is, without closures the inventive productivity would be diminished, but in the long term the closures as such do not enhance inventive productivity beyond the effects of accumulating individual-level knowledge capital stocks.

Regarding the routinization of inventive activities, both collaboration closure formation and internal brokerage are candidates for analytically interesting and operationalizable repeating processes. The processes are complementary, but the mechanisms that support each one may have conflicting features. Closure formation provides grounds for what Henderson and Cockburn (1994) call "component competence." Internal brokerage, in turn, is related to "architectural competence" (Henderson and Clark, 1990). Component competence is largely based on routine, even tacit, problem-solving strategies, the development of which requires close interaction among researchers for an extended period of time. A research organization that supports the accumulation of component knowledge is likely to have a relatively stable internal structure with well-defined areas of expertise and responsibility. Architectural competence is related to the organization's ability to combine its component competences in a flexible manner according to current needs, perhaps with little regard for established communication channels and decision-making routines. For a research organization, the initiation and successful completion of projects that require cross-disciplinary contributions are "core architectural competence." The present analysis suggests that knowledge capital measures, and not those of network closure, account for component competence at the individual level. However, internal brokerage seems to provide benefits distinct from knowledge capital, thus supporting conceptualization in which combinatory or architectural ability is also examined separately at the individual level.

The role of absorptive capacity is somewhat puzzling, at least with regard to the ways in which it has been operationalized in previous studies (e.g., Cockburn and Henderson, 1998). Specifically, neither academic outputs and stocks of academic (scientific) knowledge capital nor external ties from co-authorship relations are associated with increases in inventive productivity. This is in stark contrast to previous studies that have elaborated on the importance of scientific norms and external collaboration networks as sources of inventive productivity and overall competitive advantage. The present study, to the best of the author's knowledge, is the first one to simultaneously examine the two types of knowledge capital (academic/scientific and inventive/technological) in detail at the individual level. Nevertheless, it seems rather implausible that the conflicting results merely reflect more precise operationalization, but rather indicate that a contingency argument of the sources of inventive productivity is in order. That is, the previous studies have focused on biotechnology and the pharmaceutical industry, whereas the current study falls under the domain of communications, including Internet, technologies. In the pharmaceutical sector, patent protection is especially strong since a single patent can, in many instances, almost completely prevent "inventing around." Moreover, patent protection in pharmaceuticals is relatively broad, since a single patent can offer protection to a marketable end-product. In contrast, there are often multiple ways to implement a given feature in complex communications systems, which makes "inventing around" of other patents feasible. Furthermore, communications products are typically protected by a number of complementary patents, and in this sense the protection offered by a given patent is rather narrow. Taken together, there is a fundamental difference in how "systemic" or "atomistic" the technology in these different industries is (cf. Teece, 1992).

Based on variance in the systemic nature of technology, a contingency argument for the mechanisms of absorptive capacity can be proposed. If the technology is comparatively atomistic, the firms should align their internal structure and decision-making mechanisms to correspond to the external sources of knowledge, and cultivate extensive networks of external collaborative ties. Thus, the internal structure is optimized to process (i.e., seek innovations from inventions based on external knowledge) the wide variety of knowledge from external sources, whereas most knowledge generation and exploration is performed within the external networks. Internal structure is comparatively static, corresponding to component competences, whereas norms supporting external collaboration provide flexibility via access to diverse sources of knowledge, and changing network configurations external to the firm provide variation in knowledge, i.e., potential inventive opportunities. However, if the technology is comparatively systemic, the mechanisms for absorptive capacity must respond to different challenges. Specifically, many inventions involve re-combinations across component competences, and singular inventions offer less innovation potential. Thus, the evaluation of innovation potential related to external knowledge requires coordination across component competences, and even in cases where high innovation potential is perceived, complementary inventions may be required. Therefore, internal flexibility is required in ways that the external networks are unable to provide.

These arguments should not be interpreted as suggesting that absorptive capacity is of little value to firms inventing and innovating with systemic technologies. Rather, there are boundary conditions for mechanisms of absorptive capacity that have received little attention in previous research, one of which is the systemic nature of technology. As indicated by the strong positive effects of technological knowledge capital on inventive productivity in this study, the specific skills and knowledge of individual researchers are of importance whether or not the technology is systemic. Thus, instead of facilitating the flow of knowledge across organizational boundaries, firms can facilitate the flow of individuals across both internal and external organizational boundaries in order to promote invention and innovation. Indeed, empirical surveys have found that in many engineering-oriented industries, firms rate access to skilled personnel, not specific pieces of knowledge, as the major benefit provided by universities (Freeman and Soete, 1997).

The present study suggests several practical considerations for managers of research organizations. First, collaboration closure formation and internal brokerage were presented as recurring processes that contribute to the organization's inventive output. The rate at which the processes operate can be examined based on the collaborative relationships implied by inventive and academic outputs. By being aware of these processes and by tracking longitudinal changes in them, managers can be mindful of trends in organizational inventiveness resulting from factors such as changes in the organization's size, maturity, or norms supporting collaboration. Second, the results related to absorptive capacity should alert managers to the contingencies that systemic or atomistic technologies set to the utilization of external innovation networks. Specifically, the guidelines and recommendations for collaboration provided, based on experiences from firms in domains of atomistic technologies, should not be carelessly applied to firms involved with systemic technologies. It was also argued that internal flexibility – even at the cost of efficient routines – is at a premium when inventions in systemic technologies are pursued. In addition, firms can invent in both systemic and atomistic technologies or technological domains can gradually change their nature as a result of factors such as the establishment of de facto standards. In these cases, firms must either facilitate simultaneously different kinds of innovation processes or dynamically change their routines according to the changes in technological domains.

Although the longitudinal data used in this study partly mitigates the challenges of validity, a replication of the study at multiple organizations would be valuable. In addition, the study could be extended to examine the characteristics of inventions in more detail. Several authors have noted that technological inventions tend to be generated by "local search" (e.g., Stuart and Podolny, 1996; Rosenkopf and Nerkar, 2001). It seems rather plausible that the localness of search is a function of internal and external collaboration practices. Thus, research that combines data of technological trajectories with network measures related to the inventors' collaborative positions might provide insights into trade-offs between incremental and radical inventive activities (cf. Trajtenberg et al., 1997; Hansen et al., 2000; Ahuja and Lampert, 2001).

Lastly, the study raises some issues that are relevant to innovation policy. The implications of the atomistic vs. systemic nature of technologies were discussed from the perspective of firms involved in inventive activities. Similar contingency arguments, however, are likely to apply to research carried out at universities. Moreover, policy instruments, e.g., technology programs, that aim to encourage university–industry cooperation are subject to these considerations. Further research is required before attempting to draw any policy conclusions. In the mean time, however, policymakers should remain alert to differences in the dynamics of inventive activities between technological domains.

Notes

1   See, for example, Fifth Framework Programme FP5 at http://www.cordis.lu/fp5/.
2   http://www.tekes.fi

References

Ahuja, G. and Lampert, C. M. 2001: Entrepreneurship in the large corporation: a longitudinal study of how established firms create breakthrough inventions. *Strategic Management Journal*, 22 (June–July Special Issue), 521–43.

Allen, T. J. 1977: *Managing the Flow of Technology: Technology Transfer and the Dissemination of Technological Information with the R&D Organization*. Cambridge, MA: The MIT Press.

Amabile, T. 1988: A model of creativity and innovation in organizations. *Research in Organizational Behavior*, 10, 123–67.

Amit, R. and Schoemaker, P. J. 1993: Strategic assets and organizational rent. *Strategic Management Journal*, 14 (1), 33–46.

Ancona, D. G. and Caldwell, D. F. 1992: Bridging the boundary: external activity and performance in organizational teams. *Administrative Science Quarterly*, 37 (4), 634–65.

Borgatti, S. P., Everett, M. G., and Freeman, L. C. 1999: *UCINET 5.0 for Windows Version 5.34: Software for Social Network Analysis*. Natick: Analytic Technologies.

Brown, J. S. and Duguid, P. 1991: Organizational learning and communities-of-practice: toward a unified view of working, learning, and innovation. *Organization Science*, 2 (1), 40–57.

Brown, J. S. and Duguid, P. 1998 Invention, innovation, and organization. Working paper.

Buderi, R. 2000: Funding central research. *Research–Technology Management*, 43 (4), 18–25.

Burt, R. S. 2000: The network structure of social capital. In R. I. Sutton and B. M. Staw (eds.), *Research in Organizational Behavior*. Greenwich: JAI Press, 345–423.

Burt, R. S. 2001: Structural holes versus network closure as social capital. In N. Lin, K. S. Cook, and R. S. Burt (eds.), *Social Capital: Theory and Research*. New York: Aldine de Gruyter, 31–56.

Burt, R. S. 2002: The social capital of structural holes. In M. F. Guillen, R. Collins, P. England, and M. Meyer (eds.), *The New Economic Sociology: Developments in an Emerging Field*. New York: Russell Sage Foundation, 148–90.

Cockburn, I. M. and Henderson, R. M. 1998: Absorptive capacity, coauthoring behavior and the organization of research in drug discovery. *Journal of Industrial Economics*, 46 (2), 157–83.

Cohen, W. M. 1995: Empirical studies of innovative activity. In P. Stoneman (ed.), *Handbook of the Economics of Innovation and Technological Change*. Oxford: Blackwell, 182–264.

Cohen, W. M. and Levinthal, D. A. 1989: Innovation and learning: the two faces of R&D. *The Economic Journal*, 99 (397), 569–96.

Cohen, W. M. and Levinthal, D. A. 1990: Absorptive capacity: a new perspective on learning and innovation. *Administrative Science Quarterly*, 35 (1), 128–52.

Debackere, K. and Rappa, M. A. 1994: Technological communities and the diffusion of knowledge: a replication and validation. *R&D Management*, 24 (4), 355–71.

Dosi, G. 1982: Technological paradigms and technological trajectories: a suggested interpretation of the determinants and directions of technical change. *Research Policy*, 11 (3), 147–62.

Dosi, G. 1988: Sources, procedures and microeconomic effects of innovation. *Journal of Economic Literature*, 26 (3), 1120–71.

Freeman, C. and Soete, L. 1997: *The Economics of Industrial Innovation*, 3rd edn. London: Pinter.

Freeman, L. C. 1979: Centrality in social networks: conceptual clarification. *Social Networks*, 1, 215–39.

Grant, R. M. 1996: Prospering in dynamically-competitive environments: organizational capability as knowledge integration. *Organization Science*, 7 (4), 375–87.

Hansen, M. T., Podolny, J. M., and Pfeffer, J. 2000: *So many ties, so little time: a task contingency perspective on the value of social capital in organizations*. Paper presented at OSWC 2000.

Hargadon, A. and Sutton, R. I. 1997: Technology brokering and innovation in a product development firm. *Administrative Science Quarterly*, 42 (4), 716–49.

Hausman, J., Hall, B. H., and Griliches, Z. 1984: Econometric models for count data with an application to patents–R&D relationship. *Econometrica*, 52 (4), 909–38.

Henderson, R. M. and Clark, K. B. 1990: Architectural innovation: the reconfiguration of existing product technologies and the failure of established firms. *Administrative Science Quarterly*, 35 (1), 9–30.

Henderson, R. M. and Cockburn, I. 1994: Measuring competence? Exploring firm effects in pharmaceutical research. *Strategic Management Journal*, 15 (Winter, special issue), 63–84.

Henderson, R. M. and Cockburn, I. 1996: Scale, scope and spillovers: the determinants of research productivity in drug discovery. *Rand Journal of Economics*, 27 (1), 32–59.

Katz, R. 1997: Organizational socialization and the reduction of uncertainty. In R. Katz (ed.), *The Human Side of Managing Technological Innovation*. New York: Oxford University Press, 25–38.

Katz, R. and Allen, T. J. 1997: Managing dual ladder systems in RD&E settings. In R. Katz (ed.), *The Human Side of Managing Technological Innovation*. New York: Oxford University Press, 472–86.

Kline, S. J. and Rosenberg, N. 1986: An overview of innovation. In R. Landau and N. Rosenberg (eds.), *The Positive Sum Strategy: Harnessing Technology for Economic Growth*. Washington, DC: National Academy Press, 275–305.

Kreiner, K. and Schultz, M. 1993: Informal collaboration in R&D: the formation of networks across organizations. *Organization Studies*, 14 (2), 189–209.

Leonard-Barton, D. 1992: Core capabilities and core rigidities: a paradox in managing new product development. *Strategic Management Journal*, 13 (Summer, special issue), 111–25.

Liebeskind, J. P., Oliver, A. L., Zucker, L., and Brewer, M. 1996: Social networks, learning and flexibility: sourcing scientific knowledge in new biotechnology firms. *Organization Science*, 7 (4), 428–43.

Mokken, R. 1979: Cliques, clubs and clans. *Quality and Quantity*, 13, 161–73.

Mowery, D. C. 1983: Industrial research and firm size, survival and growth in American manufacturing, 1921–1946: an assessment. *Journal of Economic History*, 43 (4), 953–80.

Mowery, D. C. 1990: The development of industrial research in US manufacturing. *American Economic Review*, 80 (2), 345–9.

Nahapiet, J. and Ghoshal, S. 1998: Social capital, intellectual capital and the organizational advantage. *Academy of Management Review*, 23 (2), 242–66.

Nelson, R. R. and Winter, S. G. 1982: *An Evolutionary Theory of Economic Change*. Cambridge, MA: The Belknap Press of Harvard University Press.

Nokia Corporation 1996: *Annual Report 1995*.

Nokia Corporation 2001: *Annual Report 2000*.

Nonaka, I. 1994: A dynamic theory of organizational knowledge creation. *Organization Science*, 5 (1), 14–37.

Nonaka, I. and Konno, N. 1998: The concept of "ba": building a foundation for knowledge creation. *California Management Review*, 40 (3), 40–54.

Pake, G. E. 1986: Business payoff from basic science at Xerox. *Research Management*, 29, 35–40.

Patel, P. and Pavitt, K. 1995: Patterns of technological activity: their measurement and interpretation. In P. Stoneman (ed.), *Handbook of the Economics of Innovation and Technological Change*. Oxford: Blackwell, 14–51.

Penrose, E. 1959: *The Theory of the Growth of the Firm*. Oxford: Oxford University Press.

Pisano, G. P. 1990: The R&D boundaries of the firm: an empirical analysis. *Administrative Science Quarterly*, 35 (1), 153–76.

Pisano, G. P. 1991: The governance of innovation: vertical integration and collaborative arrangements in the biotechnology industry. *Research Policy*, 20 (3), 237–49.

Polanyi, M. 1958: *Personal Knowledge: Towards a Post-Critical Philosophy*. Chicago: University of Chicago Press.

Powell, W. W., Koput, K. W., and SmithDoerr, L. 1996: Inter-organizational collaboration and the locus of innovation: networks of learning in biotechnology. *Administrative Science Quarterly*, 41 (1), 116–45.

Prahalad, C. K. and Hamel, G. 1990: The core competence of the corporation. *Harvard Business Review*, 68 (3), 79–91.

Rappa, M. A. and Debackere, K. 1992: Technological communities and the diffusion of knowledge. *R&D Management*, 22 (3), 209–20.

Rosenkopf, L. and Nerkar, A. 2001: Beyond local search: boundary-spanning, exploration and impact in the optical disk industry. *Strategic Management Journal*, 22 (4), 287–306.

Salmenkaita, J.-P. 2001: Organizational Learning in Industrial Research: Inventive Productivity vs. Emergence of Technological Programs. *Strategic Management Society 21st Annual Conference*, October 21–24, 2001.

Scherer, F. M. and Harhoff, D. 2000: Technology policy for a world of skew-distributed outcomes. *Research Policy*, 29, 559–66.

Schumpeter, J. A. 1934: *The Theory of Economic Development.* Cambridge, MA: Harvard University Press.

Schumpeter, J. A. 1942: *Capitalism, Socialism and Democracy.* New York: Harper.

Stuart, T. E. and Podolny, J. M. 1996: Local search and the evolution of technological capabilities. *Strategic Management Journal*, 17 (Summer, special issue), 21–38.

Teece, D. J. 1992: Competition, cooperation and innovation: organizational arrangements for regimes of rapid technological progress. *Journal of Economic Behavior and Organization*, 18 (1), 1–25.

Teece, D. J. 1998: Capturing value from knowledge assets: the new economy, markets for know-how and intangible assets. *California Management Review*, 40 (3), 55–79.

Teece, D. J., Pisano, G., and Shuen, A. 1997: Dynamic capabilities and strategic management. *Strategic Management Journal*, 18 (7), 509–33.

Trajtenberg, M., Henderson, R., and Jaffe, A. 1997: University versus corporate patents: a window on the basicness of invention. *Economics of Innovation and New Technology*, 5, 19–50.

Wasserman, S. and Faust, K. 1994: *Social Network Analysis: Methods and Applications.* Cambridge: Cambridge University Press.

Wernerfelt, B. 1984: A resource-based view of the firm. *Strategic Management Journal*, 5 (2), 171–80.

Zucker, L. G., Darby, M. R., and Brewer, M. B. 1998: Intellectual human capital and the birth of US biotechnology enterprises. *American Economic Review*, 88 (1), 290–306.

CHAPTER 10

The Impact of Intangible Resources: How Organization Reputation and Employee Know-How Affect Performance

Alice C. Stewart and Henry Y. Zheng

Keywords: Intangible resources, resource-based view, data envelopment analysis, reputation, know-how.

Abstract

In this study we examine two related intangible resources, reputation and employee know-how. We also provide a measurement methodology, data envelopment analysis (DEA), which evaluates the relative productivity associated with the intangible resource employee know-how. The measures created from the DEA are entered into a structural equations model. Using various models, we provide support for the hypothesis that the relationship between the intangible resource, reputation, and firm performance is mediated by employee know-how. Our results suggest that the effect of reputation is strengthened in firms where the employee know-how and the management processes occurring to create it are maximized. This supplies support for the conventional wisdom of managers that suggests that firms must "walk the talk" to be successful. The results suggest that consistency in reputation and practice increase organizational performance.

Introduction

In a recent debate on the value of the resource-based view (RBV) of the firm, Priem and Butler (2001) suggested that the primary contribution of RBV was in the assertion that the sustainability of competitive advantage hinged on difficulty of imitation. Resources that are easily imitated are not a source of sustainable competitive advantage. Intangible resources are difficult to imitate; thus resources that are both valuable and intangible will provide sustainability as well as advantage.

While theoretically appealing, the idea of ensuring competitive advantage through development of intangible resources creates challenges for both researchers and practitioners. Resources are often described as intangible when they refer to value adding elements of a firm that, while affecting more quantifiable elements, cannot themselves be quantified (Barney, 1991). In essence, only the effect of the intangible asset can be observed. A common financial measure of intangibility is "goodwill" (Reilly, 2001). Goodwill is defined as the total value of the firm in excess of the fixed assets and net working capital. This is a collective measure of all of the intangible value created as a result of the firm's operations. One limitation of this method is that it does not allocate the component values of each intangible asset present within the firm. An assumption in accounting literature is that the organizational value created from intangibility, while important, is not possible to allocate operationally within the firm.

To empirically evaluate the impact of particular intangible resources the relationship between intangibility and its management must be examined. If a resource is completely intangible, as under the financial accounting assumption, can it be managed? If the resource can be managed strategically and if value-added can be assessed, is it truly intangible? If it cannot be managed, the idea of creating "strategy" associated with its development is problematic.

Following the lead of Hall (1992), Hall and Andriani (1998), and Sanchez et al. (2000), we suggest that strategically valuable resources can exist along a continuum, with some resources relatively more intangible than others. The most intangible of resources result from the path-dependent, socially complex, activities of the organization. These resources may be created as a result of luck or be unexplainable due to causal ambiguity. Firms with these types of intangible resources (and there may be very few) may thrive, but may deal with a substantial amount of uncertainty associated with their source of competitive advantage.

Somewhat farther away from this un-measurable intangibility are other resources that are intangible on their face, but are more likely to result from identifiable, yet very complex, management practices. At this position in the intangibility continuum exist embedded organizational routines, policies, and procedures that may not be explicitly articulated, but are critical to the success of the organization. "Well-managed" companies develop the consistency of strategy, structure, and organizational processes that create a virtuous cycle of performance. This performance can be tracked and evaluated, but simultaneously, the processes that create it are difficult for outsiders to imitate. For these firms, the intangibility of the resource comes from collective tacit knowledge created through a combination of individual experience and organization processes repeated over time (Hitt et al., 2001; Nonaka, 1991; Ambrosini and Bowman, 2001).

This type of intangibility may not be directly imitable, but its effects can certainly be observed. For example, there may be two organizations with similar inputs and very different performance outcomes. The difference is often attributable to some "intangible" difference in managerial process or managerial competence. In this form, intangibility may actually exist, but its effects and relative value can be discerned. In this form, intangibility is the result of management practice and strategic choice, and less a result of luck. This suggests that for these types of intangible resources, their relative value can be identified. While direct measurement of intangible processes is beyond our ability, measuring the effectiveness of the process is possible by examining the relative inputs and subsequent performance outcomes associated with it. Measuring stakeholders' perceptions of firm outcomes is also an indication of the value of the intangible process.

In this paper we examine intangible resources that are of each of the two types described above. We identify two commonly described intangible resources that are often mentioned as a potential source of competitive advantage for firms. We hypothesize about the relative value of these resources and about the relationship between the two. We use organization-level data to measure the relative strength of each of two intangible resources. We then use a structural equations model to determine the impact of each resource on performance and to establish a causal link between the two intangible resources. Thus the contribution of this research is to provide a model of how to empirically relate outcomes attributed to intangible organizational processes to organizational performance.

Theory Development and Hypotheses

Reputation and know-how as intangible resources

In this paper, we examine the relative effect of two commonly discussed and important resources, organization reputation and employee know-how. Both types of resources have been discussed as a potential source of competitive advantage (Weigelt and Camerer, 1988; McGuire et al., 1990). In an empirical study, McGuire et al. (1990) found that an organization's reputation has a positive and statistically significant relationship with performance. Employee know-how also has been linked empirically to the strategic advantages created by process capabilities (Amit and Schoemaker, 1993; Coff, 1997). According to Michalisin et al. (1997), employee know-how has an additional effect due to its influence on the selection, creation, and management of other firm resources.

Both resources, reputation and employee know-how, have been identified as intangible and valuable (Hall, 1993; Petrick et al., 1999). As such, they are more difficult to imitate and theoretically can be significant sources of competitive advantage. As such, hypothetically, each should be able to independently attract economic rent from the marketplace (Barney, 1991). However, there are differences in the nature of these two intangible resources.

Reputation is often the result of competitive processes that evolve over long periods of time (Petrick et al., 1999). Of the two, reputation is arguably the more difficult to imitate. It is "a fragile resource; it takes time to build, cannot be bought, and can be easily damaged" (Petrick et al., 1999: 60). A favorable reputation is a rare corporate asset and thus potentially quite valuable. Reputation "depends on specific, difficult-to-duplicate historical settings" (Barney, 1991: 115). It is based on socially complex relationships with large numbers of stakeholders that have been reinforced over time (Michalisin et al., 1997). Finally, reputation is based, to some extent, on perception. Owing to the historical nature of reputation, the actual activities and relationships for which it proxies may be different in the present than in the past or in the future. Thus the effect of reputation, favorable or unfavorable, may exceed the effect of the *current* activities upon which it is based (Hall and Andriani, 1998).

Employee know-how is potentially easier to imitate than reputation. Employee know-how is the knowledge, experience, insight, skill, etc., of the organization's employees. While this collective resource is also socially complex and historically determined to some extent, it is potentially more easily imitated or substituted than reputation. Hall and Andriani (1998) describe know-how as a functional type of intangible resource. Functional resources reflect how things are done. Human resource procedures can hire necessary employees from successful competitors, thus creating an acquisition mechanism for employee know-how. Training can increase the skill set of existing employees and thus increase employee know-how. Technology can be applied to organizational processes and result in a net increase in employee know-how via enterprise reporting systems and the strategic use of internal information and knowledge. Because employee know-how is more reflective of functional and managerial capability within the organization, there is less likely to be inconsistency between the current activities of the organization and actual employee know-how. Thus, to summarize, reputation is powerful, long-lived, but difficult to manage. Employee know-how is a powerful, "real time" value-adding and more manageable resource.

According to the RBV perspective, the relative inimitability of a valuable resource is directly related to its ability to provide a source of sustainable competitive advantage. The extent to which a resource can provide competitive advantage should be directly related to its ability to extract economic rent from the market. Thus, ceteris paribus, resources with greater inimitability should have a stronger effect on performance. This leads to the following hypotheses:

> Hypothesis 10.1. Ceteris paribus, reputation will have a greater direct effect on performance than employee know-how.

The interaction of intangible resources

One area that has only rarely been explored is the interplay between multiple intangible organizational resources. If two resources are both intangible, have socially complex components, and are in some way related to the strategic activities of the organization, then there may be some interdependence between them. What is not clear is if the less intangible resource reinforces the value of the more intangible resource or vice versa.

The nature of the relationship between resources is an under-explored area in strategy. Hall (1993) surveyed CEOs in an attempt to understand the relative importance of intangible resources, but his work did not speak to the interplay between them. Strategy literature in the 1970s, however, did tend to believe that there was a benefit to having consistency among organizational processes (Miles and Snow, 1978; Porter, 1980). Synthesis between the current RBV approach and traditional typologies of the past might suggest that certain resources reinforce each other. Thus, the effectiveness of one resource may be associated with the quality of another. Generally, we suggest that if two resources have different levels of relative intangibility, it is likely that the less intangible of the two will mediate the relationship between the more intangible resource and performance. In the case of reputation and employee know-how, this suggests that the relationship between reputation and performance is mediated by the employee know-how present within the organization.

Employee know-how may interact with reputation in some way to affect firm performance. With the extensive literature describing the value of each of these intangible resources, there is very little which theorizes about the nature of the relationship between the two and how they might be theoretically linked. Both are described as intangible, both are described as valuable. Employee know-how may be more immediately manageable within the organization, but researchers also suggest that managers attend to reputation management activities as well (Bontis et al., 1999). Traditionally, reputation is identified as an outcome of organizational processes (Petrick et al., 1999); but the strategic activities of the employees and how they apply their know-how may be instrumental in creating reputation. Once the reputation exists, however, it may allow firms to more easily attract employees with higher levels of skill and ability, ultimately enhancing know-how, a second-order outcome. Thus we suggest the following competing hypothesis:

Hypothesis 10.2. The effect of reputation on organizational performance is mediated by employee know-how.

Methodology

One area where employee know-how and reputation are key to the ability to attract external rents is higher education. In the past decade universities have come under increasing pressure to justify federal and state support as well as tuition increases. Often, this justification has resulted in the creation of data regarding university reputation, student quality, and faculty productivity. Though most Research I universities participate in knowledge creation through publication and patenting, and participate in knowledge distribution through teaching, there is significant variance in both institution reputation and faculty productivity.

For this study, we focus on the assessment of America's public Research I universities as defined by the Carnegie Classification System before its revision in 2000. The restriction of the research scope to only public Research I universities is an attempt to control for variation in mission, organizational structure, human capital, and financial capacity. Complete data was available for 42 public Research I universities. This represents a very high percentage of all public Research I universities.

We use structural equations modeling to test the hypotheses described above. In the structural equations model, we create two latent variables, know-how and reputation. We then evaluate two models to understand the relationship between these variables and organization performance. To create the latent variables, we must measure observed variables in a way that serves as a reasonable proxy for their intangible effectiveness. We create two observed measures of organization process effectiveness associated with employee know-how using a relatively new technique, data envelopment analysis (DEA).

Establishing the relative performance of resources across organizations

While measuring process efficiency using DEA is rare in the strategy field, it is not completely absent. In a study of the telecommunications industry Majumdar (1998) used DEA analysis to show empirically that the effectiveness of resource utilization was important to understand the competitive advantages of firms within this industry. Majumdar (1998) argued that analyzing efficiency in resource use with the DEA was a reasonable proxy for the intangible management skill that was the true source of competitive advantage. Other fields have begun to use this analysis tool as a way of determining the relative performance of organizational functional units (Cook and Hababou, 2001; Sola and Prior, 2001) or decision-making units (De Lancer Julnes, 2000).

Over the last two decades, Data Envelopment Analysis (DEA) has emerged as a truly multidimensional approach to assessing the overall performance of organizations (Seiford, 1990). While some of the earliest applications of DEA were in the area of elementary and secondary education (Charnes et al., 1978; Desai, 1986, 1992), its application to measure the performance of university faculty (Walters et al., 1997; Walters et al., 1998), of university departments (Johnes and Johnes, 1993; Sinuany-Stern et al., 1994; Tomkins and Green, 1988), and of institutes of higher learning as a whole (Kao, 1994; Sarrico et al., 1997) is fairly recent.

This analysis tool is also appearing in public-sector organizational literature and emerging in business and management literature. Sarkis (1997) reported the use of DEA to model a decision to invest in flexible manufacturing systems. Garbaccio and Hermalin (1994) used DEA to compare efficiency among firms in the savings and loan industry. Ozcan (1993) explored the possibility of using DEA to measure the technical efficiency of hospitals.

More recently, Banker et al. (2002) used DEA to evaluate the impact of information technology in the public accounting industry. DEA was used to assess the productivity of five accounting offices over several periods both before and after implementation of an information technology system. Brockett et al. (2001) used DEA as a way to empirically determine best practices in the computer industry. DEA provided the analytic tool to examine and integrate a vast array of benchmark data and identify star performers in the computer industry.

Because of its relative newness, studies of organizational effectiveness through the use of the DEA technique are mostly exploratory in nature. The small number of studies that have been done show clear promise that DEA, when coupled with reliable data, can be used to map the relative performance of organizations with similar missions and contextual constraints. Minimally, a DEA study can serve as an attention-focusing device to help managers identify opportunities for organizational improvement. One nice feature of DEA analysis is that its findings can be used to locate the optimal resource allocation strategies for organizations with different missions. Organizations can be compared on how well they managed their resources or on how well they achieve their outcomes.

Data envelopment analysis – a description of the analysis tool

DEA has its origins in economic theories of production and linear programming. Linear programming is concerned with the general problem of allocating limited resources among competing activities in the best possible way. Hillier and Lieberman, in their classic textbook, refer to linear programming as being "among the most important scientific advances of the mid-twentieth century" (1984: 29). Charnes et al. (1978) are credited with the development of DEA based on the principles of linear programming.

Through DEA analysis, efficiency can be examined in two ways: maximizing the outputs given a certain level of inputs or minimizing the inputs given a certain level of outputs. For example, consider the performance of a university that, for the sake of simplicity of exposition, uses two resources, R1 and R2, to achieve two outcomes, O1 and O2. Using data on these four variables from a number of universities, we can obtain figures similar to those shown in Figure 10.1.

In the figure on the left, the axes measure two resources R1 and R2 being used to produce the desired outcomes. The data from the different universities would yield a scatter plot. Assuming that the three universities, A, B, and C, represent the minimal combinations of resource utilization they identify the best-observed practice. The frontier, RABCR, denotes a "best practice frontier." We may thus construct performance measures based on the distance from the frontier.

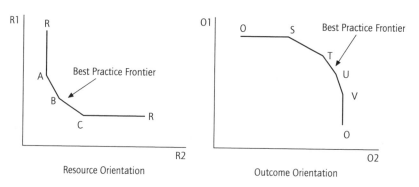

Figure 10.1   The best practice frontiers

Similarly, the universities can be compared on the basis of outcomes achieved. Assuming more to be preferable in this instance, universities S, T, U, and V would define the best practice frontier in the figure on the right. Thus, distance from the frontier OSTUVO could be used to define outcome performance measures. As in the case of resource utilization, universities on the frontier (that is, S, T, U, and V) are deemed to be effective while institutions "below" the frontier lag behind in their performance.

To provide a simplified illustration of DEA, let us use the benchmarking of university research performance as an example. In this example, only three variables are used: research expenditure, number of faculty, and number of journal publication. From these three variables, we create two ratio-based measures: number of publication per $1,000 research spending and number of publication per faculty FTE. The graph below shows how DEA can be used in such a simple case to identify best practices and identify performance gaps. The graph in Figure 10.2 illustrates how DEA maps the performance frontier.

Points A, B, C, and D connect to form a curved line that defines the frontier of research outcome. College A represents the highest possible publication numbers per $1,000 research expenditure whereas College D represents the highest possible publication per faculty FTE. Any school that lies on the line ABCD is considered effective and a best practice school. Any school that lies under the best practice frontier is considered less effective. A university that is not on the best practice frontier would look for best-practice schools that have similar resource allocation structures to improve its performance. For example, E has the same publication outcome per $1,000 investment as C but lower publication per faculty input. E also has the same publication numbers per faculty input as B but lower publication outcomes per $1,000 research expenditure than B. Either B or C can be E's best-practice reference. E needs to emulate B's or C's practices in order to improve its research effectiveness.

Of course, this illustration oversimplifies the issues. Performance assessment in higher education is far more complex than this example and requires the simultaneous

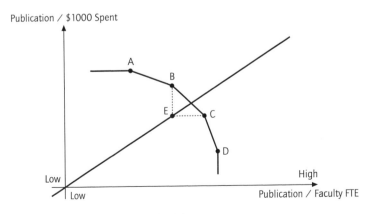

Figure 10.2 The simple case of research productivity

consideration of multiple variables that reflect not only the inputs and outputs, but also process and environmental factors. However, DEA models with more than two variables cannot be easily represented graphically. In such cases, we need to formulate the problem in mathematical terms.

The mathematical formulation of the problem is not restricted to two dimensions. The best practice frontiers as well as the performance indices can be readily constructed using multiple resources and multiple outcomes. In the context of university performance, the index corresponds to the following verbal statement:

Find a multivariate ratio, which (1) characterizes each university in terms of its outcomes and resources and (2) provides an ordering, from best to worst, of universities with similar outcomes, resources, and environmental constraints. A multivariate ratio that meets the above description can be expressed, for each of *n* universities, as:

$$\sum_r u_{rj} Y_{rj} \Big/ \sum_i w_{ij} X_{ij} \qquad\qquad \text{(Equation 10.1)}$$

Where: Y_{rj} = Level of outcome r of university j;
X_{ij} = Amount of resource i being used by university j;
and u_{rj} and w_{ij} are weights assigned to the outcomes and the inputs.

In creating DEA, Charnes et al. (CCR) (1978) proposed a mathematical programming formulation of this problem which simultaneously solves for the weights and provides a relative ordering of these programs. Ideally, each university would want to maximize Equation 10.1. Since the weights are to be "objectively" assigned so as to maximize this ratio, there is no upper limit to the size of the ratio. Hence, in order to bind the maximum value that any program could obtain, CCR proposed an upper bound of 1. CCR operationalized the above verbal statement of the problem as follows:

Find weights u_{rj} and w_{ij} such that the ratio of virtual outcomes to virtual inputs is maximized subject to the constraint that no ratio exceeds unity.

Thus, we have in Equation 10.1 the ratio of a *virtual* outcome to a *virtual* resource, where the virtual outcome is the weighted linear combination of outcomes and the virtual resource is the weighted linear combination of resources. Given this characterization of the university's activity, the computational issue to be addressed is a) how should the weights *u* and *w* be obtained and subsequently, b) how should the ordering of these programs be achieved.

Hence the performance score, h_k, for the kth university given n universities using m resources to result in s outcomes is obtained by solving a fractional program (Desai, 1999). Charnes and Cooper (1962) developed a transformation that yields a linear equivalent for this fractional program, thereby considerably simplifying the computation of h_k. Hence, the computation of these performance scores entails solving the following linear program:

Maximize: $\quad h_k = \sum_{r=1}^{s} \mu_{rk} \Upsilon_{rk}$

Such that:

$$\sum_{r=1}^{s} \mu_{rk} \Upsilon_{rj} - \sum_{i=1}^{m} v_{rk} X_{ij} \leq 0; \, j = 1, \ldots, n$$

$$\sum_{i=1}^{m} v_{ik} X_{ik} = 1$$

$$\mu_{rk} \geq 0; \, r = 1, \ldots, s$$
$$v_{ik} \geq 0; \, i = 1, \ldots, m$$

The performance score for each university is obtained by solving n such linear programs, one for each university in the study sample. Computer software for efficiently solving such complex mathematical programs is readily available. This research project used Frontier Analyst provide by Banxia Corp.

Data collection

Data used for this study comes from databases available from the Integrated Higher Education Data Systems (IPEDS) of the US Department of Education, WebCaspar from the National Science Foundation, and the Institute for Scientific Information (ISI). We believe that these organizations have by far the most consistent and accurate data resources for this type of analysis. The fact that ISI, IPEDS, and WebCaspar have multiple years of data affords us the opportunity to examine organizational performance not only for a particular point in time but also longitudinally. Additional data on reputation and student selectivity came from U.S. News and World Report Guide to Colleges and Universities.

Measures

Dependent variable
The dependent variable used for the study was market share of all federal and state grants and industrial contracts averaged over a four-year period from 1995 to 1999. Market share is defined as the percentage of the total of all federal, state, and industrial contracts obtained by a university. Because the absolute amount of available funding changes each year, market share was used to standardize the relative amount of funding received across the time period. This data was acquired from WebCaspar, which is the funding database from the National Science Foundation. The use of a four-year average instead of data for a particular year is to avoid bias caused by possible fluctuations in a university's funding performance. This is especially true in assessing research outcomes where an occasional dip or jump may or may not represent the normal state.

Grants and contracts represent substantial economic rents coming into a university. This variable includes all grants and contracts. While federal grants and contracts are usually the most prestigious, a substantial amount of research funding is available through state, industry, and foundation sources. These grants and contracts are awarded based on competition among primarily Research I universities (the population of interest). Also, because the measure is based on market share we believe this measure is a good proxy for competitiveness in university performance.

Independent variables (observed)
In this study, reputation is a latent variable that is created by combining measures of academic reputation and student selectivity. Academic reputation is the reputation score of each individual university in the *US News and World Report Guide to Universities.*

Student Selectivity is measured by two items, the percentage of the entering freshmen class in the top 10 percent of their high school class and the percentage of the entering freshmen in the top 25 percent of their high school class. This data was obtained from the *US News and World Report Guide to Universities.* These data are routinely provided to the public so that students and other stakeholders can assess the relative "quality" of universities. Universities with better reputations generally attract a higher percentage of excellent students, thus these measures are part of the creation of the latent independent variable, reputation.

Independent variables (calculated from DEA)
Though know-how is discussed as an intangible resource, it is often proxied in empirical research as quantity of patents and licenses held by an organization (Henderson and Cockburn, 1994). We suggest that it is possible to provide a better empirical proxy for the relative quality of employee know-how in universities through capturing relative faculty research productivity with data envelopment analysis (DEA).

The DEA model was used to create two scores: publication efficiency (Effscore) and research funding efficiency (Resscore). To create the publication efficiency score, the following input and outcome measures were entered into the model:

- Faculty FTE (input was measured as the number of full-time equivalent faculty in each institution).
- Financial expenditure (input was the total of direct research, academic support, and plant and equipment expenditures).
- Total number of publications (output was measured by counting the total number of publications attributed to each university from the Institute for Scientific Information, an organization that tracks and organizes this information).
- Total number of citations (output was measured by counting the total number of citations attributed to publications from each university). This data came from the Institute for Scientific Information.

The Publication Efficiency score for each university in the sample is provided in Table 10.1.

Table 10.1   Performance scores for public Research I universities

University	Publication Efficiency	University	Research Funding Efficiency
Indiana University	100.00	Indiana University	100.00
Louisiana State University	100.00	University of Michigan	100.00
Purdue University	100.00	University of Minnesota	100.00
Texas A & M University	100.00	University of North Carolina	100.00
University of Alabama	100.00	University of Pittsburgh	100.00
University of Michigan	100.00	University of Washington	100.00
University of Minnesota	100.00	University of Wisconsin	100.00
University of Nebraska	100.00	Georgia Institute of Tech	100.00
University of North Carolina	100.00	Texas A & M University	95.29
University of Pittsburgh	100.00	The University of Texas at Austin	91.85
University of Virginia	100.00	Ohio State University	88.65
University of Washington	100.00	University of Mo-Columbia	85.08
University of Tenn-Knoxville	99.76	University of New Mexico	83.19
University of Arizona	98.17	University of Illinois at Chicago	82.88
University of Wisconsin	95.58	Purdue University	82.25
University of Utah	95.53	University of Alabama	81.83
University of Mo-Columbia	95.23	University of Arizona	81.52
The University of Texas at Austin	93.14	University of Connecticut	80.66
University of Florida	89.03	University of Virginia	79.65
Iowa State	88.91	University of Illinois at Urbana	79.55
Oregon State University	86.96	University of Florida	78.72
University of Iowa	86.14	University of Iowa	77.52
Georgia Institute of Tech	82.35	University of Cincinnati	74.82
Colorado State	81.74	Louisiana State University	74.43
Pennsylvania State	81.28	Pennsylvania State	74.16
University of Connecticut	81.05	Michigan State	71.10
Ohio State University	80.55	University of Utah	70.92
University of Hawaii	79.14	University of Tenn-Knoxville	67.56
North Carolina State	76.98	Arizona State	65.74
University of Illinois at Urbana	75.61	Wayne State University	65.07
University of Maryland	72.83	North Carolina State	64.94
University of Kentucky	71.51	University of Nebraska	64.70
University of Kansas	71.00	University of Maryland	64.39
University of New Mexico	70.94	Iowa State	63.33
Virginia Polytechnic Inst	67.38	Virginia Polytechnic Inst	58.87
University of Illinois at Chicago	67.14	Utah State University	58.74
University of Cincinnati	66.79	Colorado State	58.31
Wayne State University	61.52	New Mexico State	56.81
Michigan State	59.53	University of Kansas	56.77
Arizona State	58.58	University of Georgia	54.54
University of Georgia	55.48	Oregon State University	52.91
Florida State	50.22	University of Kentucky	52.70
New Mexico State	47.80	University of Hawaii	48.47
Utah State University	47.54	Florida State	48.31

To create the research funding efficiency score, we used the following data:

- Faculty FTE (input was measured as the number of full-time equivalent faculty in each institution). This data was obtained from IPEDS.
- Total number of publications (input was measured by counting the total number of publications attributed to each university from the Institute for Scientific Information, an organization that tracks and organizes this information).
- Total number of citations (input was measured by counting the total number of citations attributed to publications from each university). This data came from the Institute for Scientific Information.
- Research grant dollars (output was measured by the total amount of federal research dollars acquired by each university). This data came from the National Science Foundation.

The Research Funding Efficiency score for each university in the sample is provided in Table 10.1.

Each variable was calculated using a three-year average instead of data for a particular year to avoid bias caused by possible fluctuations in a university's performance.

Once the DEA analysis was done, the scores for each university were included with the other data and were entered in a structural equations model using the AMOS software program. Given its use in the past as a proxy for know-how, growth in patents was entered in the model as a control variable.

Analysis and Results

Table 10.2 presents the means, standard deviations, and correlation matrix for the variables in the structural equation model. The variables seemed to have acceptable levels of variance. While some collinearity exists, the nature of a structural equations models where latent or endogenous constructs are the items of interest means that some collinearity is to be expected among the observed or exogenous variables. Interestingly, all variables in the model have a significant correlation with the dependent variable. For the growth rate in patenting (CGMSPAT), however, the correlation to market share of federal funding is negative.

Table 10.2 Means, standard deviations and correlations of key variables

	Descriptive Statistics		
	Mean	Std. Deviation	N
PR1SHR	1.8417	1.28883	42
EFFSCORE	82.6230	16.68996	44
RESSCORE	75.8234	16.23535	44
TOPTEN	34.1860	15.03845	43
TOP25PCT	65.2093	16.49034	43
REPUTATI	3.4159	.50160	44
CGMSPAT	55.1495	100.70703	44

Table 10.2 (*cont'd*)

Correlations

		PR1SHR	EFFSCORE	RESSCORE	TOPTEN	TOP25PCT	REPUTATI	CGMSPAT
PR1SHR	Pearson Correlation	1	.463**	.648**	.419**	.463**	.596**	-.319*
	Sig. (2-tailed)	.	.002	.000	.006	.002	.000	.040
	N	42	42	42	41	41	42	42
EFFSCORE	Pearson Correlation	.463**	1	.630**	.175	.125	.493**	-.109
	Sig. (2-tailed)	.002	.	.000	.262	.425	.001	.482
	N	42	44	44	43	43	44	44
RESSCORE	Pearson Correlation	.648**	.630**	1	.262	.253	.651**	-.255
	Sig. (2-tailed)	.000	.000	.	.089	.102	.000	.095
	N	42	44	44	43	43	44	44
TOPTEN	Pearson Correlation	.419**	.175	.262	1	.931**	.694**	-.244
	Sig. (2-tailed)	.006	.262	.089	.	.000	.000	.115
	N	41	43	43	43	43	43	43
TOP25PCT	Pearson Correlation	.463**	.125	.253	.931**	1	.701**	-.255
	Sig. (2-tailed)	.002	.425	.102	.000	.	.000	.099
	N	41	43	43	43	43	43	43
REPUTATI	Pearson Correlation	.596**	.493**	.651**	.694**	.701**	1	-.243
	Sig. (2-tailed)	.000	.001	.000	.000	.000	.	.111
	N	42	44	44	43	43	44	44
CGMSPAT	Pearson Correlation	-.319*	-.109	-.255	-.244	-.255	-.243	1
	Sig. (2-tailed)	.040	.482	.095	.115	.099	.111	.
	N	42	44	44	43	43	44	44

**. Correlation is significant at the 0.01 level (2-tailed).
*. Correlation is significant at the 0.05 level (2-tailed).

Publication Efficiency (EFFscore) has a fairly high mean, but substantial variance. For this measure 12 of the 42 universities appeared to be equally efficient and scored 100 percent. A score of 100 percent indicates that all of these universities performed on the efficiency frontier. For Research Funding Efficiency (RESscore) the mean is lower, but the standard deviation is still large. Fewer universities (eight) scored on the efficiency frontier and the range of scores was more widely distributed. Only six of the universities (Indiana, Michigan, Minnesota, North Carolina, Pittsburgh, and Washington) maximized efficiency in both measures. For the purposes of this study, these results suggest that the highest levels of employee know-how reside within these six schools.

To test hypothesis 10.1, a structural equation model was specified. The initial path analysis (Figure 10.3) tested a direct effect of reputation and employee knowledge

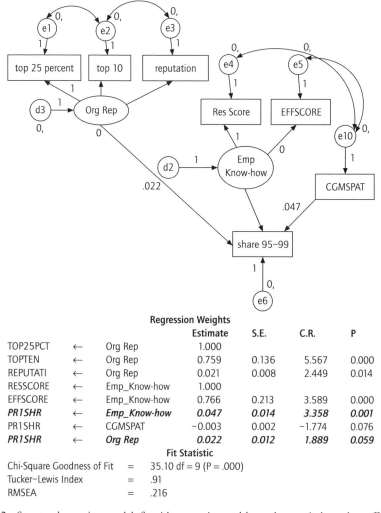

Regression Weights

			Estimate	S.E.	C.R.	P
TOP25PCT	←	Org Rep	1.000			
TOPTEN	←	Org Rep	0.759	0.136	5.567	0.000
REPUTATI	←	Org Rep	0.021	0.008	2.449	0.014
RESSCORE	←	Emp_Know-how	1.000			
EFFSCORE	←	Emp_Know-how	0.766	0.213	3.589	0.000
PR1SHR	←	*Emp_Know-how*	*0.047*	*0.014*	*3.358*	*0.001*
PR1SHR	←	CGMSPAT	−0.003	0.002	−1.774	0.076
PR1SHR	←	*Org Rep*	*0.022*	*0.012*	*1.889*	*0.059*

Fit Statistic

Chi-Square Goodness of Fit = 35.10 df = 9 (P = .000)
Tucker–Lewis Index = .91
RMSEA = .216

Figure 10.3 Structural equation model: fit with reputation and know-how as independent effects

on market share of grants and contracts. Regression weights reported in the model indicated a direct relationship between each latent variable and the performance variable. Reputation was positively and significantly related to performance, as was employee know-how. (All reported estimates are significant at $p > 0.05$). The standardized estimates suggest that the impact of know-how was almost twice that of reputation. Though the underlying assumption that both types of intangible resources influence performance was supported, this result directly refutes hypothesis 10.1, that reputation adds more value than employee know-how.

In addition, the model as specified did not pass standard goodness-of-fit tests used to evaluate the appropriateness of the theoretical paths. In this model, each intangible resource was assumed to have a direct effect on organization performance. Though

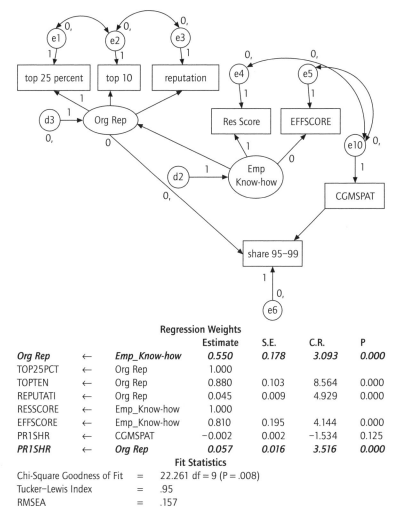

Regression Weights

			Estimate	S.E.	C.R.	P
Org Rep	←	*Emp_Know-how*	*0.550*	*0.178*	*3.093*	*0.000*
TOP25PCT	←	Org Rep	1.000			
TOPTEN	←	Org Rep	0.880	0.103	8.564	0.000
REPUTATI	←	Org Rep	0.045	0.009	4.929	0.000
RESSCORE	←	Emp_Know-how	1.000			
EFFSCORE	←	Emp_Know-how	0.810	0.195	4.144	0.000
PR1SHR	←	CGMSPAT	−0.002	0.002	−1.534	0.125
PR1SHR	←	*Org Rep*	*0.057*	*0.016*	*3.516*	*0.000*

Fit Statistics

Chi-Square Goodness of Fit   =   22.261 df = 9 (P = .008)
Tucker–Lewis Index   =   .95
RMSEA   =   .157

Figure 10.4 Structural equation model: fit model with employee know-how mediated by reputation (Model A)

each variable does independently predict performance based on the regression, the overall model does not pass the tests for goodness of fit. The chi-square for goodness of fit is significant, indicating that the null hypothesis, that the data fits the model, is rejected. In addition, the Tucker–Lewis index of 0.91 and the Root Mean Square Error of Approximation (RMSEA of 0.216) support the assertion that the model does not fit the data. This suggests that though reputation and know-how both affect performance, the relationships shown in the model are not the best representation of how performance is affected.

In Figures 10.4 and 10.5, we examine two alternate ways of relating reputation and know-how. In Figure 10.4 the model is drawn to suggest that the relationship

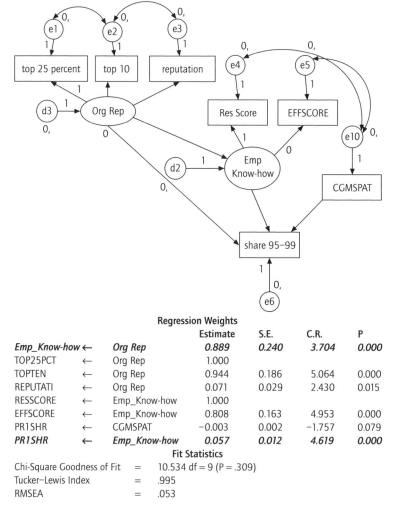

Regression Weights

			Estimate	S.E.	C.R.	P
Emp_Know-how ←	*Org Rep*		*0.889*	*0.240*	*3.704*	*0.000*
TOP25PCT	←	Org Rep	1.000			
TOPTEN	←	Org Rep	0.944	0.186	5.064	0.000
REPUTATI	←	Org Rep	0.071	0.029	2.430	0.015
RESSCORE	←	Emp_Know-how	1.000			
EFFSCORE	←	Emp_Know-how	0.808	0.163	4.953	0.000
PR1SHR	←	CGMSPAT	−0.003	0.002	−1.757	0.079
PR1SHR	*←*	*Emp_Know-how*	*0.057*	*0.012*	*4.619*	*0.000*

Fit Statistics

Chi-Square Goodness of Fit	=	10.534 df = 9 (P = .309)
Tucker–Lewis Index	=	.995
RMSEA	=	.053

Figure 10.5 Structural equation model: fit model with reputation mediated by employee know-how (Model B)

between employee know-how and performance is mediated by reputation. This model showed that in the regression weights, employee know-how does predict reputation and reputation predicts performance. Again, this is consistent with the results found in the correlation matrix. However, examination of how well this model fits the data suggests that, again, the relationships between the variables are mis-specified.

In Figure 10.5, the model is drawn to suggest that the relationship between reputation and performance is mediated by employee know-how. Reputation predicts employee know-how and employee know-how predicts organization performance. The test for goodness of fit shows, and an insignificant chi-square suggests, that we

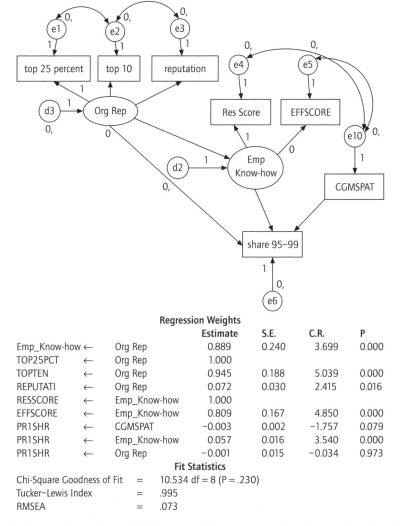

Regression Weights

			Estimate	S.E.	C.R.	P
Emp_Know-how	←	Org Rep	0.889	0.240	3.699	0.000
TOP25PCT	←	Org Rep	1.000			
TOPTEN	←	Org Rep	0.945	0.188	5.039	0.000
REPUTATI	←	Org Rep	0.072	0.030	2.415	0.016
RESSCORE	←	Emp_Know-how	1.000			
EFFSCORE	←	Emp_Know-how	0.809	0.167	4.850	0.000
PR1SHR	←	CGMSPAT	−0.003	0.002	−1.757	0.079
PR1SHR	←	Emp_Know-how	0.057	0.016	3.540	0.000
PR1SHR	←	Org Rep	−0.001	0.015	−0.034	0.973

Fit Statistics

Chi-Square Goodness of Fit = 10.534 df = 8 (P = .230)
Tucker–Lewis Index = .995
RMSEA = .073

Figure 10.6 Structural equation model: fit model with reputation mediated by employee know-how and direct effect of reputation (Model C)

can accept the null hypothesis that the model fits the data. The additional measures of the Tucker–Lewis Index and the RMSEA are also within the ranges that suggest a well-fit model. Thus this result provides support for hypothesis 10.2.

In an ad hoc analysis shown in Figure 10.6, the model is drawn to show that reputation is mediated by know-how, but also suggests a direct effect of reputation on performance in addition to the mediated effect. For this sample, there was no significant direct effect from reputation to performance. In addition, though the model shows an insignificant chi-square, the RMSEA of .07 (which exceeds the threshold of acceptability of .06) suggests that this model is not as good of a fit with the data.

Discussion

In this study we have begun to shed empirical light on the nature of intangible resources. The research findings suggest that the generally accepted maxim that intangible resources add value is accepted. However, an interesting finding from the above research is that the more *tangible* of the two resources was the more valuable. This is a contradiction of the traditional resource-based view. In RBV, inimitability suggests sustainability. In this study, reputation was posited as the more inimitable of the two resources.

Reputation is seen as a very socially complex, path-dependent resource. Reputation is something that takes a substantial amount of time to build and perception of reputation is often maintained long after the original incidents that created it occurred. Since imitability of reputation is very difficult, a good reputation should be a source of sustainable competitive advantage. In fact, in this study, reputation was strongly related to performance. However, in the modeling of the relative strength of reputation and employee know-how, the more tangible of the two, know-how, was the stronger predictor.

In the best-fitting model, know-how mediated the relationship between reputation and performance, thus employee know-how strengthens that relationship. This suggests that the more valuable reputation belongs to companies that can back it up with actual management practice. The employee know-how in this study is captured by measuring the relative efficiency of the organizations in converting inputs to desirable outcomes. Without this tangible evidence of the intangible processes for which the organization gained its reputation, there would be little impact of reputation on performance at all. Thus we can conclude that reputation is a source of sustainable competitive advantage when there is consistency between the external reputation and the internal practices. In other words, the firm must "walk the talk." While this makes intuitive sense, this study provides empirical evidence to support this intuitive understanding. In fact, in looking at the model which accounts for a possible direct effect absent the know-how effect, the model showed no significant direct relationship between reputation and performance.

One additional contribution of this research is the use of the DEA to evaluate the relative resource effectiveness across the organizations studied. Since we can now assert the importance of management practice, the DEA is of further value in that the algorithm can suggest to each individual unit exactly what magnitude of input or output change would be required to maximize efficiency. Thus in a study which supports the importance of organizational process and management practice, we have used a tool that can also be of prescriptive help in determining improvement.

Finally, this research has implications for research in the field of strategy. Often strategy research uses financial data as the measure of organizational performance. While profitability is a legitimate measure of comprehensive performance, it does not provide insight about the source of competitive advantage. Often, that advantage is inferred from other data on organizational differences. To use the data in this way, however, there is an implied assumption of homogeneity of resources and similarity of path. DEA as an analysis tool can assist researchers in evaluation of the relative strengths of each organization as they utilize different resource combinations to achieve the same competitive outcome. In doing so, research in strategy may become more consistent with resource-based logic of heterogeneity and path dependence.

This study also has limitations. The primary limitation is the type of organization studied. Though reputation and employee know-how are critical in the higher education industry, the nature of this set of organizations may be quite different than those that operate in a market-driven environment. In addition, the sample size of the study is rather small. Because of the numbers of parameters estimated, the degrees of freedom are low and results should be interpreted cautiously. Finally, the reputation variable relies heavily on a measure based on perception. This perceived reputation may not be as useful as other possible measures. Even with its limitations, replication of this study in other industries using other relevant proxies for reputation and employee know-how would be valuable.

Conclusion

These results shed light on the interesting interplay of organizational resources. In this study we examine two related intangible resources. We also provide a measurement methodology that evaluates the relative productivity associated with the intangible resource employee know-how. Finally, we test the causal relationships between the intangible variables and determine the effect on performance. This research has contributed to the literature associated with the resource-based view of the firm by examining the impact of related resources on firm performance. We believe that this is a first step toward the development of more sophisticated empirical tools to assist in the application of the RBV theories to organizational performance.

References

Ambrosini, V. and Bowman, C. 2001: Tacit knowledge: some suggestions for operationalization. *Journal of Management Studies*, 38 (6), 811–29.

Amit, R. and Schoemaker, P. J. H. 1993: Strategic assets and organizational rent. *Strategic Management Journal*, 14, 33–46.

Banker, R. D., Chang, H., and Kao, Y. 2002: Impact of information technology on public accounting firm productivity. *Journal of Information Systems*, 16, 209–23.

Barney, J. B. 1991: Firm resources and sustained competitive advantage. *Journal of Management*, 17, 99–120.

Bontis, N., Dragonetti, N., Jacobsen, K., and Roos, G. 1999: The knowledge toolbox: a review of the tools available to measure and manage intangible resources. *European Management Journal*, 17, 391–402.

Brockett, P. L., Golden, L. L., Sarin, S., and Gerberman, J. H. 2001: The identification of target firms and functional areas for strategic benchmarking. *Engineering Economist*, 46, 274–300.

Charnes, A. and Cooper, W. W. 1962: Programming with linear fractional functionals. *Naval Research Logistics Quarterly*, 9, 181–5.

Charnes, A., Cooper, W. W., and Rhodes, E. 1978: Measuring the efficiency of decision-making units. *European Journal of Operational Research*, 2, 90–119.

Coff, R. W. 1997: Human assets and management dilemma; coping with hazards on the road to resource based theory. *Academy of Management Review*, 22, 374–402.

Cook, W. and Hababou, M. 2001: Sales performance measurement in bank branches. *Omega*, 29, 4.

De Lancer Julnes, P. 2000: Decision-making tools for public productivity improvement: a comparison of DEA to cost benefit and regression analyses. *Journal of Public Budgeting, Accounting and Financial Management*, 12, 625–48.

Desai, A. 1986: *Extensions to measures of relative efficiency with an application to educational productivity*. Doctoral Dissertation, University of Pennsylvania.

Desai, A. 1992: Data envelopment analysis: a clarification. *Evaluation and Research in Education*, 6, 39–41.

Desai, A. 1999: Program evaluation, best practices and data envelopment analysis. In S. S. Nagel (ed.). *Policy Analysis Methods*. Commack, NY: Nova Science Publishers, 183–203.

Garbaccio, R. F. and Hermalin, B. E. 1994: A comparison of nonparametric methods to measure efficiency in the savings and loan industry. *Journal of the American Real Estate and Urban Economics Association*, 22, 169–94.

Hall, R. 1992: The strategic analysis of intangible resources. *Strategic Management Journal*, 13, 135–44.

Hall, R. 1993: A framework linking intangible resources and capabilities to sustainable competitive advantage. *Strategic Management Journal*, 14, 607–18.

Hall, R. and Andriani, P. 1998: Management focus: analyzing intangible resources and managing knowledge in a supply chain context. *European Management Journal*, 16, 685–97.

Henderson, R. and Cockburn, I. 1994: Measuring competence? Exploring firm effects in pharmaceutical research. *Strategic Management Journal*, 15, 63–84.

Hillier, F. and Lieberman, G. 1984: *Introduction to Operations Research*. Oakland, CA: Holden-Day Publishing.

Hitt, M., Bierman, L., Shimizu, K., and Kochhar, R. 2001: Direct and moderating effects of human capital on strategy and performance in professional service firms: a resource based perspective. *Academy of Management Journal*, 44, 13–28.

Johnes, G. and Johnes, J. 1993: Measuring the research performance of UK economics departments: an application of data envelopment analysis. *Oxford Economics Papers*, 45, 332–47.

Kao, C. 1994: Evaluation of junior colleges of technology: the Taiwan case. *European Journal of Operational Research*, 72, 43–51.

Majumdar, S. 1998: On the utilization of resources: perspectives from the US telecommunications industry. *Strategic Management Journal*, 19, 809–31.

McGuire, J. B., Schneeweis, T., and Branch, B. 1990: Perceptions of firm quality: a cause or result of firm performance. *Journal of Management*, 16 (1), 167.

Michalisin, M. D., Smith, R. D., and Kline, D. M. 1997: In search of strategic assets. *International Journal of Organizational Analysis*, 7, 1–20.

Miles, R. E. and Snow, C. C. 1978: *Organizational Strategy, Structure, and Process.* New York: McGraw-Hill.

Nonaka, I. 1991: The knowledge creating company. *Harvard Business Review*, 69, 96–104.

Ozcan, Y. A. 1993: Sensitivity analysis of hospital efficiency under alternative output/input and peer groups: a review. *Knowledge and Policy*, 5, 1–30.

Petrick, J., Scherer, R., Brodzinski, J., Quinn, J., and Ainina, M. 1999: Global leadership skills and reputational capital: intangible resources for sustainable competitive advantage. *Academy of Management Executive*, 13, 58–69.

Porter, M. E. 1980: *Competitive Strategy: Techniques for Analyzing Industries and Competitors.* New York: Free Press.

Priem, R. and Butler, J. 2001: Tautology in the resource-based view and the implications of externally determined resource value: further comments. *Academy of Management Review*, 28, 57–66.

Reilly, R. F. 2001: Valuation of intangible assets in dot.com and intellectual property intensive companies. *American Journal of Family Law*, 15, 36–46.

Sanchez, P., Chaminade, C., and Olea, M. 2000: Management of intangibles: an attempt to build a theory. *Journal of Intellectual Capital*, 1, 312–27.

Sarkis, J. 1997: Evaluating flexible manufacturing systems alternatives using data envelopment analysis. *Engineering Economist*, 43, 25–47.

Sarrico, C. S., Hogan, S. M., and Athanassopoulos, A. D. 1997: Data envelopment analysis and university selection. *Journal of the Operational Research Society*, 48, 1163–98.

Seiford, L. M. 1990: *A bibliography of Data Envelopment Analysis (1978–1990), Version 5.0.* Technical Report, Dept. of Industrial Engineering, University of Massachusetts, Amherst, MA.

Sinuany-Stern, Z., Mehrez, A., and Barboy, A. 1994: Academic departments' efficiency via DEA. *Computers and Operations Research*, 21, 543–56.

Sola, M. and Prior, D. 2001: Measuring productivity and quality changes using data envelopment analysis: an application to Catalan hospitals. *Financial Accountability and Management*, 17, 3.

Tomkins, C. and Green, R. 1988: An experiment in the use of data envelopment analysis for evaluating the efficiency of UK university departments of accounting. *Financial Accountability Management*, 4, 147–64.

Walters, L. C., Cornia, G. C., and Chabries, D. 1997: *Benchmarking university faculty performance.* Presented at the APPAM Annual Research Conference, Washington, DC.

Walters, L. C., Cornia, G. C., and Chabries, D. 1998: *A framework and national benchmarks for assessing academic productivity.* Presented at the INFORMS Research Conference, Montreal, Canada.

Weigelt, K. and Camerer, C. 1988: Reputation and corporate strategy: a review of recent theory and applications. *Strategic Management Journal*, 9, 443–54.

CHAPTER 11

The Fall of a Silicon Valley Icon: Was Apple Really Betamax Redux?

Joel West

Keywords: Network externalities, technology strategy, standards competition, computer industry, strategy implementation.

Abstract

Apple Computer was once the face of the PC revolution and Silicon Valley's first global icon. However, in the 1990s Apple was eclipsed by rivals, entering a long period of relative decline that many assumed would be terminal. The accepted truth of the Apple exemplar is that its Macintosh was "another Betamax" – that its failure to license this technology consigned Apple to a small market share and, through positive network effects, sealed its inevitable decline.

This paper examines Apple's troubled decade in light of the network theories and simpler explanations of strategic and operational error. It concludes by showing how a skeptical reading of the accepted "Betamax" wisdom might affect the strategies of other IT firms.

Introduction

The term "Silicon Valley" has been used to refer to a geographic region, a group of industries, and an archetypal organizational culture. Practices and cognitive frameworks that exemplify "Silicon Valley" have diffused both throughout the region and also to high technology firms around the world. This diffusion has been both direct through employee mobility and indirectly through mass media coverage of the region, industries, and culture.

These diffusion patterns have been defined by the Valley's exemplary companies. One of the region's earliest technology companies, Hewlett Packard, provided a corporate culture for all to imitate. Meanwhile, in the 1950s and 1960s, semiconductor companies like Fairchild and Intel gave "Silicon Valley" its name and started the process of venture-funded spin-offs that continues to this day (Rogers and Larsen 1984; Morris and Ferguson 1993; Kenney 2000).

But the valley remained little more than a geographic region, invisible to the rest of the world, until the personal computer revolution of the 1970s. A local company, Apple Computer, became both the face of the PC revolution and Silicon Valley's first global icon. Apple was the earliest and most successful of the Valley's first-generation PC companies, and, in fact, the only one to survive the end of the 8-bit era. However, in the 1990s, Apple was eclipsed by Compaq, Dell, and others, entering a long period of relative (and absolute) decline that many interpreted as a terminal slide.

The accepted explanation for Apple's decline is that its Macintosh was "another Betamax." To wit, the failure to license its technology consigned the Macintosh standard to a small share, and, through the lack of software, sealed its inevitable decline. Under this explanation, the fall of Apple is an exemplar for the positive feedback model of network externalities mediated by a supply of complementary assets (Shapiro and Varian, 1999; Ferguson, 1999).

But is it this simple? Or would a more nuanced examination of the company's fortunes suggest other explanations? In particular, did Apple's preoccupation with the Betamax allegorical tale keep it from finding a viable niche strategy until it had dissipated more than a billion R&D dollars and the majority of its market share?

The paper first reviews the Betamax-derived theories of network-based industry competition that, according to received wisdom, explain Apple's fate. It then examines the evidence for four possible explanations for Apple's fall: its failure to license its technology, its premium pricing strategy, errors in its product strategies/execution, and poor operational efficiency. It also presents Apple's contemporaneous fears that making the Macintosh standard ubiquitous (as IBM had done with the IBM PC) would not necessarily translate into company success.

The paper concludes by showing how a more skeptical reading of the accepted "Betamax" wisdom might affect the strategies of other firms engaged in IT standards competition.

Accepted Wisdom on Standards Competition

The accepted theoretical wisdom is that *de facto* information technology compatibility standards competition is a "winner take all" battle driven by two forces, positive network externalities and switching costs (e.g., Katz and Shapiro, 1985; Morris and Ferguson, 1993; Arthur, 1996). While such Betamax-derived theories are widely accepted, there have been a few criticisms of the empirical evidence and the theories themselves (Liebowitz and Margolis, 1999).

Positive-feedback models

The concept of positive consumption externalities for networks of users was originally developed for physical communication networks, in which "the utility that a subscriber derives from a communications service increases as others join the system" (Rohlfs, 1974: 16). Examples of such networks include telephones, telexes, fax machines, and e-mail systems. Katz and Shapiro (1985) extended this to the more abstract concept of a "network," in which only one of three categories corresponds to Rohlfs' (1974) physical networks.

Katz and Shapiro (1985) identified another category of goods, those that conform to a "hardware–software paradigm," in which buyers of a type of hardware (e.g., VCR) require specialized software (pre-recorded video tapes). When hardware makers rely on outside suppliers of software (rather than their own subsidiaries), then the larger the number of hardware users, the more attractive that market is to software makers to produce the specialized software. Then, the leading standard enjoys demand-side economies of scale, where every new adopter increases its advantage over rivals. Eventually producers shift to making products compatible with the dominant standard rather than sticking with an incompatible losing standard (Katz and Shapiro, 1985, 1986, 1994; Farrell and Saloner, 1985, 1986; Teece, 1986; Besen and Farrell, 1994).

A related stream of standards research examines the effect of asymmetric switching costs upon adopter decisions (David, 1985; Beggs and Klemperer, 1992). If intra-standard adoption of successive generations of products is less expensive than inter-standard adoptions, customers tend to "lock in" to one standard, as Greenstein (1993) demonstrated with US mainframe computer purchases. New adopters are presumed to calculate the net present value of a prospective switching cost, decreasing the attractiveness of a flagging standard that might eventually disappear.

This is one reason researchers have concluded that the combination of network externalities and switching costs lead to the "tipping" of the standards contest (Farrell and Saloner, 1986; Arthur, 1996; Katz and Shapiro, 1994; Shapiro and Varian, 1999). Specifically, the theories make a strong and unambiguous prediction that, ceteris paribus, a virtuous cycle will inevitably "tip" a standard contest in favor of the leader, consigning the trailing standard(s) to market pressures that irrevocably force its share to zero (Arthur, 1989; West, 1999).

The most often cited example of such a tippy standards battle is that of VHS vs. Betamax (Cusumano et al., 1992).[1] And whether directly through the VCR wars or indirectly through academic theories of positive network externalities, this winner-take-all, positive-feedback model has driven standards-related decisions by adopters and producers (West, 1999). For more than a decade, producers developed aggressive strategies to improve the actual supply of software, by courting software developers and using a penetration pricing strategy to quickly establish a market share lead that would attract developers. Producers have also sought to influence the perception of software availability and market share (truthfully or otherwise) to attract both users and producers of complementary assets. Adopters have sought to reduce their likelihood of adopting a losing standard (and paying the concomitant switching costs) by handicapping standards battles based on the availability of software and perceived market share.

Limitations

Such research makes very strong predictions about the strength of network effects. Katz and Shapiro (1986: 824) assert that "the dynamics of industries subject to network externalities are fundamentally different from those of conventional industries." These models are based on strong assumptions that are questioned by other research:

- Software variety maximization: The assumption is that more adopters leads to more software variety, and more software variety is valued by users. In their pioneering model of network externalities, Katz and Shapiro predicted positive feedback in the PC standards adoption decisions:

    > . . . because the amount and variety of software that will be supplied for use with a given computer will be an increasing function of the number of hardware units that have been sold. (Katz and Shapiro, 1985: 424)

    This assumption of software variety maximization by buyers is either explicit (e.g., Besen and Farrell, 1994) or implicit (Katz and Shapiro, 1992) in all subsequent models predicting positive network effects and tipping for *de facto* standards contests.
    But is software variety maximization important for all standards contests? Other durable goods require specialized complementary assets (Teece, 1986), but differences in variety of add-on complements does not play a significant role in the selection of many other categories of durable goods. West (1999) argues the variety maximization assumption is an overgeneralization of the VCR case where unique software is regularly "consumed," but is not applicable to classes of goods where users will satisfice to a minimum level of available software.
- Adequate foresight assumes that adopters can adequately anticipate future standards sales. For example, Katz and Shapiro (1985: 426) write:

    > [A]n individual's consumption benefits will depend on the future size of the relevant networks. Consumers will base their purchase decision on *expected* network sizes.

Is this a realistic assumption for utility-maximizing consumers? Garud et al. (1997) have shown that even industry professionals have limited foresight in situations of high ambiguity such as during radical technological change, so it seems a stretch to expect accurate prognostication by even the most enthusiastic early adopter.

Empirical research that directly tests such theories has been rare. As one proponent wrote:

> Network effects, and demand-side economies of scale more generally, have been shown *in theory* to have implications for a variety of activities . . . There have not, however, been any attempts to test econometrically for the effects of networks on these phenomena. (Saloner and Shepard, 1995: 479)

Liebowitz and Margolis (1990, 1994, 1999) have asserted that these limitations mean that the theories are unproved and, in fact, demonstrably false. Others have questioned whether there are unidentified moderators, which determine whether or not network effects will apply to a given standard or standards competition (Shapiro and Varian, 1999; West, 1999). For example, West and Dedrick (2000) show that the cost advantage of the late-arriving PC standard in Japan helped it displace an incumbent, despite the latter's established software libraries and market share.

"Cloning" and Related Explanations for Apple's Decline

Founded in 1977, Apple was the most successful manufacturer of the 8-bit PC era (1975–81). Its Apple II became the standard in US K-12 education and was also long popular with home users. But the 1981 introduction of the IBM PC effectively ended business demand for 8-bit computers, leading to the inevitable decline of Apple's original product line.

After several false starts, with its 1984 introduction of the Macintosh, Apple designed a successful 16-bit PC to compete with IBM. The Mac was immediately recognized as incorporating breakthrough technology, most notably being the first with a graphical user interface for mass-market PC buyers. Given the unique capabilities of the Macintosh in its first few years, Apple had the option of commanding a large price premium for its highly differentiated product – which it exploited. Despite the premium pricing strategy, the company's revenues and market share both grew in the late 1980s and early 1990s (Figure 11.1). However, from 1995 to

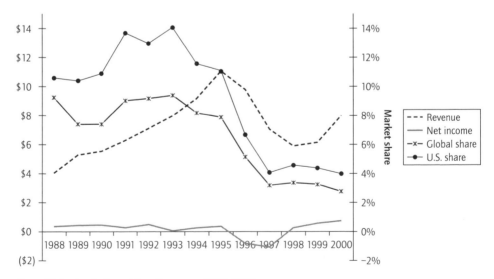

Figure 11.1 Apple financial performance, 1988–2000
Source: Company annual reports; Dataquest
Note: Revenues and net income for fiscal year ending September 30; market share is calendar year.

1998 Apple suffered from a plunge in revenues and market share that was widely expected to be fatal.

The conventional wisdom is that Apple's fall stemmed from its unwillingness to allow other firms to clone the Macintosh, which would have increased the popularity of the Macintosh platform and thus primed the pump of the positive-feedback model.[2] For example, a profile of the company by a leading financial newspaper recently wrote:

> If Apple had licensed Mac software to other computer makers in the late 1980s, it would have cut the generous gross margins the company enjoyed on Mac sales – but it also would have created a huge market for cheap Mac clones, driving Apple's market share sharply higher and letting what nearly everyone agrees was a superior machine duke it out with the PC.
>
> For whatever reason – arrogance, perfectionism, timidity, technical hurdles – Apple never did that. Instead, dozens of PC clone makers made cheap machines and delivered the keys to the computing world to Microsoft. Apple, meanwhile, continued to talk about how it was creating a computer for "the rest of us," even though that computer was much more expensive than the other choices the rest of us were offered. (*WSJ.com*, 2001)

Such analysis – both contemporaneously and retrospectively – made causal predictions linking cloning to lower prices, lower prices to higher market share, higher market share to more software availability, which in turn would further increase market share; cloning was also expected to improve the variety of hardware (Figure 11.2).

According to this argument, Apple could be successful if and only if it aggressively licensed its technology to competitors. Such licensing would "prime the pump" of adoption, which through the positive feedback model would assure Apple of an adequate supply of complementary assets and future survival.

The remainder of this section uses this causal chain as a framework to consider Apple's strategies from 1984 to 2000, the options available at the time, decisions made and their outcomes. As an alternative to the cloning hypothesis, it considers independently the effect of Apple's premium pricing strategy; the following section contrasts these explanations with more prosaic problems in the company's tactics and execution.

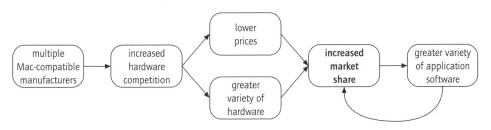

Figure 11.2   Causal chain of arguments by cloning advocates

To clone or not to clone

Both inside and outside Apple, the key question facing Apple in the late 1980s and early 1990s was whether it should license its technology to other PC makers. The question divided Apple from its developers and users, as well as groups within Apple. Some within and outside Apple felt that Apple needed to make the Macintosh platform ubiquitous to assure an adequate supply of software. Otherwise, they feared, the Mac would repeat the same death spiral of declining software and share that had doomed the Betamax.

The controversy focused on two issues: whether Apple should let other firms make Macintosh "clones," and, if so, whether in fact the company should exit the hardware business entirely and become a software-only company to compete with Microsoft. The presumption was that having more than one company make Mac-compatible computers – as on the IBM PC side – would expand the total sales of the Macintosh platform, its share of the overall PC market, and also its supply of complementary assets such as software. Against this, others argued that Apple could expand market share without licensing if it reduced prices and gross profit margins.

While the software-only option was very popular with external analysts, there is little evidence that it won serious consideration within Apple due to the wrenching changes required. However, the internal debate among Apple executives (and with third-party partners) over whether to license clone makers racked Apple from 1985 to 1994 (Carlton, 1997a; Linzmayer, 1999; Malone, 1999).

Arguments for and against cloning

The cloning option was broached in a then-secret June 1985 memo to CEO John Sculley from Microsoft CEO Bill Gates. From the success of the Apple II and IBM PC, Gates anticipated the as yet unpublished theories of positive network externalities when he wrote:

> As the independent investment in a "standard" architectures grows, so does the momentum for that architecture. The industry has reached the point where it is now impossible for Apple to create a standard of their innovative technology without support from, and the resulting credibility of, other personal computer manufacturers. (Carlton, 1997a: 40–41)

Gates also contacted two PC makers, HP and AT&T, and relayed their interest in cloning to Apple.

While Apple did not follow through on Gates' proposal, it marked the beginning of the cloning debate both inside and outside Apple (Table 11.1). Co-founder Steve Jobs later described 1988–92 as the "golden window of opportunity [for Apple] to license its Macintosh operating system software" (*MacWEEK.com*, 1997).

Many industry analysts and trade journal reporters had backed clones for many years (e.g., Davis 1988). Those with a direct financial interest, not surprisingly, strongly backed licensing. This included producers of co-specialized assets (such as software and magazines) who would enjoy a larger potential market if, by licensing, Apple expanded the number of Macintosh computer buyers. Also supporting

Table 11.1 Proposals for licensing Macintosh technology for other platforms

Date	Firm	Status	Reason Apple killed it
1985	Microsoft	Proposal by Bill Gates	Unknown
1987	Apollo	Contract signed by Apollo	CEO (Sculley) changed his mind
1987	Sony, Tandy	Requested license to GUI	Unknown
1990	Sun	Merger approved by management	Dropped when Apple adopted IBM PowerPC
1992	Intel, Novell	Working prototype of Mac OS on Intel hardware	Product champion left company
1994–1997	Power Computing, Pioneer, Motorola, others	Nearly 500,000 computers sold based on PowerPC chips	Canceled by new CEO (Jobs)
1995	Gateway 2000	Contracts ready for signature	Opposition from Apple sales executives

Sources: Grindley (1995), Carlton (1997a); Linzmayer (1999), Malone (1999)

licensing were users, who anticipated greater variety or lower prices brought on by increased competition.

All of the arguments for cloning were premised on increasing the market share for the Macintosh platform. As late as 1994, the pro-licensing camp projected that the standard would need at least 20 percent of the global PC market to maintain the Macintosh as a viable platform. One of the most forceful advocates was then-CFO Joseph Graziano, who backed cloning:

> If you are asking, What does the Macintosh platform share have to be to be a sustainable business proposition? – which I think is the big strategic question – then I think that it has to be well over 20 percent. . . . That doesn't mean that it has to be all Apple. That's where this licensing thing comes in. (Lach, 1994)

However, there were real concerns about the economic feasibility of allowing clones. Those opposed to licensing contended that there was no guarantee that expanding the market for the platform would make up for Apple's smaller share of that market. In particular, they noted that the OS sales accounted for a small proportion of the PC revenues, and that without hardware sales it would not have the revenues and profits it needed to support its R&D (Carlton, 1997a: 50–53). Through 1996, Apple's R&D intensity (R&D/sales ratio) was twice that of Compaq and roughly four times that of Dell Computer, the two other major PC-only companies.

Some Apple execs worried that by splitting the Mac market with rivals, they would be unable to profit from a shared standard: from 1986 onward, they need only witness IBM's ever-smaller minority share of the "IBM compatible" market. There would also be questions whether other PC makers would be willing to depend on a PC rival for their essential technology, as IBM found trying to license

OS/2 (Grove, 1996: 48). Both factors proved major problems during Apple's brief 1994–97 cloning era.

Finally, others doubted that even with cloning Apple would make significant inroads into the market share of Microsoft's operating system – particularly after the 1990 introduction of Windows 3.0. To Graziano's 20 percent market share goal, analyst Jonathan Seybold replied "It is really, really difficult to find a scenario where Apple gets even 20 percent" (Lach, 1994).

After the "golden window of opportunity" had closed, in late 1994 Apple started licensing new PC entrants to make clones, and the first clones were sold beginning in May 1995.[3] Although Apple refused to license strong competitors such as Gateway 2000, within two years the clones captured about 20 percent of the Mac market. In the summer of 1997, Apple unilaterally ended cloning, buying out the largest clone maker. Acting CEO Steve Jobs justified the decision by the failure of clones to attract new users and a net decline in overall Mac unit sales during the period clones were sold (*MacWEEK.com*, 1997).

The OS focus strategy

Some had proposed that Apple should spin off its hardware business to become a software-only company analogous to Microsoft. To support this, they noted that Apple's differentiation (particularly since 1984) had come from its software, and that hardware itself was rapidly becoming a commodity. A second argument pointed to the potential conflicts of selling operating systems to rival PC makers while Apple continued to make PCs.

Related to such arguments were theories of industry transformation advanced beginning in the early 1990s. Successful firms in the computer industry in the period 1964–81 had been vertically integrated makers of proprietary systems, typified by IBM, Digital Equipment, and Apple. However, since the emergence of the IBM PC, the most successful computer companies had been those who adopted high-volume, low-cost standardized components that they shared with their rivals. Similarly, to achieve economies of scale and thus reduce unit costs, suppliers of components and operating systems needed to make their standards ubiquitous, including avoiding conflicts with potential customers (Morris and Ferguson, 1993; Grove, 1996; Moschella, 1997).

To give Apple more options to become a software-only company, and to allow the combined Macintosh platform to enjoy the economies of scale already enjoyed by the vast majority of the PC industry, many had suggested that Apple must adapt its operating system to work on the same Intel-based hardware as used by the rival MS-DOS and Windows operating systems. In a videotaped 1992 speaking appearance, CEO John Sculley told a group of business executives:

> I wish we had started moving our technology to the Intel processor years ago so we had more options. Because most of the industry is taking advantage of the tremendous price drops that are going on the Intel world, and we can't because we're not on it, and it's very difficult to move our technology over to that processor very quickly. So I wish we had that option, which we don't. 'Cause we could have come out and done a Windows, and that would be very . . . you could build a whole company around that. (Yoffie, 1992)

Sculley and other Apple executives appear to have for many years overestimated the difficulty of moving the technology to Intel-based chips, as later that same month a small team of engineers began a project to get the Mac OS running on Intel-based technology. After three months they had a working prototype, but the project was abandoned when its internal sponsor left Apple (Carlton, 1997a: 170–71).

But, technical issues aside, the shift from being an integrated computer company would have been traumatic due to sheer differences in scale. In 1988, at the beginning of the "golden window of opportunity," Apple had revenues more than 6× as large as Microsoft and net profits 3× as large. This allowed it to sustain a larger R&D budget than Microsoft until 1994; if it were a software-only company with the second most popular OS, it would have had a far smaller R&D budget than Microsoft during this period.

Also, by being vertically integrated, Apple had been able to make sure that products were available to showcase its technologies. The history of the computer industry shows that those firms dropping hardware to concentrate on software – Novell, Daisy Systems, NeXT, and Palm – must first produce an integrated product to establish their platform standard with a large market share. Such share both provides sufficient scale to support a software-only company and also enough demand by hardware firms to support the software. Despite highly visible technological differentiation, NeXT failed to grow after its 1993 decision to exit the hardware business.

Appropriating success from the standard

Another key question is if Apple were licensing to others, whether it would be able to profit from its technological innovation. In a general sense, such profit depends on both the legal protectability of an innovation, as well as how much an innovator must share economically with suppliers, customers, and developers of complementary assets (Teece, 1986).

Apple was known for being aggressive in maximizing its share of the profits from its innovation. The 1983 Lisa (an unsuccessful high-priced predecessor to the Mac) had included a complete suite of business applications, forestalling the need for many third-party packages. With the Macintosh Apple aggressively courted third-party software developers, but waited three years before it openly embraced suppliers of complementary hardware peripherals such as printers and expansion cards.

Apple succeeded in appropriating the returns from its Macintosh standard, where IBM did not. IBM had used its mainframe computer-derived reputation to establish a PC standard, but by 1988 it was playing an ever-declining role in the "IBM PC" compatible market (Moschella, 1997). In purchasing its processor and operating system from outside vendors IBM knew it was enabling rivals, but had expected to prevent 100 percent compatible computers by copyrighting the software ROM chips embedded in its systems. However, a series of court rulings allowed rivals to make "clones" if they followed specified procedures in designing the ROMs for competing systems, eliminating any legal barriers that might protect IBM and its innovation from direct competition.

Apple would not face the same risk because, as the operating system vendor, Apple enjoyed superior *de jure* and *de facto* protections against imitation. However, from a business standpoint Apple's hardware division (if still integrated) might have had difficulty competing with clone makers, and Apple would have faced practical limits on how much it could charge clone makers for its software to pay for the required R&D. Both turned into actual problems for Apple when clones were sold from 1995 to 1997: its hardware did have difficulty competing, and it was eventually unable to convince clone makers to pay the OS royalties it sought to support R&D.

Pricing and market share strategies

With or without cloning, Apple's pricing strategy had a major impact on its market share. Its choices were a high-price strategy that maximized profitability, or a high-volume strategy that would produce higher revenues, economies of scale, and, it was hoped, higher gross profits. Licensing competitors to clone the Mac would have increased competition and thus reduced hardware profit margins, but Apple could have chosen to unilaterally initiate its own price reductions. So while a high-priced strategy ultimately lowered market share, an absence of cloning was not inherently a low-share, cream-skimming strategy – even though this is the approach Apple adopted from 1984 to 1990.

In its initial 1979 plans, Apple had intended the Macintosh as a mass-market consumer appliance priced at $500. When the designers switched to a 16-bit processor and added a built-in display, the price shot up to $1,500. In 1983, the eventual design had a cost of goods of $500 per unit, which under standard Apple markups would be sold for $1,995. However, to support a $15 million product launch advertising campaign, new CEO John Sculley argued for a price of $2,495 (Linzmayer, 1999: 67–76). Sculley's strategy won, in part, because it would not cost any sales – Apple knew it would be unable to manufacture enough computers to meet demand for the first six months (Sculley, 1987: 170).

Upon its January 1984 introduction, the Mac won rave reviews for its innovative graphics and user interface, but with limited utility (no color, hard disk, few applications) few were willing to pay $2,500 for a product derided by many as a "toy." Thus Apple sold only 250,000 of the 450,000 units originally forecast for 1984, and 45 percent of these were discounted sales to universities and software developers (Kawasaki, 1990: 20–21). So instead of seeking to capture the mass market as originally conceived, Apple used its unique differentiation to support a premium price niche strategy that emphasized profits over market share (Figure 11.3).

Software developers, industry analysts, users, and many Apple employees argued that a lower-priced product line would increase Apple's market size and thus its total profits. But from 1987 to 1989, they were overridden by top executives, who often pointed to Apple's high R&D expenses compared to those of MS-DOS-based PC makers like Compaq. At times, the resistance of key executives to cutting prices came across as arrogance, as when R&D head Jean-Louis Gasée said "We don't want to castrate our computers to make them inexpensive . . . We make Hondas, we don't make Yugos" (Levy, 1994: 233; Carlton, 1997a). With increasing memory prices and overhead costs and seeking to increase gross profit margins from 51 to 55 percent, in the fall of 1988 Apple raised prices up to 29 percent – unprecedented

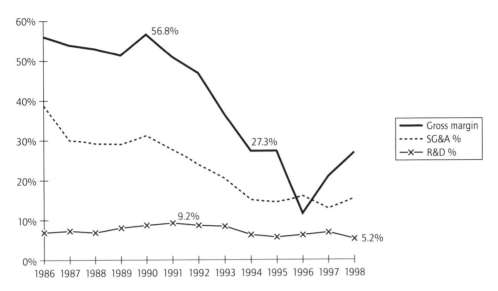

Figure 11.3 Changes in key operating ratios, Apple Computer, 1986–98
Source: CompuStat
Note: In fiscal year 1996, excludes $375 million write-off for discontinued R&D

in an industry which had grown accustomed to annual price cuts for constant performance. Apple eventually reversed its price increase, but sales plummeted in the crucial Christmas quarter (Carlton, 1997a: 79–81).

In 1989, severely chastened and with a new CFO, Apple abandoned profit margins to chase market share, privately seeking to raise its global PC share from 7.5 to 10.5 percent. A combination of administrative cost cutting and more aggressive product design allowed it in October 1990 to release a family of lower-cost products, including its first sub-$1,000 Macintosh – bringing it to rough price parity with top-tier MS-DOS rivals (Carlton, 1997a: 135–42).

As for market share, after the Apple II faded away in the late 1980s, Apple's share of the global PC market (as driven by Macintosh sales alone) never reached 10 percent of the units sold according to IDC and Dataquest estimates. In the US, its unit share stayed above 10 percent from 1985 onward, peaking in 1993 at around 14 percent. In dollar volume, McKinsey credits Apple's higher-priced PCs as claiming 9.9–10.1 percent of the market in 1988–91 (and 1994), and more than 11 percent of the market in 1992–93 (e.g., see McKinsey, 1994).

Apple's positive-feedback downward spiral

Arguments for both the cloning and low price strategies were framed in terms of increased market share that would assure a greater supply of complementary assets. As the editor of one Macintosh-specific computer magazine wrote:

The most important potential benefit of Mac clones would be the proliferation of great software, which would increase to accommodate a larger installed base of computers. Developers would have many more machines for which to sell their software, thus enjoying increased revenue and being able to reinvest more money into producing even better software. Better software is not only great for us computer users, but it is also what will keep the Macintosh ahead of the competition in the long run. (Davis, 1988)

Instead, through fiscal year 1990, Apple forbade clones and kept profit margins high, limiting its market share. By the time it changed both strategies, it was already facing pressure from declining industry prices and the rise of Windows as a rival user interface.

The virtuous positive feedback cycle predicted by Arthur (1996) to accrue to a winning standard becomes a vicious self-reinforcing cycle for a standards loser. After the release of Windows 95 – which eliminated Apple's perceived ease-of-use advantage – Apple faced a downward spiral of bad news:

- plummeting market share worldwide and in key markets in 1996 and 1997, from a 1993 US high of 14% to less than 5%;
- losses in fiscal 1996 and 1997 totaling nearly $2 billion;
- delays and cancellations of development projects for its long-awaited next-generation operating system;
- the forced resignation of Apple CEOs in January 1996 and July 1997.

The bad news brought a demonstrable reduction in the available third-party software. While the Macintosh began the decade with far more menu-driven, GUI-based software packages than Windows, the advent of Windows 3.1 and Window 95 attracted a wide range of packages. Meanwhile, from 1997 onward, many third-party software developers either cut spending on future Macintosh versions or withdrew from the Mac market entirely to concentrate on developing Windows software.

Are network externalities the explanation?
The collapse of Apple's sales from 1995 to 1997 fits the textbook example of a "bandwagon" effect. However, the software-mediated positive network externalities model is but one example of such effects (Farrell and Saloner, 1985). In retrospect, it appears that the self-reinforcing nature of the bad news was as much psychological as economic; the decisions of computer buyers and third-party software developers were driven more by the trend in Apple's market share and profits than the actual level. In particular, both groups (fueled by the press) reacted to the Apple's expected future market share and the increasing doubts about Apple's survival. When the market share leveled out from 1997 to 2000 at 1997 levels, a limited number of Mac users and developers who defected during the 1996–97 collapse returned (e.g., Alsop, 2001).

Also, despite the inexorable decline of Macintosh market share, it did not exactly emulate the Betamax pattern. The Beta format was actually introduced a year ahead of its VHS rival and held 100 percent of the market (Cusumano et al., 1992); by comparison, the Mac was more than two years behind the IBM PC and never held more than 20 percent share, even in the US. The US share of Beta format fell continuously from 85 percent to 1 percent in 11 years (Redmond, 1991). Meanwhile, 15 years after its introduction, from 1997 to 2000 the Macintosh maintained a stable level of about 4 percent in the US and 3 percent worldwide. One possible explanation is that the VCR utility model is tied to novelty seeking and the consumption of new complementary assets – as compared to computer users, who generally require a small but stable supply of complementary assets (West, 1999).

So while many of the mechanisms that eliminated the Betamax VCR also hurt the Macintosh, the VCR-derived theories seem inadequate to explain all the result. If falling behind in a standards war is not inevitably fatal, then perhaps it is relevant to examine some of the other strategic and operational decisions made by Apple during the relevant period.

Simpler Explanations for Apple's Fall

Strategic errors

Over the past decade, criticisms of Apple's strategy have focused on the licensing debate or (to a lesser extent) its pricing policies. But, as one author observed, "the company's fundamental problem was its dearth of effective leadership almost from the outset" (Carlton, 1997b). Apple had an opportunity to differentiate itself and establish a solid market position prior to the introduction of Windows 95, but was unable to do so because of weak leadership in the Sculley era, and rapid turnover after his departure (Table 11.2).

Resources squandered on failed diversification efforts
The belief within Apple (and the industry) that Apple was doomed to follow the "Betamax" model created tremendous pressures on the company, pressures that at times created strategic paralysis within the company and other times sent it far afield in search of the "next big thing."

Table 11.2   Chief executives of Apple Computer, Inc. 1977–2000

Who	Joins	Leaves	Title	Period
Mike Markkula	1/77	8/97*	Chairman	1/77–3/81
Steve Jobs	1/77	9/85	Chairman	3/81–4/83
John Sculley	4/83	9/93	President/CEO	4/83–11/87
			Chairman/CEO	11/87–6/93
Michael Spindler	9/80	1/96	CEO	6/93–1/96
Gil Amelio	1/96	7/97	Chairman/CEO	1/96–7/97
Steve Jobs	12/96		Interim CEO	7/97–1/00
			CEO	1/00–

* Removed from board of directors.
Sources: Rose (1989), Apple (1995), Carlton (1997a), news reports

As co-founder Steve Jobs wrote to employees in a memo justifying the buying out of the license of Apple's largest cloner:

> It is widely believed that Apple missed a golden window of opportunity to license its Macintosh operating system software to clone manufacturers in the 1988–1992 time frame, and that, had Apple done so, today the Mac OS might rival Windows as the personal computer operating system standard. We will never know. Unfortunately, the perception that Apple missed such a huge opportunity has haunted the company ever since, and finally drove Apple to make the poor business decisions resulting in the existing Mac OS licenses. (*MacWEEK.com*, 1997)

Convinced that they had "lost" the PC standards battle, Apple executives pursued an aggressive and risky product development strategy – attempting both to leapfrog operating system software, and to diversify out of PCs. Both strategies were failures.

The company's technological advantage withered away as it focused on radical innovations (which were subsequently abandoned) rather than incremental improvements to its core technology. Apple incurred tremendous self-inflicted opportunity costs, as it squandered money and technological talent on colossal failures (Taligent, Kaleida, Newton, PowerTalk, QuickDraw GX, Copland, OpenDoc, CyberDog) which far overshadowed its modest successes (QuickTime, PowerPC, System 7). Carlton (1997a: 86) estimated that engineering expenditures during 1987–97 for canceled or failed technologies totaled $1.5 billion.

In its diversification efforts, Apple stumbled on one potential success when in 1991 Sculley began promoting the personal digital assistant (PDA). But the Newton, introduced in August 1993, failed in the market as Apple refused to compromise on its original vision. Instead, the smaller, simpler, and cheaper Palm was released in April 1996 and captured the majority of the market, leading to the February 1998 termination of the Newton.

Shopping the company, not solving its problems

A major source of strategic indecision for Apple was its ongoing attempts at mergers and acquisitions. After turning down opportunities to buy Compaq (1984) and Sun (1985), the efforts by Sculley (and later Spindler) concentrated on selling Apple to a Fortune 500 parent. At the same time, long-term fixes were deferred for a decade as both men focused on selling the company, and Apple's board extended their respective tenures in the futile hope that they would soon conclude a deal. The effects became particularly acute from 1993 to 1996, when negotiations reached a fever pitch.

The attempts to sell Apple were potentially damaging with its existing customers, given that the future of the Macintosh and its users under a potential acquirer was far from certain. Apple succeeded in keeping the negotiations largely secret, before Spindler's desperate efforts to find a buyer became highly public shortly before being fired in January 1996.

With lifelong roots on the East Coast, Sculley had originally viewed his move to California as a temporary one lasting no more than five years. As that milestone came and went, and the threats to Apple became greater, he increasingly focused on selling the company. In his last two years, Sculley negotiated with both Kodak and

AT&T. The Kodak negotiations foundered due to cultural differences, while AT&T pulled back to digest its purchase of NCR and planned purchase of McCaw Communications.

When Sculley was forced out, Spindler took over and won successive offers from IBM, Canon, and Philips. The IBM offer marked the high point of nearly 15 years of merger efforts; in October 1994 IBM offered Apple $40 a share, a slight premium to the market price, but overconfident demands by Apple for $60/share and golden parachutes prompted IBM to break off negotiations.

Three subsequent purchases of Apple also failed. In April 1995, Apple rejected a $54/share buyout offer from Canon at a premium of 50 percent over the market price. In late 1995, Apple found acceptable a proposed $36/share buyout by Philips Electronics, but the deal was not approved by Philips' board of directors; soon after, Spindler was fired and Apple's days as an independent company appeared numbered. In January 1996, Sun Microsystems made a $23/share offer, at a 27 percent discount to the market price, which was rejected by Apple's board (Quinlan and Scannell, 1996; Schlender, 1996; Carlton, 1997a: 51–57, 292–98, 355–58; Linzmayer, 1999: 184–88; Malone, 1999: 346–47).

Weak products

As Liebowitz and Margolis (1994, 1999) have observed, results in the competition of standardized products are often more simply explained by differences in important product attributes. In its initial years, the Macintosh had major weaknesses when compared to the IBM PC, while in later years its relative ease-of-use advantage had declined. These two periods bracketed its one period of market share success, from 1991 to 1993 (Figure 11.1).

Delays in supplying key features
Despite its many user interface innovations, the Macintosh trailed the IBM PC in the introduction of most other significant hardware innovations (Table 11.3). These were the innovations that were highly valued by computer users, particularly among business buyers who were more performance-oriented and less price-sensitive than buyers in Apple's two strongest markets, consumers and K-12 education.

As one magazine wrote upon the Mac's initial roll-out: "The engineering is compact and elegant, and the machine is perhaps the first moderately priced computer that is easy to use. But Mac has some drawbacks. It is difficult to expand, has a small memory and does not have a color monitor" (*Time*, 1984). Such deficiencies were cited at the time as the major reasons why many firms found the Macintosh not to be a "serious" business computer, despite its ease of use and strength in preparing advertising and technical manuals. Two years later, Apple added hard disk support, but it was not until 1987 that it provided expansions slots and color monitors.

The failure of Apple to address deficiencies relative to the IBM PC were attributed to the strong opinions of the lead product design executives for the Macintosh – first Jobs, then later Jean-Louis Gasée, VP of research and development. Both were unwilling to compromise their vision of an "insanely great" design, no matter how much such perfectionism ignored the utility lost by omitting important features.

Table 11.3   Introduction of key innovations by competing PC standards

Innovation	IBM PC compatible	Macintosh	Mac lead/lag
16-bit computer	1981 (IBM PC)	1984 (Macintosh 128)	−3 years
Color computer	1981 (IBM PC)	1987 (Macintosh II)	−6 years
Expansion slots	1981 (IBM PC)	1987 (Macintosh II)	−6 years
Portable PC	1982 (Compaq)	1989 (Macintosh Portable)	−7 years
Hard disk	1983 (IBM XT)	1986 (Macintosh Plus)	−3 years
Laptop PC	1984 (HP-110)	1991 (PowerBook)	−7 years
Graphical user interface	1990–1995 (Windows 3.0, 95)	1984 (Macintosh 128)	+6 to +11 years
Mouse	1985 (Microsoft Mouse)	1984 (Macintosh 128)	+1 years
RISC-based CPU	(none)	1994 (Power Macintosh)	n/a
First server OS	1993 (Windows NT)	1999 (OS X Server)	−6 years

Such uncompromising, strong-willed leadership meant Apple was late to portable computing, with its first portable Macintosh coming in 1989, seven years after the first Compaq transportable computer. Unfortunately, the Macintosh Portable was a huge flop due to its size (too big to use on an airplane) and weight (16 pounds). The delays and failures of Apple's portable computing design efforts were widely attributed to bad decisions made by Gasée, who was fired after the Portable debacle (Carlton, 1997a: 104–05; Levy, 1994: 256–57).

This illustrates a key point: were Apple's failures (such as in laptops) due to management error, or due to inherent limitations of its go-it-alone strategy? A case could be made for both. On the one hand, under new R&D management Apple eventually developed a successful notebook product line, its 1991 PowerBook series. On the other hand, it should be noted that three early breakthrough MS-DOS portable computers came from three separate companies – which did not include IBM, the inventor of the architecture. In this case, the comparatively open architecture of the IBM PC allowed multiple design centers, which encouraged innovation.

The fact that IBM did not share in the success of this new product category (at least until its 1992 ThinkPad product line) also showed that, in the case of the IBM PC architecture, the success of the architecture did not accrue to the architecture's creator. So while cloning could have improved product offerings for the Macintosh platform, it would not have helped Apple's corporate future.

Declining product advantage

Belatedly, Apple recognized that the core of falling sales in the late 1990s was its failure to maintain a perceived advantage over its competition, particularly in the area of ease of use. For the first time, in 1997 its 10-K acknowledged this reality:

> The Company believes that the Mac OS, with its perceived advantages over Windows, and the general reluctance of the Macintosh installed base to incur the costs of switching platforms, have been driving forces behind sales of the Company's personal computer

hardware for the past several years. Recent innovations in the Windows platform . . . have added features to the Windows platform that make the differences between the Mac OS and Microsoft's Windows operating systems less significant. The Company's future consolidated operating results and financial condition is substantially dependent on its ability to maintain continuing improvements on the Macintosh platform in order to maintain perceived functional advantages over competing platforms.

Apple certainly had adequate time to prepare for this challenge. While Windows had been announced in 1983, the first usable implementation was 1990's Windows 3.0, which was widely adopted after the bugs were fixed in the 3.1 release (1991). The Windows 3.1 solution still lacked the technical elegance of the Macintosh, as most users found that the MS-DOS text-only interface was never very far away.

But the long-awaited and long-delayed Windows 95 was something else. Early technical evaluations correctly predicted that the easier-to-use Windows 95 would give PC compatibles rough parity with the Macintosh. As important as the technical triumph of Windows 95 was the marketing triumph, a testament to Microsoft's new role as the industry leader. Backed by a $200 million marketing campaign, the August 24, 1995 introduction of Windows 95 became the computer industry's largest media event ever (e.g., see Goldberg, 1995).

Apple fanatics claimed that the Mac retained an advantage, but it was too minuscule for most new users to notice as the trade press proclaimed the ease-of-use war now a draw. As one newspaper columnist later noted:

> When Microsoft Corp. introduced Windows 95, Apple enthusiasts gloated, "Windows 95 equals Apple 89." But it was a little like whistling as they passed the graveyard. What they didn't say is that "Apple 95 equals Apple 89, and Apple 96 equals Apple 89," says Dan Kusnetzky, a director with consulting firm International Data Corp. (Browser, 1997)

That Apple's 1997 technology was nearly equivalent to its 1991 version could be traced back to the investment of engineering resources in Pink/Taligent, Copland, and other new software technologies that were later abandoned. Apple had started several next-generation operating systems that never shipped. In 1988, it split its OS development group into two groups, with the "pink" team attracting the most talented engineers (at least by their own estimation). In 1991, the "pink" team became the core of the Taligent joint venture with IBM, which was killed in 1995 without ever shipping an operating system (Carlton, 1997a; Hagedoorn et al., 2001).

Around 1994, Apple began development of its "Copland" next-generation operating system that was intended to address the Windows 95 challenge. Two years behind schedule, it was canceled in the summer of 1996, earning it the #2 spot on a list of the century's "10 biggest software flops" (*PC World*, 1999). Even if Taligent or Copland had succeeded, both would have required programmers to totally rewrite their applications for new Macs, highly risky because it would render obsolete Apple's existing library of third-party software.[4]

The 1996 purchase of NeXT led to announcement of a "Rhapsody" operating system, but those plans were delayed and scaled back. Finally, Apple announced

OS X, which after, several delays, finally shipped in March 2001 – more than a decade after the challenge of Windows 3 and Apple's formation of its "pink" next-generation OS team.

Other factors

Errors in execution and operations
From the beginning, Apple had suffered from a dysfunctional organizational culture that hindered its execution (Moritz, 1984; Malone, 1999). Within the company, it was expressed by a joke: "What's the difference between Apple and a Boy Scout troop? The Boy Scouts have adult supervision."

Its numerous failures to deliver major OS revisions killed its credibility and support from third-party software developers; as one executive said, "In the last few years it was impossible for any developer to work with them. We couldn't rely on anything they said. . . . We were absolutely convinced they were going to die" (Kirkpatrick, 1998). In hardware, it was late to enter the laptop segment. While its PowerBook models were eventually successful, they lost momentum after a 1995 fire in one model was caused by a faulty battery.

The company also faced a chronic mismatch of supply and demand, with too few of popular products and bloated inventories of unpopular ones. One reason was that in 1991, the head of Apple's sales laid off five of seven demand forecasters, raising the margin of error from ±5 percent to ±50 percent (Carlton, 1997a: 329). More generally, until 1997–99 reorganizations, Apple lagged leading PC makers such as Dell and Gateway in the management of its inventory and other finances. "The company lost $1 billion in 1997 mainly as a result of asset problems, such as being too long on inventory," said Apple's subsequent senior VP of operations, Tim Cook. "We had five weeks of inventory in the plants, and we were turning inventory 10 times a year" – as contrasted with 40 turns a year for Dell (Bartholomew, 1999).

Clearly the Macintosh as a platform was less robust to tactical errors as a single company than was the larger "Wintel" alliance. When top 10 US PC makers such as IBM, Compaq, Packard Bell, and AST faltered in their product design efforts during the late 1980s and early 1990s, other companies like Dell and Gateway rushed in to pick up the slack. In these cases, sales (or market share) of the individual PC makers fell, but the Windows/Intel standard continued to grow as existing suppliers and customers easily switched their loyalties to rival hardware makers.

Unilateral strategic hostage-taking
In its role as both a PC maker and an operating system vendor, Apple had an Achilles' heel: its dependence on Microsoft for key technologies. Its Apple II was built around the Basic programming language licensed from Microsoft, while the most popular business software for the Macintosh was sold by Microsoft, and in both cases Apple lacked leverage to prevent unilateral actions by Microsoft.

In the late 1980s, Apple had vested its differentiation hopes versus Windows with intellectual property law, in particular a 1988 lawsuit accusing Microsoft of copying the Macintosh user interface with Windows 2.0. While direction of copyright protection for user interfaces was uncertain (as exemplified by cases such as *Lotus v. Paperback Software*), Apple in the end lost its lawsuit – not on the legal precedents, but on a loophole it left in a 1985 GUI license granted to Microsoft. Microsoft had, in turn, won the license in 1985 as one of the conditions for renewing the Basic license for the Apple II (Carlton, 1997a; Linzmayer, 1999).

A decade later, in January 1997, Microsoft released its Office 97 for Windows but not for the Macintosh. While Mac users had wrestled with the problem of collaborating with Office 95 co-workers, the problem was exacerbated with the Office 97 file formats, which were not readable on the Macintosh. Individual Mac owners who worked in an increasingly Windows workforce also faced this problem, as it became even more difficult to use their home Macintosh to bring work home. In fact, contemporaneous field interviews and press accounts suggested that Macintosh incompatibility with Office 97 was a crucial factor for organizations dropping Mac support throughout 1997. Doubts about the future availability of Office upgrades for the Macintosh plagued Apple until August 1997, when Apple and Microsoft's CEOs jointly announced Microsoft's commitment to develop versions of Office for the Mac (Kawamoto et al., 1997). The file format compatibility problems were finally resolved with the March 1998 release of Microsoft Office 98.

Just how crucial Microsoft Office was to Apple's survival was revealed later during the *US v. Microsoft* trial. In November 1998, an Apple executive testified that before the August 1997 agreement, Microsoft threatened to withhold future Macintosh development of Office to gain Apple's cooperation on unrelated standards issues. Barring such cooperation, "Microsoft would take any necessary action to drive Apple out of business," testified Avadis Tevanian, Apple's senior vice president for software engineering. Without this software, Apple's CFO said, "we were dead," so Apple felt it had no choice but to agree to Microsoft's terms (Grimaldi, 1998; Brinkley, 1998).

So two path-dependent decisions – the timing of the Apple II Basic license renewal and the emergence of Microsoft as the leading supplier of Macintosh business software – at crucial times left Apple without free rein to pursue its platform competition with Microsoft.

Conclusions

What explains Apple's fall?

Based on theories of network externalities, the conventional wisdom is that after refusing to allow Macintosh clones, Apple's fall was as inescapable as gravity itself.

Any *post hoc* analysis is limited by the lack of a true counterfactual, and thus it is impossible to prove what might have happened at Apple given a different strategy. However, the evidence in the preceding sections suggests that Apple's downward path was less inevitable than accepted theory would predict. Even if the theory was partially right but other factors contributed to its decline, then Apple's strategic myopia – obsessed with the cloning debate and with "leapfrog" radical innovation – may have prevented it from improving the execution of its existing standards strategy. When in 1997 Apple accepted that strategy and focused on improving pricing, product development, and operations, it quickly stabilized its market share and improved financial performance.

Does the evidence support licensing theories?
The simple argument – that Apple's decision to forbid clones sealed its fate – subsumes seven causal predictions (Figure 11.2). Based on Apple's actions from 1984 to 2000, as well as other industry trends, one can retrospectively judge the accuracy of these linkages:

- *Multiple vendors → increased competition* is supported both by the Wintel-based industry experience and also the brief period when Mac clones were available (1995–97).
- *Increased competition → greater hardware variety* is also supported; this greater variety would have improved the Mac's market share for laptop buyers (particularly from 1995 to 2000) when greater variation was evident in Wintel designs; but significant impact on desktop buyers seems doubtful, given the standardization of the PC dominant design around a few basic features.
- *Increased competition → lower prices* is supported, but not the converse: Apple possessed the option to independently lower prices without clones, as its 1990 price cuts and 1998 iMac introduction demonstrated.
- *Lower prices → increased market share* is consistently supported by the patterns of the PC industry and related industries such as PC software and handheld computers.
- *Increased share → more software → increased share*, the core prediction of network externality theory, is supported during some periods but not others.

The difficulty of such network externality theories explaining Apple's fate is that their strong predictions don't allow for the effect of product differentiation. Against the Macintosh, the IBM PC platform enjoyed a two-year head start, larger installed base, and a larger variety of software applications: all should have provided a self-perpetuating positive-feedback loop that (according to theory) would have caused the Mac's share to fall towards zero. Instead, Apple succeeded in improving the worldwide and US market share for the Macintosh through 1993 – through 1995 in Japan against an even larger head start. During this period of increasing market share, the results show that Apple maintained enough differentiation to be competing in a different market segment from the IBM PC, albeit one that drew from an overlapping pool of potential buyers.

As its differentiation disappeared with Windows 3.1 and Windows 95, then from 1995 to 1997 Apple faced the self-fueling market share collapse exactly as predicted. But from 1997 onward, Apple has enjoyed an essentially stable share despite a dramatic disadvantage in software variety. This supports the view of West (1999) that PC buyers differ from VCR buyers in the utility they derive from software libraries: in making a standards decision, PC buyers may satisfice to a minimum level of co-specialized complementary assets (web browser, e-mail, word processor, MP3 player) while the VCR buyers of Cusumano et al. (1992) consumed complementary assets and, thus, chose based on maximizing the variety of complementary assets.

Assessing Apple's strategic alternatives
Implicit in the cloning argument is the suggestion that Apple Computer would have done better if it had licensed competitors. But this was only one of several options available to Apple during the late 1980s and early 1990s. As noted earlier, the four major options were:

- *Licensing Macintosh clones*, and even spinning off PC hardware sales, in hopes of making the Macintosh standard ubiquitous. This would have helped platform sales but, as IBM's experience showed, platform success would not necessarily accrue to the originating company.
- *Penetration pricing*, lowering product prices and gross margins towards the same goal, with less market power to promote platform success but more likelihood that any success would accrue to Apple.
- *Focus on new product development* rather than long-shot efforts to achieve "next big thing" that could have maintained competitive advantage (or reduced disadvantage) in laptop sales and OS updates, particularly in reducing the impact of Windows 95.
- *Improved operational execution*, particularly in product forecasting and supply chain management, would have reduced or eliminated the severe losses of 1996–97 that shook confidence in the company and fueled the largest round of platform defections.

It should be noted that some of these options might have resulted in a more competitive industry structure and greater consumer welfare – but not necessarily better corporate performance for Apple. With doubts about the survival of Apple and its PC standard, Apple's global market share declined 62 percent from 1990 to 2000 (Table 11.4). With the success of the IBM PC standard guaranteed and IBM's corporate future more assured, IBM's share still dropped 43 percent.

However, consideration of these four strategic options raises questions about the accepted wisdom that, after losing a PC standards war, Apple's only choice was to

Table 11.4 Market share leaders in PC industry, 1990 and 2000

Company	1990 share	Company	2000 share
IBM	11.9%	Compaq	12.8%
Apple	7.4%	Dell	10.8%
Commodore	7.1%	HP	7.6%
NEC	5.6%	IBM	6.8%
Compaq	3.9%	NEC	4.3%
Toshiba	3.7%	Gateway	3.8%
Atari	3.1%		
Epson	2.5%	Apple	2.8%
Packard Bell	2.2%		
Olivetti	1.9%		

Source: Dataquest

do what Sony did and abandon its standard. As Apple's fortunes began to ebb in early 1996, no less an analyst than Bill Gates weighed in on the side of execution rather than network effects:[5]

> Business professors love to talk about strategy, and as Apple has declined, the basic criticism seems to be that Apple's strategy of doing a unique hardware/software combination was doomed to fail. I disagree. Like all strategies, this one fails if you execute poorly. But the strategy can work if Apple picks its markets and renews the innovation in the Macintosh. (Schlender, 1996)

Evidence from the Jobs II era

Apple's improved financial performance success since 1998 suggests that with a clearer strategy and operational focus, Apple's 1990s performance could have been considerably better. Among the changes made by Steve Jobs after he took over in July 1997:

- ended cloning to no longer divide the small Macintosh market with rivals;
- designed and released innovative new product designs, including the iMac and a series of improved laptop computers;
- increased the product run rate to provide economies of scale for internal R&D and independent software developers;
- reduced inventory from more than a month to less than a week;
- developed a new OS X fully upward compatible with its existing OS 9, linked to open software and the installed base of Unix and Linux systems;
- stabilized global market share at around 3% from 1997 to 2000; and
- achieved three consistently profitable years after the two years of massive losses.[6]

Apple's performance in the Jobs II era suggests that it is not following the network externality path to extinction, but instead to a small and stable niche. Is a Macintosh more like a Betamax or a BMW? A *Forbes* reporter was among the first to draw the contrast: "Can Apple continue to survive as a high-end player? In automobiles, BMW does, even though its revenues are only 11% those of General Motors" (Morgenson, 1990).

Could the company's fortunes have been turned around before 1997? At this point, it would be difficult to determine whether Apple's 3 percent share is the size of a niche insensitive to software variety, or whether correcting Apple's operational errors earlier would have stabilized Apple's share at a higher level. Apple's attempts starting in May 2001 to grow share through direct retail distribution were an attempt to prove the latter. Either way, Apple's share stabilized at a fraction of the 20 percent figure top Apple executives in 1994 predicted would be necessary to support a viable business model.

At the opposite extreme, there is clearly a minimum efficient scale for an R&D intensive company such as operating system developer. As Arthur (1996) predicts, software companies face high fixed R&D costs and low marginal product costs. Thus far Apple has remained above the minimum efficient scale, and its recent use of open source technologies undoubtedly lowers the cost and thus the required scale

(West, 2003). However, at some point a declining Mac share would bring to an end Apple's attempts to use software innovation to provide differentiation.

Implications for high-tech companies

Should firms worry about becoming "another Betamax"?
The failure of Apple's Macintosh to follow the trajectory of Sony's Betamax into oblivion raises important questions about so-called "old truths" on competitive strategy in network industries.

In particular, the predictions of the Betamax-derived theories assume that buyers derive additional utility from each marginal increase in software variety. This is consistent with entertainment that is "consumed," such as movies (rented to watch only once) or video games (which are generally used for a few weeks or months before being abandoned). A slightly weaker form of this prediction is that there are decreasing returns to additional software variety – as for example with general-purpose computers, where the most popular software is attractive to a wide audience, but additional packages reach an increasingly smaller niche.

But for other types of goods, a minimum set of software could be adequate – as when a PC is used for Internet access or a handheld computer is used for time and contact management. This pattern would be consistent with the increasing number of special-purpose computer technologies. It would also be consistent with an industry environment driven by cross-platform electronics standards, in which users derive equivalent utility from all products that support a small number of open standards, as with Apple's support for the Internet, MP3, and DVDs.

Finally, the predictions of tipping derived from the supply of co-specialized complementary assets assume that there is a significant cost of supporting multiple standards. This has traditionally been applicable to video games (which maximize performance by closely designing for system characteristics) or for GUI-based business applications (due to a lack of cross-platform programming interfaces). Tipping would not be expected if different platforms can be easily supported with co-specialization that is minor compared to the overall effort – as has been true for Unix workstations since the mid-1980s. Another example is the provision of streaming audio and video content on the Internet, where the cost of producing the master recording is comparatively high, but the cost of converting to multiple formats is very low.[7]

In cases where tipping is not expected, then market share is no longer the *sina qua non* for standardized products, and firms can concentrate on traditional product differentiation, marketing, and price leadership strategies. As with non-network industries, such strategies can include niche strategies if such niches are large or munificent enough to be profitable.

More recent standards competitions
Like Apple, the experience of other technology pioneers suggests that market share without innovation is hardly enough. And, as Apple found out, any advantage of share and innovation must be weighed against differences in market power:

- *Web browsers.* Netscape held a virtual monopoly on graphical web browsers until the 1995 release of Windows 95 and Internet Explorer. Some have attributed the rise of Microsoft's browser to its market power, as when the US government alleged that Microsoft illegally used its OS quasi-monopoly to promote its web browser. Others have contended that Netscape's decline stemmed from its failure to innovate its products, in part due to limitations in its software architecture (Liebowitz and Margolis, 1999; Ferguson, 1999).
- *Handheld computers.* Palm Computing established an early lead and managed from 1996 to 2000 to retain the majority of the market despite aggressive licensing by Microsoft. Usage patterns thus far suggest that software complementary assets are relatively unimportant, while most hardware needs would be met by a dozen add-on modules – thus obviating network adoption pressures. While market share for the Palm OS has since declined, this decline has been attributed more to limitations in OS capabilities and the failure of Palm and licensees to innovate rather than either differences in complementary assets or Microsoft's market power from its desktop Windows OS (LeToq, 2001).
- *Streaming multimedia.* Another innovative startup, Real Networks, established an early lead but was later challenged by Microsoft's Windows Media format. On the one hand, the provision of streaming multimedia from Internet servers to individual clients conforms to the physical networks models of Rohlfs (1974), for which tipping would be expected. On the other hand, the relatively small cost for both content suppliers and consumers to support both standards obviates any strong impetus to tip in favor of either standard. Hence, in 2001 Microsoft developed new strategies to leverage its operating system and web portal position market power to encourage both consumers and producers to favor its format.

Such examples of contemporary standards battles suggest that innovation is as important in information technologies as with any technology-driven industry. If there is any benefit to reconsidering the "Betamax" metaphor, it comes from a renewed focus by IT firms on establishing and maintaining innovation strategies.

Notes

1 Note that Liebowitz and Margolis (1994, 1999) question the premise of the Betamax story – that a superior product was doomed due to lack of software – citing the format's inadequate recording capacity for time-shifting.
2 The term "platform" is customarily used in the computer industry to refer to an architecture of related standards, on which complementary assets such as software can be built (Morris and Ferguson, 1993; West and Dedrick, 2000).
3 The most successful licensee was a startup clone-maker, Power Computing. Other licensees included small makers of computer peripherals, and larger companies (Motorola and Pioneer) that had not entered the PC market during its first two decades.
4 The importance of PC application software compatibility is illustrated by the fate of NEC's PC-98. In 1991, the PC-98 platform held nearly 60 percent of the Japanese

PC market but six years later NEC abandoned the standard after Windows rendered obsolete the vast PC-98 software library (West and Dedrick, 2000).

5   In 1999, the whole question of whether network effects provide Microsoft an insurmountable lead became the central issue in *U.S. v. Microsoft* (e.g., see Liebowitz and Margolis, 1999). Gates and Microsoft were accused during that trial of exaggerating the strength of the competition provided by Apple and others. But despite this, there is little reason to suspect Gates was being disingenuous three years earlier in the *Fortune* article, as his response is consistent with the confidence and bluntness of other interviews.

6   In the face of slowing global PC demand and product development delays, Apple reported unprofitable quarters during its 2001 and 2002 fiscal years, but unlike in 1996–97 its full fiscal years were profitable.

7   The Internet example is different from the traditional retailing of pre-recorded music and videos, where considerations and geographically convenient distribution have explained tipping to the more popular standard (Grindley, 1995).

References

Alsop, S. 2001: My old flame: the Macintosh. *Fortune*, June 25, 60.

Apple Computer, Inc. 1995: Corporate timeline – January 1976 to May 1995. Apple Computer website, June 1995. Accessed February 6, 1996.

Arthur, W. B. 1989: Competing technologies, increasing returns, and lock-in by historical events. *Economic Journal*, 99 (394), 116–31.

Arthur, W. B. 1996: Increasing returns and the new world of business. *Harvard Business Review*, 74 (4), July–August, 100–09.

Bartholomew, D. 1999: What's really driving Apple's recovery. *Industry Week*, 248 (6), March 15, 34–40.

Beggs, A. and Klemperer, P. 1992: Multi-period competition with switching costs. *Econometrica*, 60 (3), 651–66.

Besen, S. M. and Farrell, J. 1994: Choosing how to compete: strategies and tactics in standardization. *Journal of Economic Perspectives*, 8 (2), 117–31.

Brinkley, J. 1998: An Apple executive testifies of Microsoft bullying tactics. *New York Times*, October 31.

Browser, G. 1997: Apple barely hangs on with Next strategy. (Toronto) *Globe and Mail*, January 22, B15.

Carlton, J. 1997a: *Apple: The Inside Story of Intrigue, Egomania, and Business Blunders*. New York: Times Business.

Carlton, J. 1997b: They coulda been a contender. *Wired*, 5 (11), November, 122.

Cusumano, M. A., Mylonadis, Y., and Rosenbloom, R. S. 1992: Strategic maneuvering and mass-market dynamics: the triumph of VHS over Beta. *Business History Review*, 66 (1), Spring, 51–94.

David, P. A. 1985: Clio and the economics of QWERTY. *American Economic Review*, 75 (2), 332–7.

Davis, F. E. 1988: Send in the clones. *MacUser*, 4 (5), May, 13–15.

Farrell, J. and Saloner, G. 1985: Standardization, compatibility, and innovation. *Rand Journal of Economics*, 16 (1), 70–83.

Farrell, J. and Saloner, G. 1986: Installed base and compatibility: innovation, product pre-announcements and predation. *American Economic Review*, 76 (5), 940–55.

Ferguson, C. H. 1999: *High Stakes, No Prisoners: A Winner's Tale of Greed and Glory in the Internet Wars*. New York: Times Business.

Garud, R., Nayyar, P. R., and Shapira, Z. B. (eds.) 1997: *Technological Innovation: Oversights and foresights*. New York: Cambridge University Press.

Goldberg, C. 1995: Midnight sales frenzy ushers in Windows 95. *New York Times*, August 24.

Greenstein, S. M. 1993: Did installed based give an incumbent any (measurable) advantages in federal computer procurement? *Rand Journal of Economics*, 24 (1), 19–39.

Grimaldi, J. V. 1998: Microsoft trial: Apple exec paints grim picture of partnership. *Seattle Times*, October 29, A1.

Grindley, P. 1995: *Standards, Strategy, and Policy: Cases and Stories*. Oxford: Oxford University Press.

Grove, A. S. 1996: *Only the Paranoid Survive: How to Exploit the Crisis Points that Challenge Every Company and Career*. New York: Doubleday.

Hagedoorn, J., Carayannis, E., and Alexander, J. 2001: Strange bedfellows in the personal computer industry: technology alliances between IBM and Apple. *Research Policy*, 30 (5), 837–49.

Katz, M. L. and Shapiro, C. 1985: Network externalities, competition, and compatibility. *American Economic Review*, 75 (3), 424–40.

Katz, M. L. and Shapiro, C. 1986: Technology adoption in the presence of network externalities. *Journal of Political Economy*, 94 (4), 822–41.

Katz, M. L. and Shapiro, C. 1992: Product introduction with network externalities. *Journal of Industrial Economics*, 40, 1, 55–83.

Katz, M. L. and Shapiro, C. 1994: Systems competition and network effects. *Journal of Economic Perspectives*, 8 (2), 93–115.

Kawamoto, D., Heskett, B., and Ricciuti, M. 1997: MS to invest $150 million in Apple. CNET News.com, August 6. URL: http://news.cnet.com/news/0-1003-200-321146.html.

Kawasaki, G. 1990: *The Macintosh Way*. Glenview, IL: Scott, Foresman.

Kenney, M. (ed.) 2000: *Understanding Silicon Valley: The Anatomy of an Entrepreneurial Region*. Stanford, CA: Stanford University Press.

Kirkpatrick, D. 1998: The second coming of Apple. *Fortune* 138 (9), November 9, 86–92.

Lach, E. 1994: Spindler strives to restore Apple's shine. *Upside* 6 (6), June, 10–34.

LeTocq, C. 2001: Tough decisions ahead for Palm. CNET News.com, June 19. URL: http://news.cnet.com/news/0-1273-210-6320522-1.html.

Levy, S. 1994: *Insanely Great: The Life and Times of Macintosh, the Computer that Changed Everything*. New York: Viking.

Liebowitz, S. J. and Margolis, S. E. 1990: The fable of the keys. *Journal of Law and Economics*, 33 (1), 1–26.

Liebowitz, S. J. and Margolis, S. E. 1994: Network externality – an uncommon tragedy. *Journal of Economic Perspectives*, 8 (2), 133–50.

Liebowitz, S. J. and Margolis, S. E. 1999: *Winners, Losers and Microsoft: Competition and Antitrust in High Technology*. Oakland, CA: Independent Institute.

Linzmayer, O. W. 1999: *Apple Confidential: The Real Story of Apple Computer, Inc.* San Francisco: No Starch Press.

MacWEEK.com 1997: Jobs says licensing ghost haunts Apple. September 3. URL: http://www.zdnet.com/zdnn/content/mcwo/0903/mcwo0002.html.

Malone, M. S. 1999: *Infinite Loop: How the World's Most Insanely Great Computer Company Went Insane*. New York: Currency/Doubleday.

McKinsey 1994: *The 1994 Report on the Computer Industry*. San Francisco: McKinsey and Company, Inc.

Morgenson, G. 1990: Can Apple go it alone? *Forbes*, 146 (6), September 17, 196–202.

Moritz, M. 1984: *The Little Kingdom: The Private Story of Apple Computer*. New York: W. Morrow.

Morris, C. R. and Ferguson, C. H. 1993: How architecture wins technology wars. *Harvard Business Review*, 71 (2), March–April, 86–96.

Moschella, D. C. 1997: *Waves of Power: Dynamics of Global Technology Leadership, 1964–2010*. New York: AMACOM.

PC World 1999: The digital century: software and the Internet. cnn.com, November 23.

Quinlan, T. and Scannell, E. 1996: Amelio takes over struggling Apple. *InfoWorld*, 18 (6), February 5, 3.

Redmond, W. H. 1991: When technologies compete: the role of externalities in nonlinear market response. *Journal of Product Innovation Management*, 8 (3), September, 170–83.

Rogers, E. M. and Larsen, J. K. 1984: *Silicon Valley Fever: Growth of High-Technology Culture*. New York: Basic Books.

Rohlfs, J. 1974: A theory of interdependent demand for a communications service. *Bell Journal of Economics*, 5 (1), 16–37.

Rose, F. 1989: *West of Eden: The End of Innocence at Apple Computer*. New York: Viking.

Saloner, G. and Shepard, A. 1995: Adoption of technologies with network effects: an empirical examination of the adoption of automated teller machines. *Rand Journal of Economics*, 26 (3), 479–501.

Schlender, B. 1996: Paradise lost: Apple's quest for life after death. *Fortune*, 133 (3), February 19, 64.

Sculley, J. with Byrne, J. A. 1987: *Odyssey: Pepsi to Apple – a Journey of Adventure, Ideas, and the Future*. New York: Harper and Row.

Shapiro, C. and Varian, H. R. 1999: *Information Rules: A Strategic Guide to the Network Economy*. Boston, MA: Harvard Business School Press.

Teece, D. 1986: Profiting from technological innovation: implications for integration, collaboration, licensing and public policy. *Research Policy*, 15 (6), 285–305.

Time 1984: Apple launches a Mac attack. *Time*, 136, January 30, 68.

West, J. 1999: Reconsidering the assumptions for 'tipping' in network markets. *Proceedings of the 1st IEEE Conference on Standardisation and Innovation in Information Technology*, Aachen, Germany, September, 163–7.

West, J. 2003: How open is open enough? Melding proprietary and open source platform strategies. *Research Policy*, 7 (7), 1259–85.

West, J. and Dedrick, J. 2000: Innovation and control in standards architectures: the rise and fall of Japan's PC-98. *Information Systems Research*, 11 (2), 197–216.

WSJ.com 2001: Company profiles: Apple Computer Inc. February 27. URL: http://interactive.wsj.com/archive/retrieve.cgi?id=SB871936142264155000.djm.

Yoffie, D. B. 1992: Apple Computer – John Sculley, Chairman and CEO – Presentation to ISMP Participants, July 6, 1992. Videotape, catalog #793507, Harvard Business School Publishing, December 18.

Index

Page numbers in italics refer to figures and tables